BASIC MATHEMATICAL CONCEPTS

Tarrytown-on-Hudson, New York / Belmont, California

BOGDEN & QUIGLEY, INC.
PUBLISHERS

E 70

LOUIS M. WEINER
NORTHEASTERN ILLINOIS UNIVERSITY

BASIC MATHEMATICAL CONCEPTS

Copyright © 1972 by Bogden & Quigley, Inc., Publishers

All rights reserved. No part of this book may be reproduced, stored in a retrieval system, or transcribed, in any form or by any means, electronic, mechanical, photocopying, recording, or otherwise, without the prior written permission of the publisher, 19 North Broadway, Tarrytown-on-Hudson, New York 10591.

Cover design by Kenneth Roberts

Text design by Science Bookcrafters, Inc.

Library of Congress Catalog Card No.: 77–181703

Standard Book No.: 0–8005–0028–8

Printed in the United States of America

1 2 3 4 5 6 7 8 9 10—76 75 74 73 72

512.1
W423

To Todd, who wasn't here last time

PREFACE

This book assumes no previous knowledge of mathematics. It begins with the fundamentals of arithmetic and ends with a discussion of analytic geometry and functions. It is designed for use in general education introductory mathematics courses or in precalculus courses.

Instructors teaching an introductory mathematics course to students whose mathematical background is weak or nonexistent may wish to use two semesters to cover the entire book. A one-semester course, or the first semester of a two-term course, could cover the first two chapters—the real-number system and arithmetic—or possibly the first three chapters, depending upon the ability of the students. The second semester could then be devoted to the remainder of the book.

In classes in which students are familiar with the real numbers and arithmetic, most of Chapters Three through Seven could be covered in one semester. Pretests (which have been perforated so that they may be handed in) are provided in Chapters One and Two to help the instructor determine if his students need to begin with the fundamentals. If they do not, he may wish to begin his course with Chapter Three.

Each chapter ends with a cumulative review test (also perforated) designed to measure the student's mastery of the preceding material and his readiness to move ahead. Exercises at the end of each section are designed to check progress closely along the way. Answers to odd-numbered problems appear at the end of the book; even-numbered answers are available in a separate booklet, which also contains answers to the pretests and review tests.

The emphasis in this book is on mastery of the skills necessary for solving problems. Our objective is to acquaint students with the various techniques by means of numerous worked-out examples. Accordingly, the introduction of each new concept is complemented by detailed, step-by-step examples and numerous sketches and charts. The examples provide models which the student can use to work out solutions to problems on his own; the sketches and charts help him to visualize the more abstract concepts. In our emphasis on the development of skills we do not neglect to indicate why these techniques work.

Although some of the exercises require students to prove some simple facts, our primary objective is not to teach students to prove theorems. This approach has worked well with students and gives them a chance to sharpen their manipulative skills and to gain experience in problem solving. If handled properly, this approach can also develop an appreciation for the scope and applicability of mathematics.

The material in the book has been used successfully for several years in the one-semester introductory mathematics course at Northeastern Illinois University. Although the majority of these students have little or no previous mathematical background, most find the material understandable and interesting.

The author wishes to express his appreciation to his colleagues at Northeastern Illinois University for their many helpful suggestions and to the people at Bogden & Quigley for their help and encouragement. Acknowledgment is also due C. Ralph Verno (West Chester State College), Henry Korn (Westchester Community College), Marjorie Senechal (Smith College), Howard Reiter (Westchester Community College), and William L. Zlot (New York University), all of whom read and commented upon earlier versions of the manuscript. Louis Rotando (Westchester Community College) read a revised version of the manuscript and made many valuable suggestions.

Chicago, Illinois LOUIS M. WEINER

CONTENTS

Preface vii

Symbols xii

Pretest 1 3

1 / SETS AND NUMBERS 5

1.1 Addition of numbers 5
1.2 Multiplication of numbers 10
1.3 Subtraction and division of numbers 12

1.4 Properties of numbers 15
1.5 Decimal number system 19
1.6 Nondecimal number systems 33
1.7 Prime and composite numbers 41
1.8 Sets and logic 44
Review test 1 49

Pretest 2 55

2 / REAL NUMBERS 57
2.1 Integers 57
2.2 Operations on integers 61
2.3 Order properties 67
2.4 Rational numbers 69
2.5 Decimals and real numbers 88
2.6 Ratio and percent 99
Review test 2 103

3 / EXPONENTS AND LOGARITHMS 109
3.1 Properties of exponents 109
3.2 Zero, negative, and fractional exponents 114
3.3 Operations on irrational expressions 122
3.4 Computations with logarithms 128
Review test 3 140

4 / ALGEBRAIC EXPRESSIONS 145
4.1 Operations on polynomials 145
4.2 Operations on rational expressions 153
4.3 Factorization of polynomials 159
4.4 Simplification of algebraic expressions 165
Review test 4 170

5 / SOLVING EQUATIONS 177
5.1 Equations and identities 177
5.2 Linear equations in one unknown 179
5.3 Word problems 185
5.4 Quadratic equations in one unknown 191
5.5 Systems of linear equations 200
5.6 Solution of systems of equations by graphing 211
5.7 Solution of systems of equations by determinants 219
Review test 5 228

6 / GEOMETRY AND TRIGONOMETRY 235

6.1 Areas of plane figures 235
6.2 Congruence and similarity of triangles 248
6.3 Solution of right triangles 261
6.4 Solutions of arbitrary triangles 271

Review test 6 280

7 / ANALYTIC GEOMETRY 287

7.1 Straight lines 287
7.2 Conic sections 297
7.3 Translation of axes 307
7.4 Polar coordinates 315
7.5 Parametric equations 322
7.6 Functions 326

Review test 7 339

Answers to odd-numbered problems 345

Index 359

SYMBOLS

$=$	is equal to
$>$	is greater than
$\geq$	is greater than or equal to
$<$	is less than
$\leq$	is less than or equal to
$\subset$	is a proper subset of
$\subseteq$	is a subset of
$\supset$	properly contains
$\supseteq$	contains
$\cup$	union (of sets)

$\cap$	intersection (of sets)
$\times$	product set (of sets)
$\cong$	is equivalent to (for sets)
$\cong$	is congruent to (for triangles)
$\varnothing$ or $\{\ \}$	the empty set
GCD (a, b)	the greatest common divisor of the numbers a and b
LCM (a, b)	the least common multiple of the numbers a and b
$\lvert a \rvert$	the absolute value of the number a
$\sqrt{a}$	the square root of the number a
a^n	a to the nth power (the product of n a's)
$a:b$	the ratio of a to b
$\%$	percent or hundredths
$\approx$	is approximately equal to
$\begin{vmatrix} a & b \\ c & d \end{vmatrix}$	the determinant $(ad - bc)$
π	the ratio of the circumference of a circle to its diameter (approximately 3.14)
AB	the line segment running from A to B
$\sim$	is similar to (for triangles)
$\angle ABC$	the angle whose vertex is at B and whose sides are AB and BC
$\triangle ABC$	the triangle whose vertices are A, B, and C
$^\circ$	degrees (unit of measure for angles)
$\perp$	is perpendicular to
$\parallel$	is parallel to
$\therefore$	therefore
m	the slope of a line
$y = f(x)$	y is a function of x
$f(a)$	the value of $f(x)$ when $x = a$

BASIC MATHEMATICAL CONCEPTS

The Pretests that precede Chapters One and Two are to help you determine where you are now in your understanding of mathematics. The Pretests and the Review Tests, which appear at the ends of chapters, have been perforated so that you can hand them in to your instructor.

pretest 1

NAME ____________

DATE ____________

A Perform the following computations:

(1) $4 + 9$ ______ (2) $12 - 7$ ______ (3) 7×8 ______

(4) $6 + 5$ ______ (5) $18 \div 6$ ______ (6) $36 \div 4$ ______

(7) $14 - 6$ ______ (8) 9×6 ______

B Write these numbers in words:

(1) 587 ____________

(2) 12,643 ____________

(3) 2,817,364 ____________

(4) 906,002 ____________

C Write the numerals for these numbers:

(1) Seven hundred and sixty-eight ____________

(2) Twenty-nine thousand, six hundred and three ____________

(3) Twelve million, two hundred and forty-nine thousand, five hundred and twenty-seven ____________

D Add:

(1)
$$\begin{array}{r} 168 \\ 927 \\ 348 \\ +653 \\ \hline \end{array}$$

(2)
$$\begin{array}{r} 3153 \\ 692 \\ 481 \\ +73 \\ \hline \end{array}$$

E Subtract:

(1)
$$\begin{array}{r} 87 \\ -53 \\ \hline \end{array}$$

(2)
$$\begin{array}{r} 3729 \\ -1453 \\ \hline \end{array}$$

F Multiply:

(1)
$$\begin{array}{r} 153 \\ \times 72 \\ \hline \end{array}$$

(2)
$$\begin{array}{r} 9625 \\ \times 386 \\ \hline \end{array}$$

G Divide:

(1) $6\overline{)8629}$

(2) $563\overline{)37289}$

H Find the results:

(1) $8 + (6 \times 9)$ ____________

(2) $7(2 + 9)$ ____________

(3) $(8 \times 4) - (4 \times 3)$ ____________

(4) $(42 + 6) \div (9 + 3)$ ____________

(5) $(16 - 4) \div 3$ ____________

(6) $(12 + 5)(9 - 5)$ ____________

I If you have $0.85 and buy one thing for $0.39 and something else for $0.17, how much do you have left?

J If you work for 28 hours at $2 per hour, how much should you receive in wages?

K If you travel at 45 miles per hour, how long will it take you to travel 360 miles?

L Factor these numbers into primes:

(1) 156 ____________________

(2) 360 ____________________

(3) 2196 ____________________

(4) 5280 ____________________

M If you were told that all boys drink milk and that everyone who drinks milk is healthy, what could you conclude?

N If you were told that everyone who smokes is short and that no short people are smart, what could you conclude?

chapter one

SETS AND NUMBERS

1.1 / addition of numbers

Most people think of mathematicians as people who work with numbers. It would be natural then for someone who wants to learn mathematics to want to know something about numbers. What are they? And what are they used for? As a partial answer, it might be said that numbers are used to measure things. They are used to tell us the size of an object, to specify temperature, to measure quality in various situations, to count objects, and so on. It is the last application, counting objects, that leads us to the simplest of our numbers, the natural numbers—1, 2, 3, . . . (or counting numbers as they are sometimes called), and which introduces the notion of a set as the collection of objects to be counted.

Sets are useful not only in providing objects for us to count; by using the language and notation of sets we are able to describe many of the abstract concepts of mathematics in a simple and compact manner.

One of the reasons that a mathematician works with numbers is to solve problems. For example, if a man works for 3 hours and earns $4 per hour, how much money does he earn altogether? If we have 3 rows of objects with 4 objects in each row, how many objects are there altogether? If one tennis team has 3 members, and another has 4 members, and we wish to schedule some singles matches, each one matching a member of the first team with a member of the second team, how many combinations are possible?

In each case it is necessary to realize that in order to solve the problem, we must multiply 3 by 4, and once this is realized, it is necessary to be aware of the fact that 3 times 4 is 12. This is why pupils are required to memorize multiplication tables in elementary school. Most problems in mathematics are not this simple, but they generally follow the same pattern. Given a problem, we must first know how to set up the solutions to the problem (i.e., what computations must be performed in order to get the answer), and then we must know how to perform those computations.

Mathematical computations are usually performed to solve problems about sets of objects. To know how to solve such problems, it is necessary that we know something about sets of objects. Accordingly, part of Chapter One will concern itself with sets and the way in which they give rise to the counting numbers and arithmetic. In Chapter Two we will see that there are different kinds of numbers and that the system of counting numbers can be enlarged step by step through the integers and rational numbers until we arrive finally at the real-number system, which is basic to all of mathematics.

Arithmetic deals with numbers, but to master arithmetic it is necessary not only to be familiar with numbers but also to be able to perform certain operations with numbers. The four basic operations of arithmetic are addition, multiplication, subtraction, and division. For most people arithmetic means memorizing the basic number facts about these operations; for example, $4 + 3 = 7$, or $8 \times 5 = 40$. Memorizing the addition, multiplication, subtraction, and division facts is a very important part of arithmetic, but, unfortunately, if this is all you do in arithmetic, it can become a very dull subject.

You will find that arithmetic can be made much more interesting if you consider such questions as: What do we mean when we say $4 + 3 = 7$? Let us look at an example that may raise some doubts as to whether $4 + 3$ really does equal 7. Look at Figure 1.1, in which we have a set of letters in a square and a set of letters in a circle. To be more specific, we have a set of 4 letters in the square and a set of 3 letters in the circle. Yet altogether we have only 5 letters. So why don't we say that $4 + 3 = 5$?

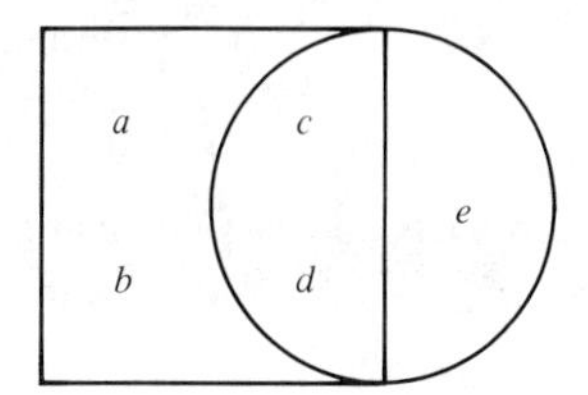

figure 1.1

To answer this question, it is necessary that we have a clear definition in our mind of what we mean by 4 + 3. Most of you will realize that the reason why we cannot use this example to conclude that 4 + 3 = 5 is because the two sets that we are putting together have letters *c* and *d* in common, so when we count up all the letters we have, we get *a*, *b*, *c*, *d*, and *e*, which totals 5 rather than 7. This common part of the two sets (the letters *c* and *d*) is called the *intersection* of the two sets, whereas the set of all the letters (the letters *a*, *b*, *c*, *d*, and *e*) is called the *union* of the two sets.

You probably realize by now that if you want to do addition by taking the union of two sets, you must be sure that they have no intersection, or as we shall say, that their intersection is the *empty set*. When this is the case, we say that the two sets are *disjoint*.

We can now state very briefly that the example shown in Figure 1.1 is not a proof that 4 + 3 = 5, because the two sets used were not disjoint.

To determine the sum 4 + 3 the proper way, we use the sets shown in Figure 1.2. The set of letters in the square and the set of letters in the circle are disjoint, so we may say that, because we have 4 letters in the square, 3 letters in the circle, and 7 letters altogether, then 4 + 3 = 7. Notice that we could also have used Figure 1.3 to show that 4 + 3 = 7.

The set of letters in the square of Figure 1.2 contains the same number of letters as the set of letters in the square of Figure 1.3. We say in this case that these two sets are *equivalent*. Thus the set of letters in the circle of Figure 1.2 is equivalent to the set of letters in the circle of Figure 1.3.

One of the nice features of mathematics is that, by adopting certain notations, long statements may be written in concise form. We will use capital letters to de-

figure 1.2

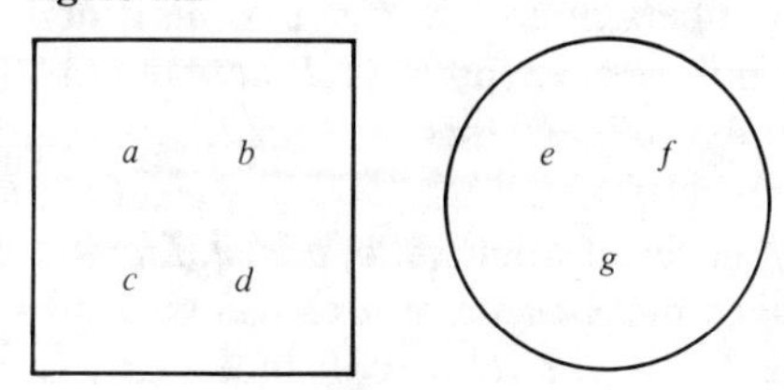

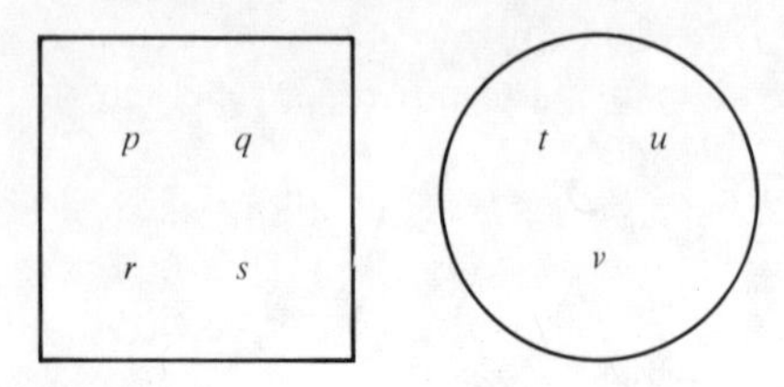

figure 1.3

note sets, and when we wish to write, "A is the set consisting of the elements a, b, c, d, and e," we will write $A = \{a, b, c, d, e\}$; when we wish to write, "B is the set consisting of the elements 1, 2, 3, 4, 5, and 6," we will write $B = \{1, 2, 3, 4, 5, 6\}$, and so on. If A is the empty set, we will write $A = \{\ \}$ or $A = \varnothing$.

Observe that if $A = \{1, 2, 3, 4, 5\}$ and $B = \{4, 5, 6, 7\}$, then the union of A and B (which we will henceforth denote by $A \cup B$) is the set $\{1, 2, 3, 4, 5, 6, 7\}$, and the intersection of A and B (which we will henceforth denote by $A \cap B$) is the set $\{4, 5\}$.

When two sets, A and B, contain identical elements, we say that the two sets are equal and write $A = B$. Thus if $A = \{a, b, c, d\}$ and $B = \{a, b, c, d\}$, we may write $A = B$. By contrast, when two sets, A and B, are equivalent we will write $A \cong B$. Thus if $A = \{a, b, c, d\}$ and $B = \{w, x, y, z\}$, we may write $A \cong B$ but not $A = B$. To see whether you understand these two concepts, determine which of the following two statements is correct:

$$\text{If } A = B, \text{ then } A \cong B.$$

$$\text{If } A \cong B, \text{ then } A = B.$$

We see that in order to explain what we mean by $4 + 3$, it is helpful to talk about sets of elements. We can now formulate the rule that, to find the sum $4 + 3$, we take two disjoint sets, one containing 4 elements and the other containing 3 elements, and look at the union of these sets. Since the union contains 7 elements, we conclude that $4 + 3 = 7$. Notice that if the sets were not disjoint, the union would not have contained 7 elements, and we would get the wrong answer.

We may now formulate a general rule for the addition of numbers:

> To find the sum $m + n$ of the numbers m and n, find two disjoint sets, A and B, such that A contains m elements and B contains n elements. Then $m + n$ is the number of elements in $A \cup B$.

As an example, to find the sum $7 + 5$, let $A = \{a, b, c, d, e, f, g\}$, and let $B = \{p, q, r, s, t\}$. We have A and B disjoint, A contains 7 elements, B contains 5 elements, and $A \cup B$ contains 12 elements. We therefore conclude that $7 + 5 =$

12. Thus when we say that $7 + 5 = 12$, we mean that if we take two disjoint sets, one with 7 elements and one with 5 elements, and put them together, then we have a set with 12 elements.

This procedure of taking unions of disjoint sets would allow us to find the sum of any two numbers. Can you imagine, however, how long it would take you to find the sum $458 + 873$ by this method? Thus it is necessary to devise a way of writing numerals to denote our numbers (which we call a *system of numeration*) and a corresponding rule for addition that will make the process of addition of any numbers short and convenient. Our decimal system of writing the counting numbers and the rule for addition that derives from it will take care of this; these are described later in this chapter.

exercises 1.1

1 State which facts of addition may be proved by the figures shown.

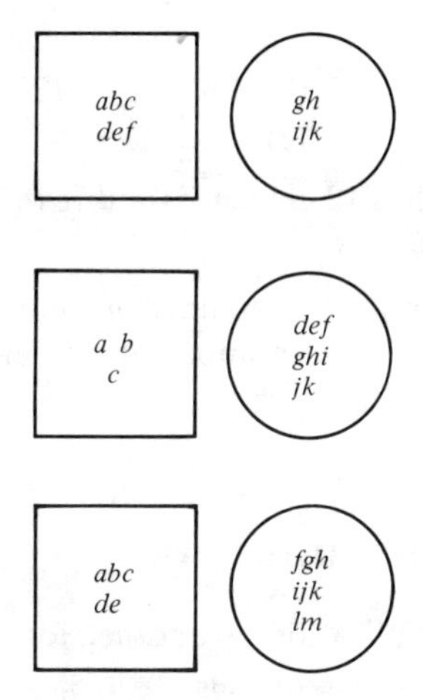

2 How many elements are there in the empty set?

3 What fact of addition is proved by the figure shown?

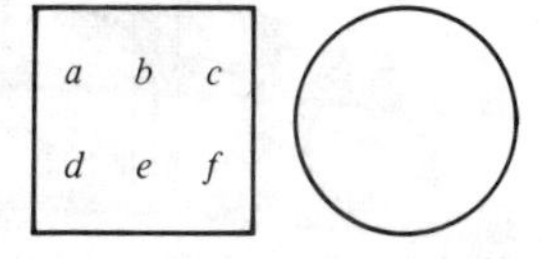

4 Can you state a rule about the sum of any number plus 0?

5 Describe a set that is equivalent to the set of fingers on your right hand.

6 State the number of elements in each of the following sets:
(a) The tables in your classroom
(b) Your feet
(c) The books on your desk
(d) The windows in your room

7 If $A = \{1, 2, 3, 4, 5, 8\}$, $B = \{2, 4, 6\}$, and $C = \{1, 3, 4, 7, 8\}$, describe the sets:
(a) $A \cup C$ (b) $A \cap B$ (c) $B \cup C$ (d) $B \cap C$

8 If $A = \{a, b, c, d, e\}$ and $B = \{d, e, f, g\}$, how many elements are there in the set $A \cap B$? How many elements are there in the set $A \cup B$? If you add the number of elements in A to the number of elements in B, what do you get for an answer?

9 Give the following sums:
(a) $4 + 5$ (b) $7 + 6$ (c) $9 + 5$
(d) $8 + 5$ (e) $9 + 3$ (f) $8 + 6$

10 Fill in the blanks:
(a) $5 + \_ = 11$ (b) $7 + \_ = 12$ (c) $6 + \_ = 14$
(d) $3 + \_ = 12$ (e) $4 + \_ = 9$ (f) $9 + \_ = 16$

11 If A has 8 elements, B has 12 elements, and $A \cap B$ has 5 elements, how many elements are there in $A \cup B$?

12 Write a general formula for the number of elements in $A \cup B$ in terms of the number of elements in A, the number of elements in B, and the number of elements in $A \cap B$.

1.2/multiplication of numbers

Many people think of multiplication as another form of addition. For example, 4×3 means the sum of four 3's, in other words, $3 + 3 + 3 + 3$. It could also mean the sum of three 4's, in other words, $4 + 4 + 4$. In either case we see that the product $4 \times 3 = 12$. You might prefer to think of 4×3 as the result of taking 4 rows, each containing 3 objects; that is,

$$4 \times 3 = \begin{matrix} | \, | \, | \\ | \, | \, | \\ | \, | \, | \\ | \, | \, | \end{matrix}$$

and when you count up the number of objects, you will see again that $4 \times 3 = 12$.

Following this example we would say that, for any two numbers m and n, $m \times n$ means the sum of m n's, or the sum of n m's, or the result of taking m rows each containing n objects, or the result of taking n rows each containing m objects; the answer is the same in each case.

Let us now examine another way of looking at the meaning of multiplication. Suppose that we have two tennis teams, A and B. Tennis team A has 4 members, a, b, c, and d, and tennis team B has 3 members, x, y, and z. Each member of tennis team A must play each member of tennis team B. How many matches must be played?

To solve this problem, we list all the matches that must take place. To do this systematically, we match a with every member of team B, and so on, and we get the following list of matches:

(a, x)	(b, x)	(c, x)	(d, x)
(a, y)	(b, y)	(c, y)	(d, y)
(a, z)	(b, z)	(c, z)	(d, z)

Thus we see that 12 matches must be played. This is another way of determining that $4 \times 3 = 12$.

When we have a set $A = \{a, b, c, d\}$ and a set $B = \{x, y, z\}$, the set of all pairs where we match each element of A with each element of B, as we did above, is called the *product set* of A and B and is denoted by $A \times B$. Notice that when A contains 4 elements and B contains 3 elements, $A \times B$ contains 12 elements. We can say, therefore, that to multiply 4×3, we take a set A containing 4 elements and a set B containing 3 elements. The product 4×3 is then the number of elements in the set $A \times B$, which will be 12.

We may now formulate an alternative rule for multiplication as follows:

> To find the product $m \times n$, find two sets A and B such that A contains m elements and B contains n elements. Then $m \times n$ is the number of elements in the set $A \times B$.

In the case of multiplication, the sets do not have to be disjoint as they do for addition.

By following any one of these rules for multiplication, you should be able to find the product of any two numbers m and n, assuming that you know how to add. If, however, you were asked to find a product such as 285×492, it would take far too long to do it by this method, so again we must find a procedure for multiplying large numbers in a short time; this will be discussed later.

In order to make use of the rules for adding and multiplying large numbers, however, you will have to know the sums and products of small numbers up to $9 + 9$ and 9×9. You can do this either by memorizing the addition and multiplication tables given later, or, if you work a lot of problems using addition and multiplication of small numbers, you will get to memorize these facts almost without trying.

exercises 1.2

1 If A has 7 elements and B has 9 elements, how many elements are there in $A \times B$?

2 If you line up 8 rows of cards, each containing 6 cards, how many cards do you have altogether?

3 If you drive for 5 hours at 20 miles per hour, how far will you drive?

4 Write the products of these pairs of numbers:

(a) 4×8 (b) 7×9 (c) 6×8
(d) 7×5 (e) 8×3 (f) 9×8

5 Fill in the blanks:

(a) $9 \times \_ = 54$ (b) $7 \times \_ = 56$ (c) $3 \times \_ = 24$
(d) $6 \times \_ = 42$ (e) $\_ \times 5 = 45$ (f) $\_ \times 8 = 48$

6 If $A = \{1, 2, 3\}$ and $B = \{4, 5, 6, 7, 8\}$, list the elements of $A \times B$.

1.3 / subtraction and division of numbers

There are at least two ways of looking at subtraction; either way you prefer is all right as long as you have a clear idea of what subtraction means. Let us consider the example $7 - 3$. This could mean starting with a set of 7 elements and taking away 3 of these elements. We would then have 4 elements left, so $7 - 3 = 4$.

We could do the same thing visually by making 7 marks and crossing out 3 of them, as in Figure 1.4. We see that we would have 4 marks left, so $7 - 3 = 4$.

Another way of looking at the meaning of $7 - 3$ is to ask: What number must be added to 3 in order to give 7? The only number that will work is 4, so we say that $7 - 3 = 4$.

By following this example, you should be able, either by trial and error, assuming that you know your addition facts, or by making marks on a piece of paper, to determine the difference, $m - n$, of any two numbers m and n. The two rules for subtraction, either one of which is acceptable, may be formulated as follows:

1. To subtract n from m (which means to take the difference $m - n$), take a set with m objects and remove n of these. The number of objects left will be $m - n$; that is, $m - n$ is the number of elements in the set that remains.

figure 1.4

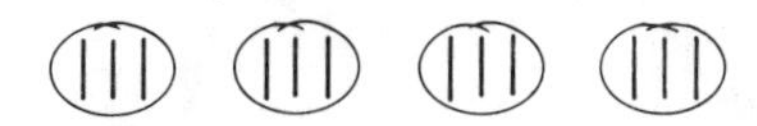

figure 1.5

2. The difference $m - n$ is the number that must be added to n to give m as an answer.

Notice that in either case m must be bigger than n; otherwise, you cannot subtract n from m. Here, too, we will need a separate rule for subtracting large numbers so that it will not take too long to do this; however, you will again have to know the differences of small numbers to use this rule.

Suppose that we start with a number such as 4, then add 3, and then subtract 3. We wind up with 4 again; in other words, subtraction is the opposite of addition in that it undoes addition. Similarly, if we start with 4, subtract 3, and then add 3, we wind up with 4 again. We say that addition and subtraction are *inverse operations.*

Multiplication also has an inverse operation—division. If we start with 4, multiply by 3, and then divide by 3, we wind up with 4 again. As with subtraction, there are two ways of looking at division. When we ask how much 12 divided by 3 (written $12 \div 3$) is, we are asking: If we start with a set of 12 objects, how many sets of 3 objects each can we get from this? To answer this question we can make 12 marks on a piece of paper and then circle groups of 3 to see how many groups we get, as in Figure 1.5. Since we get 4 groups of 3 each, we can say that $12 \div 3 = 4$.

Another way of looking at this problem is to say that $12 \div 3$ means: What number must 3 be multiplied by to give 12? Through trial and error, we see that the only number that works is 4, so $12 \div 3 = 4$.

Suppose that we consider the problem $14 \div 3$. According to this last approach, this means: What number times 3 equals 14? If you know your multiplication facts, you know that there is no such number, so we would say that 14 cannot be divided by 3. According to the first approach, however, if we put 14 marks on a piece of paper and circle groups of 3, we find that we get 4 groups and 2 marks left over (see Figure 1.6). Thus we say that when you divide 14 by 3 you get a *quotient* of 4 and a *remainder* of 2. This means that if we multiply 4×3 and then add 2, we get 14.

Many people say that division means repeated subtraction. The problem $14 \div 3$ means: How many times can you subtract 3 from 14? If you start subtracting 3's from 14, you will see that you can subtract four 3's and then you will have 2 left over. This means again that 14 divided by 3 gives a quotient of 4 and a remainder of 2.

figure 1.6

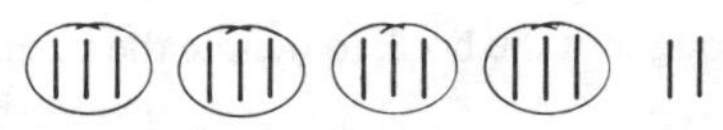

examples

1 $13 \div 5$

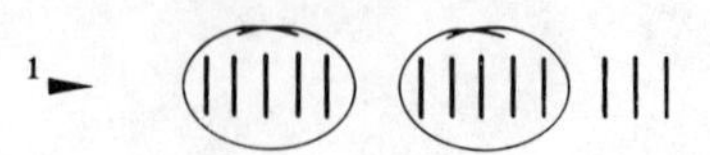

We see that $13 \div 5$ gives a quotient of 2 and a remainder of 3.

2 $10 \div 6$

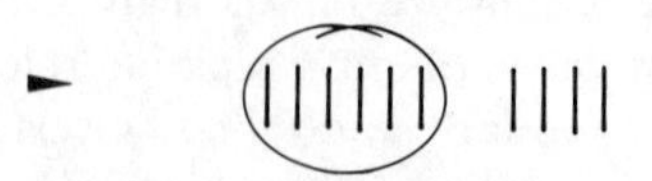

We see that $10 \div 6$ gives a quotient of 1 and a remainder of 4.

3 $10 \div 5$

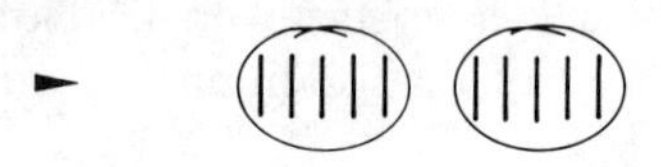

We see that $10 \div 5$ gives a quotient of 2 and no remainder, so we write $10 \div 5 = 2$.

When a number m is divided by a number n and there is no remainder, we say that *n divides m* or that m is a *multiple* of n. For example, 3 divides 9, 5 divides 20, 6 divides 18, etc.

If you know your multiplication facts for small numbers well, you will know your division facts for small numbers. For example, since $4 \times 3 = 12$, we know that $12 \div 4 = 3$ and $12 \div 3 = 4$. To do division problems for large numbers, we use the process of long division, which will be explained later.

exercises 1.3

1 Subtract:

(a) $8 - 3$ (b) $9 - 6$ (c) $15 - 8$
(d) $14 - 6$ (e) $19 - 7$ (f) $18 - 9$

2 Divide:

(a) $56 \div 8$ (b) $72 \div 9$ (c) $54 \div 6$
(d) $36 \div 4$ (e) $48 \div 6$ (f) $18 \div 3$

3 Find the quotient and remainder for each of the following:

(a) $15 \div 6$ (b) $23 \div 7$ (c) $19 \div 4$
(d) $37 \div 7$ (e) $52 \div 9$ (f) $43 \div 12$

[1]The device ► is used in the Examples throughout the book to denote the beginning of the solution.

4 Fill in the blanks:
(a) $18 - \_ = 11$ (b) $\_ - 7 = 6$ (c) $\_ - 9 = 6$
(d) $13 - \_ = 9$ (e) $\_ - 8 = 7$ (f) $13 - \_ = 5$

5 Fill in the blanks:
(a) $18 \div \_ = 6$ (b) $24 \div \_ = 4$ (c) $54 \div \_ = 9$
(d) $\_ \div 7 = 9$ (e) $\_ \div 8 = 5$ (f) $\_ \div 9 = 5$

6 If you have \$0.83 and spend \$0.58, how much do you have left?

7 If you divide \$0.48 equally among 4 people, how much will each one get?

8 If you divide \$0.53 equally among 6 people, how much will each one get, and how much will be left over?

9 If you have \$0.58 and you need \$0.25 for carfare, how much can you spend?

10 How many hours must you work at \$0.60 per hour in order to earn \$3.00?

1.4 / properties of numbers

Before we discuss the rules for adding, multiplying, subtracting, and dividing large numbers, let us pause a moment to summarize some of the properties of our number system. We know that when we take the union of two sets, A and B, it does not matter which set comes first; in other words, $A \cup B = B \cup A$. This means that when we add two numbers m and n, it does not matter which one comes first; in other words, $m + n = n + m$. This fact is known as the *commutative property for addition.*

Similarly, when we take the product of two sets, A and B, we know that $A \times B$ includes all pairs of elements consisting of an element of A and an element of B. On the other hand, $B \times A$ consists of these same pairs, except that we write the elements of B first. We see that $A \times B \cong B \times A$; that is, $A \times B$ has the same number of elements as $B \times A$. This means that when we multiply two numbers m and n, it does not matter which one comes first; in other words $m \times n = n \times m$. This fact is known as the *commutative property for multiplication.*

Let us agree here that when we are using letters such as m and n to stand for numbers, and we want to multiply them, we will not bother writing the $\times$ sign. Instead of writing $m \times n$, we will write mn. We may then write the commutative law for multiplication in the form $mn = nm$. We could not do this with regular numbers, because if we tried to write 56 instead of 5×6, this would mean fifty-six instead of 5 times 6; however, with letters we do not have this problem, so it is all right to write mn instead of $m \times n$.

Suppose now that we wish to take the sum of three numbers, say $4 + 3 + 8$. Since addition is a *binary* operation, which means that we add only two numbers at a time, we could say $4 + 3 = 7$ and then $7 + 8 = 15$. To show that we added the 4 and 3 first we write $(4 + 3) + 8$; that is, we have put parentheses around $4 + 3$ to show that we took this sum first.

We could also have done this problem by taking the sum $3 + 8$ first to get 11 and then saying $4 + 11 = 15$. If we do it this way, we would write $4 + (3 + 8)$ to show that we added the 3 and the 8 first. Whichever way we do it, the answer is the same. We see that

$$(4 + 3) + 8 = 4 + (3 + 8)$$

The same thing is true whenever we add any three numbers; it does not matter whether we add the first two first or add the last two first. This fact is known as the *associative property for addition*. If we call the three numbers m, n, and p, the associative property for addition may be written $(m + n) + p = m + (n + p)$.

We know also that when we multiply three numbers, we will get the same answer whether we multiply the first two or the last two first. For example, $(4 \times 3) \times 8 = 12 \times 8 = 96$, and $4 \times (3 \times 8) = 4 \times 24 = 96$. This fact is known as the *associative property for multiplication*.

The associative property for addition tells us that when we add three numbers, we do not have to put in any parentheses, because we will get the same answer no matter which way we do it. The associative property for multiplication tells us the same thing for multiplication.

Let us look now at the problem

$$4 \times 3 + 8$$

If we group the numbers like this,

$$(4 \times 3) + 8$$

we would say that $4 \times 3 = 12$ and then that $12 + 8 = 20$. However, if we group the numbers like this,

$$4 \times (3 + 8)$$

we would say that $3 + 8 = 11$ and then that $4 \times 11 = 44$, so we would get a different answer. We see that it does make a difference where we put the parentheses here. The associative law does not work here, since the problem involves both addition and multiplication. The associative law holds only when we have a problem that involves only addition or one that involves only multiplication.

We see from this problem that when we have a problem with more than two numbers that does not involve addition alone or multiplication alone, it is necessary to insert parentheses to indicate which numbers should be combined first. For example, the problem

$$18 - 6 \times 2$$

will give us 24 if the parentheses are inserted like this,

$$(18 - 6) \times 2$$

and it will give us 6 if the parentheses are inserted like this,

$$18 - (6 \times 2)$$

We will agree that when we have a number multiplied by something inside parentheses such as

$$4 \times (3 + 8)$$

we will not bother writing the $\times$ sign; we will just write

$$4(3 + 8)$$

since this will not cause any confusion.

/examples/

1 $4(8 - 5) = 4(3) = 12$

2 $12 - (5 \times 2) = 12 - (10) = 2$

3 $15 - (7 - 3) = 15 - (4) = 11$

4 $(15 - 7) - 3 = (8) - 3 = 5$

Notice from Examples 3 and 4 that the associative law does not hold for subtraction.

/examples/

5 $7(3 + 2) = 7(5) = 35$

6 $(7 \times 3) + (7 \times 2) = (21) + (14) = 35$

Notice from Examples 5 and 6 that when we multiply 7 by $3 + 2$, we get the same answer as when we multiply 7 by 3, then multiply 7 by 2, and add the results. Let us try this with different numbers.

/examples/

7 $4(5 + 1) = 4(6) = 24$

8 $(4 \times 5) + (4 \times 1) = (20) + (4) = 24$

Again in Examples 7 and 8, when we multiply 4 by $5 + 1$, we get the same answer as when we multiply 4 by 5, then multiply 4 by 1, and add the results.

This property is true for any numbers: When we multiply a number by a sum of two numbers, we get the same answer as when we multiply it by the first one, then multiply it by the second one, and add the results. This fact is known as the *distributive property*.

> If we call the three numbers m, n, and p, then the distributive property may be written $m(n + p) = mn + mp$.

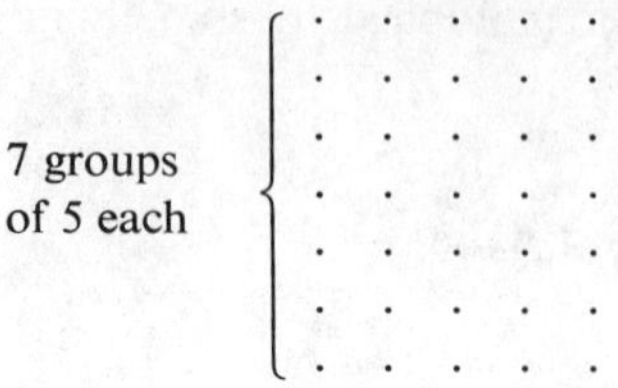

7 groups
of 5 each

figure 1.7

If we think of multiplication as repeated addition, then the distributive property tells us that 7 groups of 5 each is the same as 7 groups of 3 each plus 7 groups of 2 each. That this is true may be seen by looking at Figures 1.7 and 1.8.

We will conclude this section by summarizing the properties of the number system that we have just discussed:

1. Commutative property for addition:

$$m + n = n + m$$

2. Commutative property for multiplication:

$$mn = nm$$

3. Associative property for addition:

$$(m + n) + p = m + (n + p)$$

4. Associative property for multiplication:

$$(mn)p = m(np)$$

figure 1.8

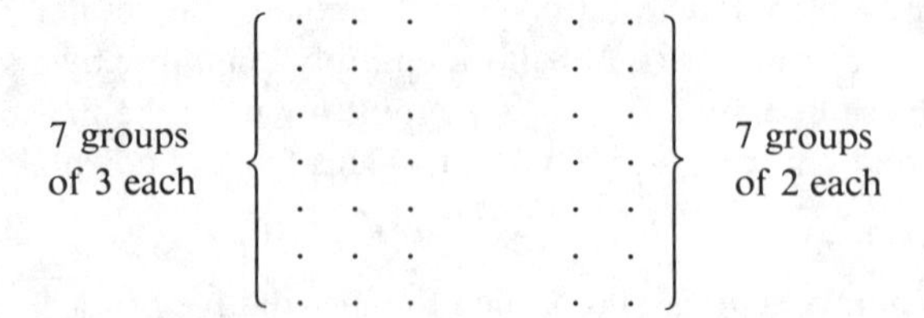

5. Distributive property:

$$m(n + p) = mn + mp$$

exercises 1.4

1 Find the answers to the following problems:

(a) $(7 + 4) - 3$ (b) $7 + (4 - 3)$
(c) $8(12 - 5)$ (d) $9 - (3 \times 2)$
(e) $(9 + 7) \div 2$ (f) $(21 \div 3) + 5$
(g) $(9 - 2) \times 6$ (h) $6(4 + 8)$
(i) $56 \div (5 + 2)$ (j) $9 + (15 - 8)$

2 Find the answers to the following problems:

(a) $(8 - 5) \times (6 + 2)$ (b) $(9 + 7) \div (10 - 2)$
(c) $(5 \times 3) - (4 \times 2)$ (d) $(4 \times 2) + (1 \times 5)$
(e) $8 + (5 \times 2) - 3$ (f) $(7 \times 6) \div (5 - 3)$
(g) $(12 - 4) - 3$ (h) $12 - (4 - 3)$

3 State which property of our numbers each of the following illustrates:

(a) $(5 \times 3) \times 2 = 5 \times (3 \times 2)$ (b) $6 \times 4 = 4 \times 6$
(c) $9(4 + 2) = (9 \times 4) + (9 \times 2)$ (d) $8 + 7 = 7 + 8$
(e) $(3 + 2) + 7 = 3 + (2 + 7)$ (f) $(4 + 6) \times 2 = (4 \times 2) + (6 \times 2)$
(g) $3 + 9 = 9 + 3$ (h) $(2 \times 5) \times 3 = 2 \times (5 \times 3)$

4 Is subtraction commutative?

5 Is subtraction associative?

6 Is division commutative?

7 Is division associative?

1.5 / decimal number system

Suppose that you were asked to take a fast look at Figure 1.9 and tell how many dots it contains. Of course you can see that there are 4 dots.

figure 1.9

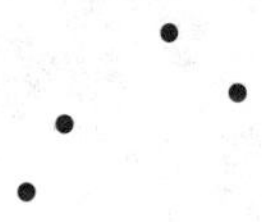

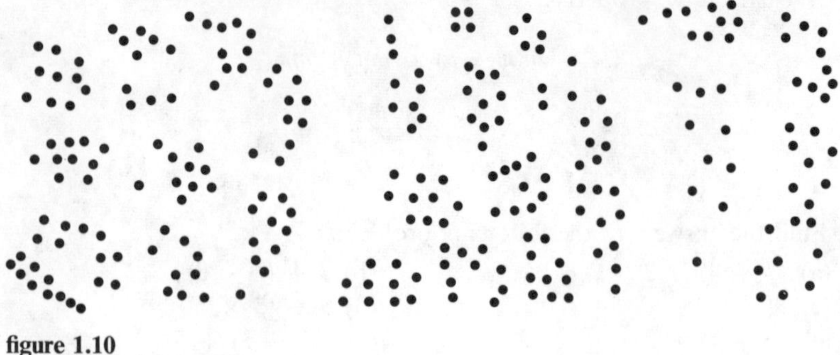

figure 1.10

Now take a fast look at Figure 1.10 and see if you can tell how many dots are in that figure.

This is much harder now because there are so many dots. To help us find out how many dots there are let us enclose each group of 10 dots with a thin line, as in Figure 1.11. Now, can you tell how many groups of 10 there are and how many are left over? This is a little easier than before, but there are still too many groups of 10, so let us enclose every 10 groups of 10 with a thicker line, as in Figure 1.12. Now we can see that we have 2 groups of ten 10's, 4 groups of 10, and 7 dots left over. Since 10 groups of 10 makes a hundred, we can now say that we have 2 hundreds, 4 tens, and 7 dots, or two hundred and forty-seven dots.

Because our number system groups things by 10's, and then groups ten 10's to make a hundred, ten 100's to make a thousand, ten 1000's to make ten thousand, etc., it is called a *decimal number system*. When we want to express a number, we group it into ones, tens, hundreds, thousands, etc. We will never have more than nine groups of any type, because whenever we get ten groups, we put these together to form one larger group. For this reason we need only the ten *digits*, 0, 1, 2, 3,

figure 1.11

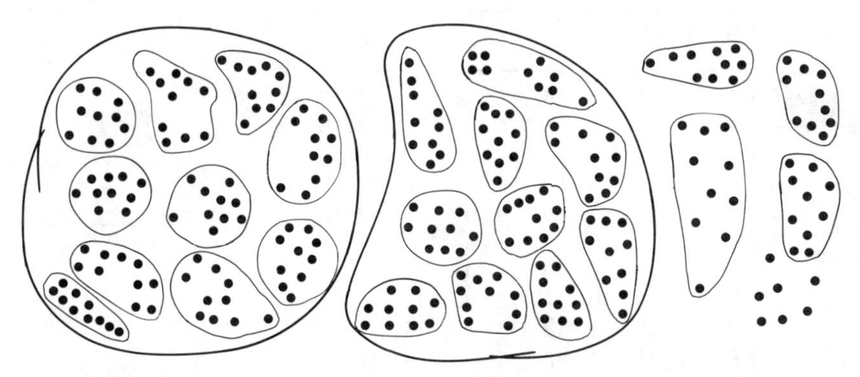

figure 1.12

4, 5, 6, 7, 8, and 9, to express any number. To express the fact that we have two hundred and forty-seven dots in Figure 1.10, we write 247, which means 2 hundreds, 4 tens, and 7 ones. The symbol 247 is the *numeral* that is used to stand for the number two hundred and forty-seven.

Because the value of each of the digits in 247 depends upon the place where it is located (the 2 stands for 2 hundreds, the 4 stands for 4 tens, and the 7 stands for 7 ones), we say that our decimal number system is a *place-value system of numeration.*

When we write a numeral to express a number, each digit of that numeral has a certain value depending upon the place where it is located. The values of these places are shown in Figure 1.13. For numbers over 1000, we usually put commas between every three digits starting from the right in order to make it easier to tell what the places stand for. The numeral

58,946,329,781

for example, is read "fifty-eight billion, nine hundred and forty-six million, three hundred and twenty-nine thousand, seven hundred and eighty-one." The number

figure 1.13

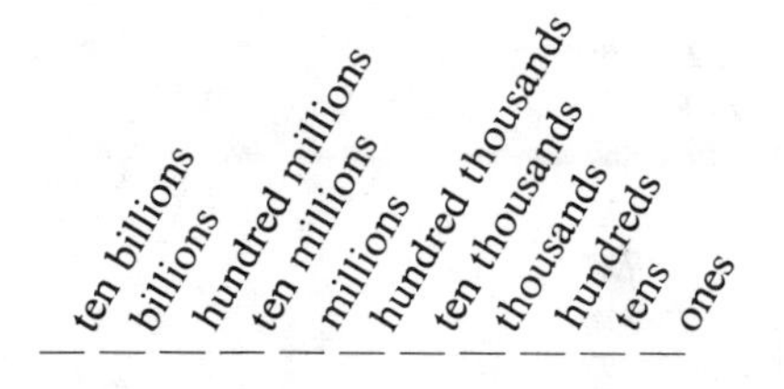

+	0	1	2	3	4	5	6	7	8	9
0	0	1	2	3	4	5	6	7	8	9
1	1	2	3	4	5	6	7	8	9	10
2	2	3	4	5	6	7	8	9	10	11
3	3	4	5	6	7	8	9	10	11	12
4	4	5	6	7	8	9	10	11	12	13
5	5	6	7	8	9	10	11	12	13	14
6	6	7	8	9	10	11	12	13	14	15
7	7	8	9	10	11	12	13	14	15	16
8	8	9	10	11	12	13	14	15	16	17
9	9	10	11	12	13	14	15	16	17	18

figure 1.14

three hundred and twenty-six million, seven hundred and twelve thousand, five hundred and three would be expressed by the numeral

326,712,503

The basic rule in doing addition, multiplication, subtraction, or division of large numbers is that we must keep track of the different place values; we must keep track of the ones, tens, hundreds, thousands, etc.

Because of the efficiency of our decimal system in representing numbers, it is not necessary to memorize the sums and products of all possible pairs of numbers; it is only necessary to memorize those sums and products involving the basic digits 0, 1, 2, 3, 4, 5, 6, 7, 8, and 9. These sums and products may be given compactly by means of an addition and multiplication table, as shown in Figures 1.14 and 1.15.

figure 1.15

×	0	1	2	3	4	5	6	7	8	9
0	0	0	0	0	0	0	0	0	0	0
1	0	1	2	3	4	5	6	7	8	9
2	0	2	4	6	8	10	12	14	16	18
3	0	3	6	9	12	15	18	21	24	27
4	0	4	8	12	16	20	24	28	32	36
5	0	5	10	15	20	25	30	35	40	45
6	0	6	12	18	24	30	36	42	48	54
7	0	7	14	21	28	35	42	49	56	63
8	0	8	16	24	32	40	48	56	64	72
9	0	9	18	27	36	45	54	63	72	81

To illustrate this, consider the following problem in addition:

ten thousands	thousands	hundreds	tens	ones
	9	5	3	7
+	4	7	2	8

If we keep track of the ones, tens, etc., and add each column separately, we get

ten thousands	thousands	hundreds	tens	ones
	9	5	3	7
+	4	7	2	8
	13	12	5	15

which means that we have 13 thousands, 12 hundreds, 5 tens, and 15 ones. Remember, however, that 15 ones means 1 ten and 5 ones, 12 hundreds means 1 thousand and 2 hundreds, and 13 thousands means 1 ten thousand and 3 thousands. Thus, if we regroup the values in our answer, it becomes

ten thousands	thousands	hundreds	tens	ones
	9	5	3	7
+	4	7	2	8
1	4	2	6	5

which is read fourteen thousand, two hundred and sixty-five. This *regrouping*, or *carrying* as it is sometimes called, can be done as we are adding up the columns:

ten thousands	thousands	hundreds	tens	ones
1	1		1	
	9	5	3	7
+	4	7	2	8
1	4	2	6	5

The 7 and 8 in the singles (or ones) place are added to give 15. Since 15 represents 1 ten and 5 singles, we enter the 5 in the singles column of the answer and *carry* the 1 ten to the tens column. This yields 1 + 3 + 2 or 6 tens, so we enter 6 in the tens column. In the hundreds column we have 5 + 7 or 12 hundreds. Since 12 hundreds means 1 thousand and 2 hundreds, we enter 2 in the hundreds column of the answer and carry the 1 thousand to the thousands column. In the thousands column we have 1 + 9 + 4 or 14 thousands, which represents 1 ten thousand and 4 thousands, so we enter 4 in the thousands column and 1 in the ten-thousands column of the answer. The numbers that are carried may be put at the top of the next column so that we do not forget them.

Here are a few addition problems for you to try:

```
                        7492
                 187     873
 5872    1974    296   54216
+2916    +352   +448   +1158
```

In the process of multiplication illustrated below, we must also keep track of the place values of the digits:

hundred thousands	ten thousands	thousands	hundreds	tens	ones
			2	8	4
		×	7	3	5
		1	4	2	0
		8	5	2	
1	9	8	8		
2	0	8	7	4	0

We first multiply 284 by the 5 in the ones column. Four times 5 is 20, so we enter 0 in the ones column and carry 2 to the tens column. The 8 tens multiplied by 5 gives 40 tens, plus the 2 that we carried yields 42 tens, so we enter 2 in the tens column and carry 4 to the hundreds column. The 2 hundreds multiplied by 5 yields 10 hundreds, plus the 4 that we carried yields 14 hundreds, so we enter 4 in the hundreds column and 1 in the thousands column.

Next we multiply 284 by the 3 in the tens column. Four times 3 tens is 12 tens, so in the second line of the answer we move one place to the left and enter 2 in the tens column and carry 1 to the hundreds column. Then 8 tens times 3 tens gives 24 hundreds, plus the 1 hundred that we carried gives 25 hundreds, so we enter 5 in the hundreds column and carry 2 to the thousands column. Then 2 hundreds times 3 tens gives 6 thousands, plus the 2 thousands that we carried gives 8 thousands, so we enter 8 in the thousands column.

Next we multiply 284 by 7 hundreds. Four times 7 hundreds is 28 hundreds, so we move over two places and enter 8 in the hundreds column and carry 2 to the thousands column. Then 8 tens times 7 hundreds is 56 thousands, plus the 2 thousands that we carried gives 58 thousands, so we enter 8 in the thousands column and carry 5 to the ten-thousands column. Then 2 hundreds times 7 hundreds gives 14 ten thousands, plus the 5 ten thousands that we carried gives 19 ten thousands, so we enter 9 in the ten-thousands column and carry 1 to the hundred-thousands column.

We now have to add up the individual columns to complete the problem. In the ones column, we have 0 and in the tens column we have 2 + 2 or 4. In the hundreds column we have 4 + 5 + 8 or 17, so we put 7 in the hundreds column and carry 1 to the thousands column. In the thousands column we have 1 + 8 + 8 or 17 plus the 1 that we carried, or 18, so we put 8 in the thousands column and carry 1 to the ten-thousands column. In the ten-thousands column we have 9 plus the one that we carried, or 10, so we put 0 in the ten-thousands column and carry 1 to the hundred-thousands column. In the hundred-thousands column we have 1 plus the 1 that we carried, so we put 2 in the hundred-thousands column, Our answer is thus 208,740, which we read as two hundred and eight thousand, seven hundred and forty.

Try the following multiplication problems for practice:

$$\begin{array}{r} 987 \\ \times 243 \\ \hline \end{array} \qquad \begin{array}{r} 1758 \\ \times 569 \\ \hline \end{array} \qquad \begin{array}{r} 18427 \\ \times 569 \\ \hline \end{array} \qquad \begin{array}{r} 4938 \\ \times 7406 \\ \hline \end{array} \qquad \begin{array}{r} 997 \\ \times 186 \\ \hline \end{array}$$

In doing subtraction problems we must again be careful to keep track of the different place values. The only difficulty that might arise is when we have to subtract a larger digit from a smaller digit in one of the places, as in the following problem:

	ten thousands	thousands	hundreds	tens	ones
	3	5	6	4	8
−	1	8	2	7	3

In the ones place 3 from 8 leaves 5, but in the tens place we cannot take 7 from 4, so we regroup the 6 hundreds into 5 hundreds and 10 tens, which, along with the 4 tens we already have, gives 14 tens. We can now take 7 tens from 14 tens and get 7 tens, which we put in the tens place:

	ten thousands	thousands	hundreds	tens	ones
	$^{2}\not{3}$	$^{1}5$	$^{5}\not{6}$	$^{1}4$	8
−	1	8	2	7	3
	1	7	3	7	5

This process is sometimes called *borrowing*.

We then continue in the hundreds place and take 2 from 5 to leave 3, which we write in the hundreds place of the answer. In the thousands place we cannot take 8 from 5, so we borrow 1 ten thousand from the 3 ten thousands to give us 15 thousands and leave 2 ten thousands instead of 3. We now take 8 thousands from 15 thousands and write 7 in the thousands place, and then take 1 from 2 in the ten-thousands place and write 1 in that place in the answer. The final answer is then 17,375, which is read as seventeen thousand, three hundred and seventy-five.

Let us look at another example:

	thousands	hundreds	tens	ones
	5	6	0	4
−	2	8	3	7

We cannot take 7 from 4, so we must borrow. However, we have no tens to borrow from; so we move over to the 6 hundreds and borrow 1 hundred, which we regroup as 10 tens. We then borrow 1 of these tens, which we add to the 4 ones to give us 14 ones. This means that we now have 5 hundreds, 9 tens, and 14 ones:

	thousands	hundreds	tens	ones
	4 ~~5~~	15 ~~6~~	9 ~~10~~	14
−	2	8	3	7
	2	7	6	7

We can now subtract 7 from 14 and write 7 in the ones column, and we can subtract 3 from 9 and write 6 in the tens column. In the hundreds column we must borrow 1 of the 5 thousands to give 15 hundreds so that we can subtract 8 from 15 and write 7 in the hundreds column, and then subtract 2 from the 4 that we have left in the thousands column and write 2 there. The answer is thus 2,767, which we may read two thousand, seven hundred and sixty-seven.

Now here are some subtraction problems for you to try:

$$\begin{array}{r} 8173 \\ -5821 \\ \hline \end{array} \qquad \begin{array}{r} 7259 \\ -4873 \\ \hline \end{array} \qquad \begin{array}{r} 15962 \\ -8147 \\ \hline \end{array} \qquad \begin{array}{r} 4062 \\ -1591 \\ \hline \end{array} \qquad \begin{array}{r} 50029 \\ -2685 \\ \hline \end{array}$$

You can check your answer by adding it to the number that was subtracted. If you get the number on top for an answer, you worked the problem correctly.

Next we look at the process of long division. Consider the problem 86,924 ÷ 279, which may also be written in the form

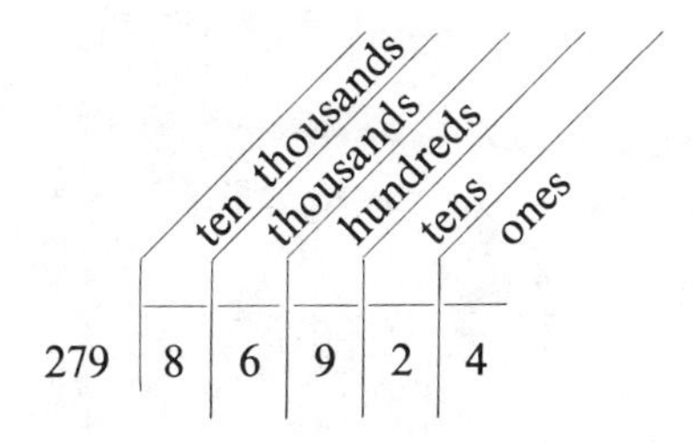

The number 86924 is called the *dividend*, and the number 279 is called the *divisor*. We take the divisor 279 and see how many digits of the dividend we must take, starting from the left, until we get something that 279 will go into. It will not go into 8, it will not go into 86, but it will go into 869 about 3 times. Since the 869 actually means 869 hundreds, 279 will go into 869 hundreds about 300 times, so we put a 3 in the hundreds place of the quotient. We then multiply the 3 (which stands for 300) by 279 and get 837 hundreds, which we put right under the 869 hundreds:

	ten thousands	thousands	hundreds	tens	ones
			3		
279	8	6	9	2	4
	8	3	7		
		3	2		

We now subtract 837 hundreds from 869 hundreds and get 32 hundreds. If we had gotten more than 279 when we subtracted, this would mean that the 3 that we put in the hundreds place was too small and we would replace it by 4 and multiply 4 by 279 and subtract again. If when we multiplied the 3 by 279, we had gotten a number larger than 869, this would mean that the 3 was too large, and we would replace it by 2 and try again. We now have 32 hundreds (which is the same as 320 tens), and we bring down the 2 tens, so that we have 322 tens:

	ten thousands	thousands	hundreds	tens	ones
			3	1	
279	8	6	9	2	4
	8	3	7		
		3	2	2	

We then say that 279 into 322 goes once, so we put a 1 in the tens place of the quotient, since the 322 stands for 322 tens. Then 1 ten times 279 gives 279 tens, which we write directly under the 322 tens, and then we subtract:

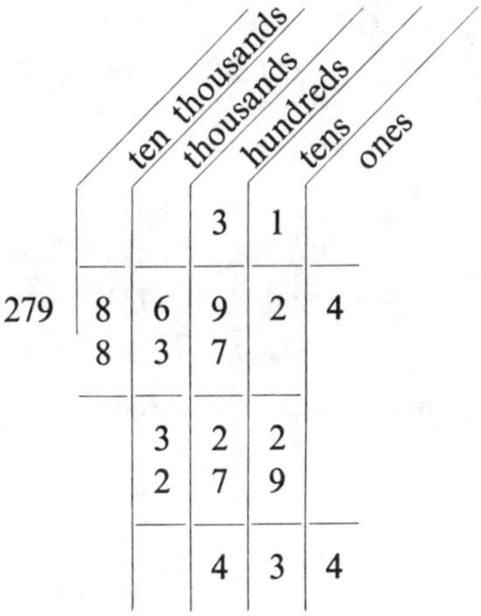

	ten thousands	thousands	hundreds	tens	ones
			3	1	
279	8	6	9	2	4
	8	3	7		
		3	2	2	
		2	7	9	
			4	3	4

This leaves 43 tens. We then bring down the 4 ones, which gives us 434 ones, and since 279 goes into 434 once, we put a 1 in the ones place in the quotient. One times 279 is 279, so we put 279 under the 434 and subtract to get a remainder of 155:

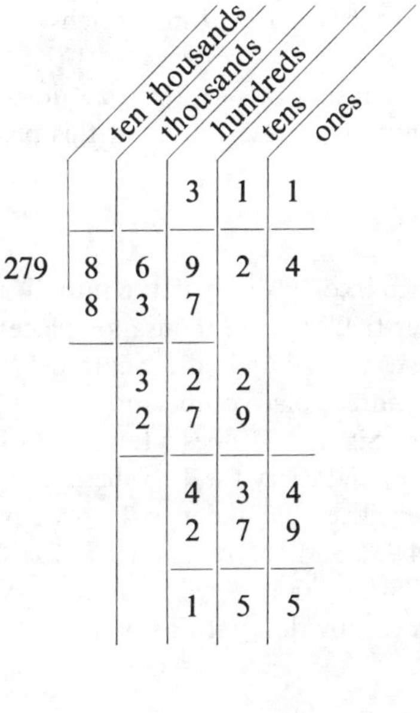

	ten thousands	thousands	hundreds	tens	ones
			3	1	1
279	8	6	9	2	4
	8	3	7		
		3	2	2	
		2	7	9	
			4	3	4
			2	7	9
			1	5	5

We are now done, since 279 will not go into 155, so we say that 86,924 ÷ 279 gives a quotient of 311 and a remainder of 155.

Here is another long-division problem worked out; see if you can follow all the steps:

```
        319
859 | 274618
      2577
       1691
        859
        8328
        7731
         597
```

The quotient is 319, and the remainder is 597.

Let us try one more:

```
        904
683 | 617453
      6147
        275
        000
        2753
        2732
          21
```

The quotient is 904, and the remainder is 21. Notice that in the tens place 683 does not go into 275, so we put a 0 in the tens place in the quotient and then continue as before.

When the divisor consists of one digit, we do not have to use long division; we can use a much shorter process. Look at this problem:

$$\begin{array}{r} 4\ 9\ 5\ 2 \\ 6\overline{)2\ 9^5 7^3 1^1 4} \end{array}$$

Since 6 does not go into 2, we say that 6 into 29 goes 4 times with a remainder of 5. We put a 4 over the 9 (in the thousands place) and carry the 5 to the hundreds place, which gives us 57 hundreds. Six into 57 goes 9 times with 3 left over, so we put a 9 in the hundreds place of the quotient and carry the 3 to the tens place, which gives us 31 tens. Six into 31 goes 5 times with 1 left over, so we put a 5 over the 1 (in the tens place) and carry the 1 to the ones place, which gives us 14 ones. Six goes into 14 twice with a remainder of 2, so we put a 2 over the 4, and we say that the quotient is 4,952 and the remainder is 2. This problem could also have been done by long division, and we would have gotten the same answer; we save a lot of writing, however, by doing it this way.

Now here are some division problems for you to try

$58\overline{)27496}$ $7\overline{)41273}$ $582\overline{)749107}$ $608\overline{)483712}$ $4872\overline{)630729}$

You can check each of your answers by multiplying the quotient by the divisor and then adding the remainder. If you get the dividend for an answer, then you worked the problem correctly.

exercises 1.5

1 Write the words for the following numbers:

(a) 968 (b) 2,573 (c) 816,436
(d) 29,562,185 (e) 40,040 (f) 1,001,001

2 Write the numerals for the following:

(a) Sixty-four thousand, three hundred and four
(b) Eight million, two hundred and forty-six thousand, three hundred and thirty
(c) Twenty-three thousand and twenty-three

3 Add:

(a) $\begin{array}{r} 5564 \\ +3297 \\ \hline \end{array}$ (b) $\begin{array}{r} 743 \\ 916 \\ 827 \\ +357 \\ \hline \end{array}$

4 Subtract:

(a) $\begin{array}{r} 5134 \\ -3816 \\ \hline \end{array}$ (b) $\begin{array}{r} 9237 \\ -6583 \\ \hline \end{array}$

5 Multiply:

(a) $\begin{array}{r} 6412 \\ \times 98 \\ \hline \end{array}$ (b) $\begin{array}{r} 9427 \\ \times 385 \\ \hline \end{array}$

6 Divide:

(a) $5\overline{)84139}$ (b) $628\overline{)291482}$

7 Check your answers to problem 4 by addition.

8 Check your answers to problem 5 by division.

9 Check your answers to problem 6 by multiplying the quotient by the divisor and then adding the remainder.

10 Work the following problems:

(a) $(4132 \times 296) - 437{,}291$
(b) $(512 \times 937) \div 2153$
(c) $(18495 - 9132) \times 58$

(d) (18429 + 1758) ÷ (2572 − 659)
(e) (1592 + 2537) − (4736 ÷ 37)
(f) (452 × 916 × 73) ÷ 229
(g) [(452 × 916) ÷ 229] × 73

11 Fill in the blank spaces:
(a) 14518 ÷ __ = 427
(b) 91642 − __ = 82143
(c) __ − 5429 = 2748
(d) __ × 392 = 4704
(e) __ + 9158 = 27381

12 Complete the following table in *less than 2 minutes*:

+	0	1	2	3	4	5	6	7	8	9
0										
1										
2										
3										
4										
5										
6										
7										
8										
9										

13 Complete the following table in *less than 3 minutes*:

×	0	1	2	3	4	5	6	7	8	9
0										
1										
2										
3										
4										
5										
6										
7										
8										
9										

1.6 / nondecimal number systems[1]

In Section 1.5 we grouped objects into tens, and when we got 10 groups of ten, we made 1 group of a hundred, when we got 10 groups of a hundred, we made 1 group of a thousand, etc. This led to our place-value decimal system, which we say uses *base 10*. There is nothing special about the number 10. We could just as well have used base 5 or base 7 or any other number for a base.

If we were using base 5, we would group things into fives, then group fives into groups of twenty-fives, then group twenty-fives into groups of one hundred and twenty-fives, etc. The place values would then be those shown in Figure 1.16. There would never be more than four groups of any kind, because as soon as we get five groups of any kind, we make one larger group. This means that in base 5, we need only the digits 0, 1, 2, 3, and 4. The numeral 413 in base 5 would mean 4 twenty-fives, 1 five, and 3 ones, so 413 in base 5 stands for 108 in base 10. Of course, you would have to be told that 413 was written in base 5 before you could figure out how much it stands for.

If we use base 7, we group things in sevens, then group sevens into groups of forty-nine, then group forty-nines into groups of three hundred and forty-three, etc. The place values in base 7 are then those shown in Figure 1.17. There would never be more than six groups of any kind, because as soon as we get seven groups of any kind we make one larger group. This means that in base 7, we use only the digits 0, 1, 2, 3, 4, 5, and 6. The numeral 5162 written in base 7 stands for 5 three hundred and forty-threes, 1 forty-nine, 6 sevens, and 2 ones, which adds up to 1808 in base 10.

The same idea carries through for any other base. As a general rule, to figure out what the values of the various places stand for in any base, the first place on the right stands for ones. To see what the next place on the left stands for, multiply by the base that you are using and keep repeating this procedure

figure 1.16

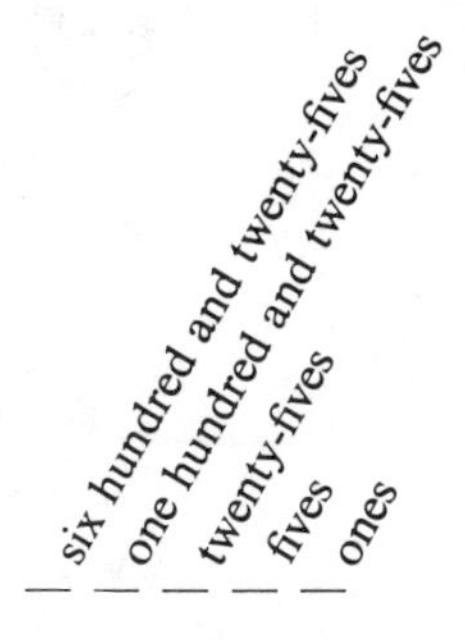

[1]This section may be omitted without disturbing the continuity of the book.

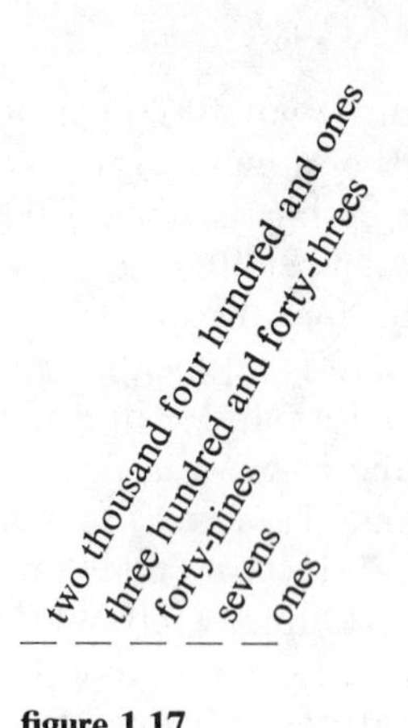

figure 1.17

until you know the values of as many places as you need. Thus in base 3, the first few places stand for those given in Figure 1.18. In base 12, the first few places stand for those shown in Figure 1.19. In base 12 we would need twelve digits, so we would have to make up two new symbols or digits to stand for 10 and 11. We could not use 10 and 11 for these, because in base 12 the numeral 10 stands for 1 twelve and no ones, which is 12, and 11 stands for 1 twelve and 1 one, which is 13. To keep things simple we will just talk about bases less than 10 in this section.

Another problem that we have to consider here is the following: When a number is written in base 10, how would you write it in another base? For example, how would you write the number 973 (written in base 10) in base 6? First, we must see what the first few places stand for in base 6; these are shown in Figure 1.20. We then start filling in the places. We cannot get any twelve hundred and ninety-sixes out of 974, so we move over to the two-hundred-and-sixteens place. We can get 4 two hundred and sixteens out of 974, so we enter 4 in that place:

4

— — — — —

figure 1.18

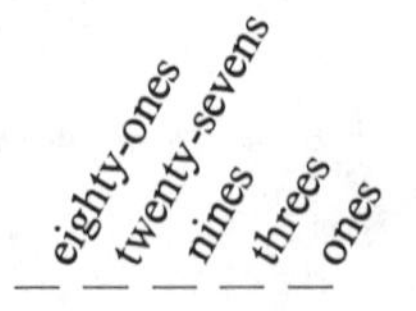

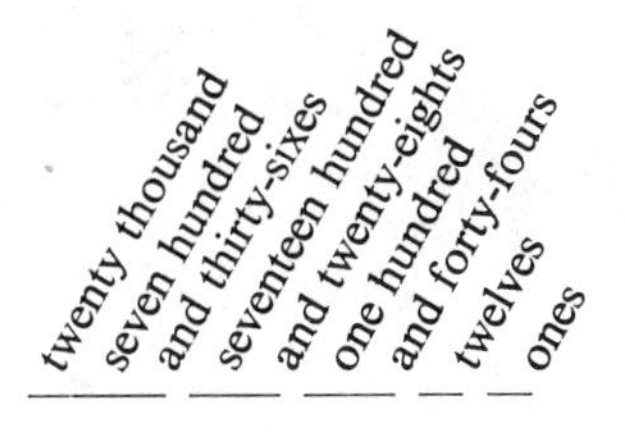

figure 1.19

Four two hundred and sixteens is 864, so we have 974 − 864 or 110 left. Then we move to the thirty-sixes place. We can get 3 thirty-sixes out of 110, so we put a 3 in that place:

4 3
— — — — —

Three thirty-sixes is 108, so we now have 110 − 108 or 2 left. We cannot get any sixes out of 2, so we put a 0 in the sixes place. Two is 2 ones, so we put a 2 in the ones place:

4 3 0 2
— — — — —

This means that the number 974 is written as 4302 in base 6.

Now let us change 394 from base 10 to base 4. The place values are those shown in Figure 1.21. We cannot get any one thousand and twenty-fours out of 394, so we move to the next place. We can get 1 two hundred and fifty-six out of 394, so we put a 1 in that place, and this leaves us 394 − 256 = 138. Next we can get 2 sixty-fours out of 138, so we put a 2 in that place; 2 sixty-fours is 128, and 138 − 128 = 10. There are no sixteens in 10 so we put a 0 there, and we still have 10 left. There are 2 fours in 10 and 2 ones left over, so we put 2 in the tens place and 2 in the ones place. This means that 394 is written as 12022 in base 4.

We can do addition, subtraction, multiplication, and division in bases other than 10 also, but in order to do this we must know how to write small numbers in that base; that is, we must know the addition and multiplication tables in that base.

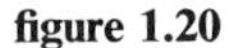

figure 1.20

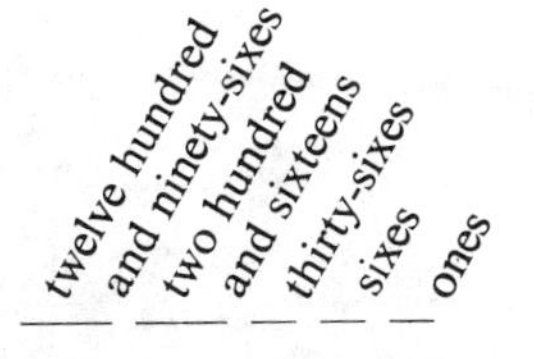

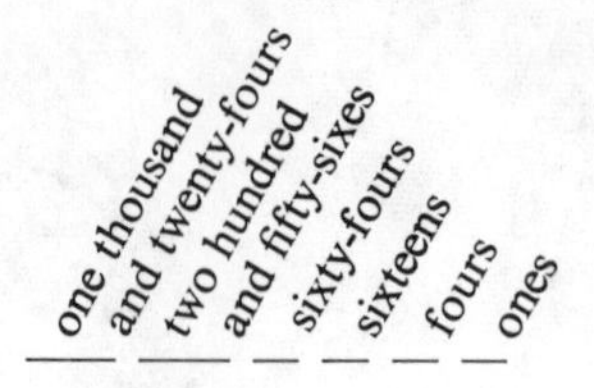

figure 1.21

In base 7, for example, $6 + 5$ is still eleven, but in base 7 we must write eleven as 14, not as 11. In base 8, for example, 6×5 is still thirty, but we must know that 30 is written as 36, not as 30.

Figures 1.22 and 1.23 give addition and multiplication tables for base 4 and base 6. After you have studied these, try to write addition and multiplication tables for bases 2, 3, 5, 7, 8, and 9 to check that you know how to express sums and products in the various bases. With the aid of an addition and multiplication table for a given base it is not too hard to add or multiply any numbers in that base. Subtraction and division are a bit harder, but they still follow the same principles given in Section 1.5.

/ examples /

1 Add in base 7

$$\begin{array}{r} 453 \\ 614 \\ 305 \\ +263 \\ \hline \end{array}$$

figure 1.22

base 4

+	0	1	2	3
0	0	1	2	3
1	1	2	3	10
2	2	3	10	11
3	3	10	11	12

×	0	1	2	3
0	0	0	0	0
1	0	1	2	3
2	0	2	10	12
3	0	3	12	21

base 6

+	0	1	2	3	4	5
0	0	1	2	3	4	5
1	1	2	3	4	5	10
2	2	3	4	5	10	11
3	3	4	5	10	11	12
4	4	5	10	11	12	13
5	5	10	11	12	13	14

×	0	1	2	3	4	5
0	0	0	0	0	0	0
1	0	1	2	3	4	5
2	0	2	4	10	12	14
3	0	3	10	13	20	23
4	0	4	12	20	24	32
5	0	5	14	23	32	41

figure 1.23

► Adding up the first column of digits on the right gives 15. In base 7, 15 is written 21, so we put down the 1 and carry the 2:

$$\begin{array}{r} {\scriptstyle 2\,} \\ 453 \\ 614 \\ 305 \\ +263 \\ \hline 1 \end{array}$$

The sum of the middle column, including the 2 that we carried, is 14, which is written 20, so we put down the 0 and carry the 2:

$$\begin{array}{r} {\scriptstyle 2\,2\,} \\ 453 \\ 614 \\ 305 \\ +263 \\ \hline 01 \end{array}$$

The sum of the last column, including the 2, is 17, which is written 23:

$$\begin{array}{r} {\scriptstyle 2\,2\,} \\ 453 \\ 614 \\ 305 \\ +263 \\ \hline 2301 \end{array}$$

The answer in base 7 is therefore 2301.

To check this, and to get some practice in changing numbers from base 7 to base 10, change all the numbers including the sum to base 10, add in base 10, and check that the sum is correct.

2 Multiply in base 6

$$\begin{array}{r} 524 \\ \times 253 \\ \hline \end{array}$$

► We first multiply by the 3. Four times 3 is 12, which in base 6 is written 20, so we put down the 0 and carry the 2. Two times 3 is 6 plus the 2 that we carried is 8, which is written 12, so we put down the 2 and carry the 1. Five times 3 is 15, plus the 1 that we carried is 16, which is written 24, so we mark down 24:

$$\begin{array}{r} 524 \\ \times 253 \\ \hline 2420 \end{array}$$

Next we multiply by the 5, which stands for 5 sixes, so we move over one place and start in the sixes column. Four times 5 is 20, which is written 32, so we put down the 2 and carry the 3. Two times 5 is 10, plus 3 is 13, which is written 21, so we put down the 1 and carry 2. Five times 5 is 25, plus 2 is 27, which we write as 43. We now have

$$\begin{array}{r} 524 \\ \times 253 \\ \hline 2420 \\ 4312 \end{array}$$

Next, we multiply by the 2, which stands for 2 thirty-sixes, so we move over and start in the thirty-sixes column. Four times 2 is written 12. Put down 2 and carry 1. Two times 2 is 4, plus 1 is 5, which is written the same way in base 6 as in base 10, namely 5. Five times 2 is written 14. We now have

$$\begin{array}{r} 524 \\ \times 253 \\ \hline 2420 \\ 4312 \\ 1452 \\ \hline \end{array}$$

We now have to add up the individual columns, but remember that you must add in base 6. This gives

$$\begin{array}{r} 524 \\ \times 253 \\ \hline 2420 \\ 4312 \\ 1452 \\ \hline 235140 \end{array}$$

Thus the answer, written in base 6, is 235140. You can check this answer also by changing 524, 253, and 235140 from base 6 to base 10 and multiplying in base 10.

3 Subtract in base 8

$$\begin{array}{r} 5537 \\ -2564 \\ \hline \end{array}$$

► In the ones place, 7 − 4 is 3. In the eights place we cannot take 6 from 3, so we take 1 of 5 sixty-fours, which is 8 eights, and move these 8 eights to the eights place. We now have 11 eights, or, in base 8, 13 eights:

$$\begin{array}{rcccc} & 5 & \overset{4}{\not{5}} & {}^{1}3 & 7 \\ - & 2 & 5 & 6 & 4 \\ \hline & & & & 3 \end{array}$$

Six eights from 11 (written 13) eights is 5 eights, so we put down 5 in the eights column:

$$\begin{array}{rcccc} & 5 & \overset{4}{\not{5}} & {}^{1}3 & 7 \\ - & 2 & 5 & 6 & 4 \\ \hline & & & 5 & 3 \end{array}$$

In the sixty-fours place we cannot take 5 from the 4 that we now have left after borrowing 1 sixty-four from the 5 sixty-fours that we had originally. We therefore borrow 1 from the five-hundred-and-twelves column, which is 8 sixty-fours. This along with the 4 sixty-fours is 12 (written 14) sixty-fours. Five sixty-fours from 12 (written 14) sixty-fours is 7 sixty-fours:

$$\begin{array}{rcccc} & \overset{4}{\not{5}} & \overset{14}{\not{5}} & {}^{1}\not{3} & 7 \\ - & 2 & 5 & 6 & 4 \\ \hline & & 7 & 5 & 3 \end{array}$$

Finally, in the five-hundred-and-twelves column, 2 from 4 is 2:

$$\begin{array}{rcccc} & \overset{4}{\not{5}} & \overset{14}{\not{5}} & {}^{1}\not{3} & 7 \\ - & 2 & 5 & 6 & 4 \\ \hline & 2 & 7 & 5 & 3 \end{array}$$

You can check this subtraction by adding 2753 and 2564 in base 8 to see that you get 5537 in base 8.

Base 2, which is referred to as the *binary* system, deserves special mention because it is so useful, especially in computers. In the binary system, the place

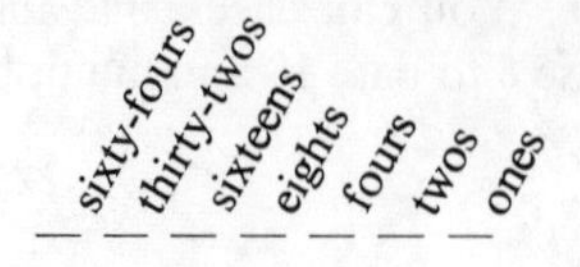

figure 1.24

values are those shown in Figure 1.24. The reason that the binary system is so practical is that it requires only two digits, 0 and 1. As an example, the number 110101 in base 2 stands for 32 + 16 + 4 + 1 or 53 in base 10. If we wish to change the number 107 from base 10 to base 2, we would come up with 1101011.

exercises 1.6

1 (a) Change 5142 from base 6 to base 10.
(b) Change 3725 from base 8 to base 10.
(c) Change 1011101 from base 2 to base 10.
(d) Change 3123 from base 4 to base 10.
(e) Change 10221 from base 3 to base 10.

2 (a) Change 5212 from base 10 to base 7.
(b) Change 927 from base 10 to base 5.
(c) Change 516 from base 10 to base 4.
(d) Change 3212 from base 10 to base 8.
(e) Change 157 from base 10 to base 2.

3 Add in base 6

$$\begin{array}{r} 514 \\ 325 \\ 431 \\ +503 \\ \hline \end{array}$$

4 Add in base 3

$$\begin{array}{r} 212 \\ 102 \\ 211 \\ +112 \\ \hline \end{array}$$

5 Subtract in base 8

$$\begin{array}{r} 15735 \\ -6253 \\ \hline \end{array}$$

6 Subtract in base 5

$$\begin{array}{r} 4133 \\ -1442 \\ \hline \end{array}$$

7 Multiply in base 4

$$\begin{array}{r} 323 \\ \times 213 \\ \hline \end{array}$$

8 Multiply in base 2

$$\begin{array}{r} 10110 \\ \times 1011 \\ \hline \end{array}$$

9 Which is larger, 5172 in base 8 or 24152 in base 6?

10 Write out an addition table and a multiplication table for base 5.

1.7 / prime and composite numbers

In this section we will discuss the prime numbers, which may be thought of as the building blocks of our number system in that every number will be seen to be a product of prime numbers. As we mentioned previously, if m and n are any numbers, and p is a number such that $n \times p = m$, then we say that n *divides* m, or that n is a *divisor* of m, or that n is a *factor* of m. We say also in this case that m is a multiple of n.

A natural number[1] p is called a *prime* number if it is different from 1 and its only divisors are 1 and p. Some examples of prime numbers are 2, 3, 5, 7, 11, 13, and 17, because they have no divisors except 1 and themselves. A natural number m is called a *composite* number if it is different from 1 and is not a prime number. Some examples of composite numbers are 4, 6, 8, 9, 12, 14, and 15.

Notice that the set of natural numbers may be broken down into three disjoint subsets: the prime numbers, the composite numbers, and the number 1.

A composite number, c, is a number distinct from 1 that may be written as a product of two numbers a and b, neither of which is 1. Both a and b must be less than c. If a is composite, it, in turn, may be broken down into two smaller factors, and similarly for b. Proceeding in this way, we see that c may eventually be written as a product of prime numbers. It can be shown that these prime numbers are unique; that is, if two people are asked to factor a number into primes, they will both come out with the same prime factors, possibly in a different order, provided that they do not make any mistakes. This may be summarized by the *fundamental theorem*[2] *of arithmetic*, which says that any natural number other than 1 may be expressed uniquely as a product of prime numbers apart from the order in which the primes are written. If the number is prime to begin with, then we say that it is a product of one prime number—itself.

To illustrate the procedure above, the number 3900 may be factored into primes as follows:

$$3900 = 39 \times 100 = 13 \times 3 \times 10 \times 10 = 13 \times 3 \times 2 \times 5 \times 2 \times 5$$

[1] The natural numbers are the numbers 1, 2, 3, 4, 5, etc.

[2] A theorem is a mathematical statement that can be proved to be true.

Suppose that we are given a natural number m and asked to determine whether or not it is a prime number. We would naturally start with 2 and see whether or not 2 divides m. If not, we would try 3, and so on.

As an example, if we wish to determine whether 143 is a prime number, we would try the primes 2, 3, 5, and 7 and find that they were not divisors of 143 but that 11 was. This means that 143 is a composite number.

To check whether 149 is prime, we would try 2, 3, 5, 7, 11, and 13 and find that none of these were divisors of 149. This is as far as we need to go to conclude that 149 is a prime number, because if 149 had a factor larger than 13, the other factor would have to be less than 13 since $13 \times 13 = 169$.

When we talk about reducing fractions in Chapter Two, it will be helpful to know how to find the largest number that is simultaneously a divisor of two given numbers. This largest divisor is defined as follows:

definition 1.1 If a and b are natural numbers, the *greatest common divisor* of a and b, denoted by GCD (a, b), is the largest number c that divides both a and b.

When we come to the process of adding fractions in Chapter Two, it will be helpful to know how to find the lowest common denominator that is really the least common multiple. The latter is defined as follows:

definition 1.2 If a and b are natural numbers, the *least common multiple* of a and b, denoted LCM (a, b), is the smallest number d that is a multiple of both a and b, that is, such that a and b both divide d.

There is a procedure known as the *Euclidean algorithm* for finding the greatest common divisor of two numbers. In order to describe this process, we must remember that if a and b are two numbers, then when we divide a by b, we get a quotient q and a remainder r that is less than b. If $r = 0$, then b divides a. If you recall the process of long division, this theorem is rather obvious, because if the remainder should come out equal or greater than b, this means that the quotient could be increased by at least 1 and the remainder reduced accordingly.

Returning now to the Euclidean algorithm, to find the greatest common divisor of two numbers, we divide the larger number by the smaller to obtain a quotient and a remainder that is less than the smaller number. Next, divide the smaller number by the remainder to obtain a second quotient and a second remainder that is less than the first remainder. Next, divide the first remainder by the second remainder to obtain a third quotient and a third remainder that is less than the second remainder. Continue in this manner until you get a remainder of 0. The last remainder you divided by is then the greatest common divisor of the two original numbers.

To illustrate this process, we find GCD (51, 189):

	2	2	2	1	3
3	6	15	36	51	189
	6	12	30	36	153
	0	3	6	15	36

We obtained 0 for a remainder when we divided 6 by 3; therefore, 3 is the greatest common divisor of 51 and 189.

It can be shown that when the greatest common divisor of two numbers is multiplied by their least common multiple, the result is the product of the numbers. This may be written

$$\text{GCD}\ (a, b) \times \text{LCM}\ (a, b) = ab$$

Accordingly, we may find the least common multiple of two numbers by dividing their product by their greatest common divisor.

In the previous example we have

$$\text{LCM}\ (51, 189) = \frac{51 \times 189}{3} = \frac{9639}{3} = 3213$$

This means that 3213 is the smallest number that has both 51 and 189 as divisors. For small numbers, such as 8 and 12, you should be able to spot their GCD without going through the Euclidean algorithm.

If a and b are two natural numbers such that GCD $(a, b) = 1$, then a and b are said to be *relatively prime*. According to our rule above, the least common multiple of two numbers that are relatively prime is their product.

We have seen that the prime numbers form the foundation of our number system in that every natural number may be expressed uniquely as a product of primes. The question naturally arises: Do we ever run out of primes? Euclid showed that we do not have to worry about this, because he proved that the number of primes is infinite. This means that no matter how many primes we may find, there are always more.

exercises 1.7

1 For each of the following pairs of numbers, state whether the first one divides the second one:

(5, 24)	(12, 36)	(7, 42)	(11, 91)	(13, 91)
(16, 42)	(14, 56)	(8, 56)	(9, 75)	(3, 81)
(7, 24)	(15, 90)	(10, 90)	(5, 12)	(17, 68)

2 State whether or not each of the following numbers is a prime number:

7, 13, 27, 37, 45, 53, 57, 61, 68, 75, 83, 97, 112, 117, 121, 127, 153, 211, 368, 373, 411, 563, 871

3 Factor each of the following numbers into its prime factors:
58, 364, 96, 3960, 256, 729, 100, 1068, 216, 782

4 For each of the following pairs of numbers, find their GCD and their LCM:

(8, 12)	(9, 15)	(56, 128)	(36, 81)
(34, 51)	(186, 306)	(51, 65)	(16, 162)

5 Which number has more prime factors, 210 or 781?

6 Find the greatest common divisor of the numbers 6, 9, and 15.

7 Find the least common multiple of the numbers 6, 9, and 15.

8 Write a list of all the prime numbers between 1 and 100.

9 Are there more prime numbers between 20 and 30 or between 30 and 40?

10 Are there more prime numbers between 60 and 70 or between 70 and 80?

1.8 / sets and logic[1]

To be a good mathematician, you have to be able to think logically, and the use of sets is very helpful in organizing our thoughts. We often have two sets where one is part of the other. For example, the set A of all men in the United States is part of the set B of all people in the United States. In a case like this, where every member of set A also belongs to set B, we say that *A is contained in B* and write $A \subseteq B$ to indicate this. Figure 1.25 is a diagram depicting the fact that $A \subseteq B$. Diagrams such as this one are called *Venn diagrams.*

Remember that when two sets are disjoint, we say that their intersection is the empty set. We use the symbol $\varnothing$ to stand for the empty set, so we write $A \cap B = \varnothing$ to indicate that A and B are disjoint sets. Figure 1.26 is a Venn diagram showing two disjoint sets, A and B.

We will now state two theorems on sets, which will be obvious from their corresponding Venn diagrams, and then we will show how these theorems may be applied to draw logical conclusions.

figure 1.25

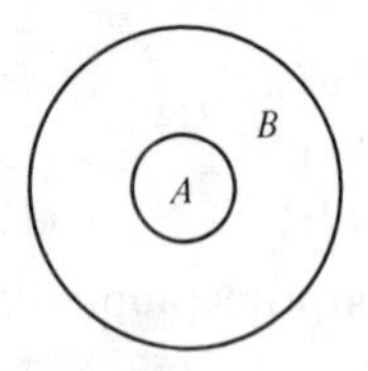

[1]This section may be omitted without disturbing the continuity of the book.

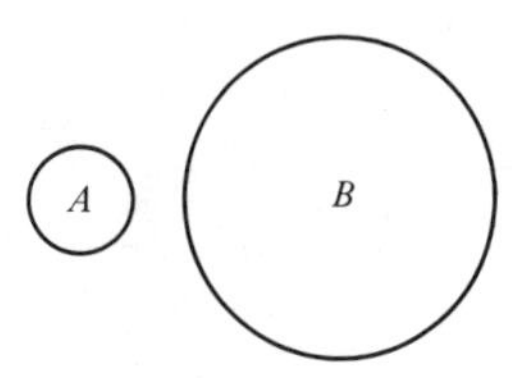

figure 1.26

theorem 1.1 *If $A \subseteq B$ and $B \subseteq C$, then $A \subseteq C$.* See Figure 1.27.

theorem 1.2 *If $A \subseteq B$ and $B \cap C = \varnothing$, then $A \cap C = \varnothing$.* See Figure 1.28.

To see how these two theorems may be applied to simple problems in logic, suppose that you are given the following statements:

All men are mortal.

Socrates is a man.

You could conclude that

Socrates is mortal.

Statement 3 follows immediately from Theorem 1 if we let A be the set consisting of the single element Socrates, B the set of all men, and C the set of all mortal creatures.[1]

Again, suppose that we are told that

All men are mortal.

No mortal creatures are smart.

figure 1.27

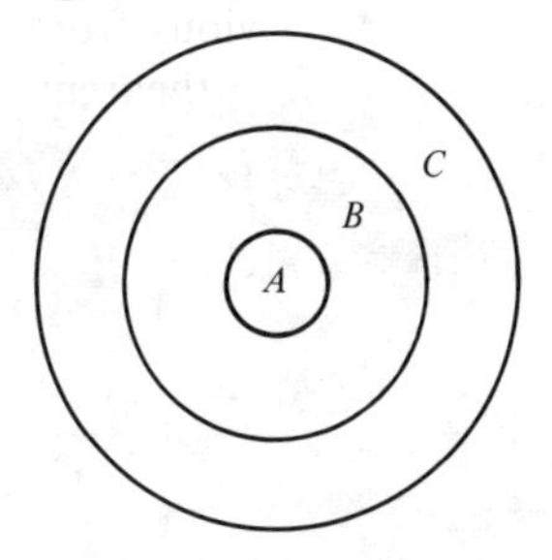

[1]Note that all that we are saying here is that statement 3 is true if statements 1 and 2 are true. We are not worried about whether statements 1 and 2 are actually true. If either one of them is false, then we could not logically conclude that statement 3 is true.

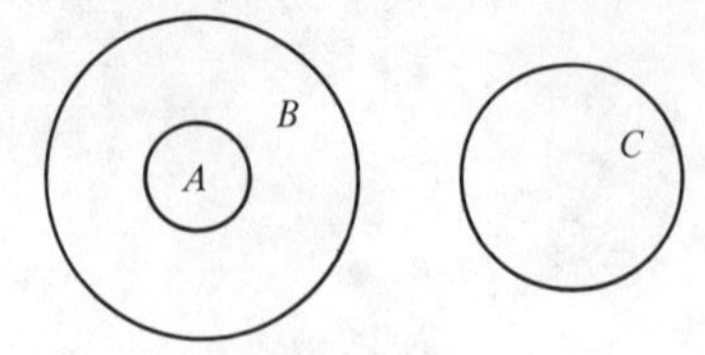

figure 1.28

It follows that

No men are smart.

This may be seen by applying Theorem 2 with A the set of all men, B the set of all mortal creatures, and C the set of all smart creatures.

Figures 1.29 and 1.30 show how these conclusions may be reached by Venn diagrams. Now see if you can draw a logical conclusion from each of the following pairs of statements; draw the Venn diagram that corresponds to each situation:

1. No cats are smart.
   Felix is a cat.
2. John is a student.
   All students are smart.
3. All students are tall.
   All scientists are students.
4. All men are teachers.
   No teachers are rich.
5. No athletes are weak.
   All students are athletes.
6. Bill is a boy.
   No boys are lazy.

The following example requires a different Venn diagram from those corresponding to Theorems 1 and 2:

Some students are smart.

All students are men.

figure 1.29

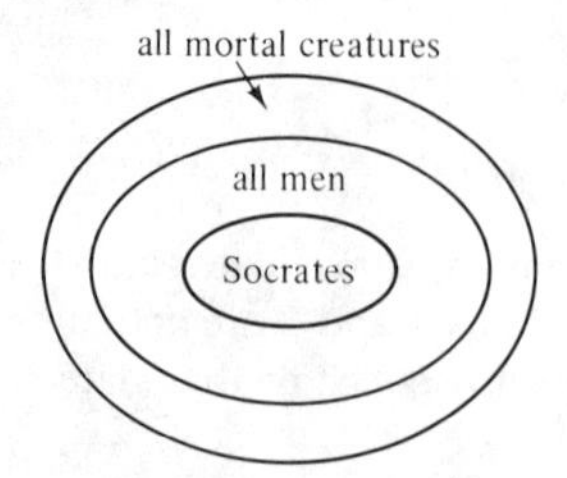

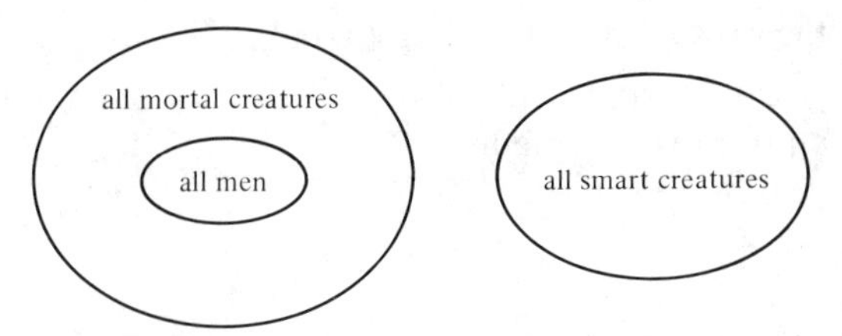

figure 1.30

The Venn diagram for this situation is shown in Figure 1.31. We can see that the conclusion is

Some men are smart.

Now try to draw logical conclusions for the following pairs of statements with the aid of Venn diagrams:

1. All horses are brown.
   Some horses are fast.
2. Some people are smart.
   All people are rich.
3. Some books are good.
   All books are made of paper.
4. All money is valuable.
   Some money is counterfeit.

exercises 1.8

1 Find a subset of each of the following sets:

(a) The students in your room
(b) The numbers from 1 to 10
(c) The fingers on your left hand
(d) The rooms in your house

figure 1.31

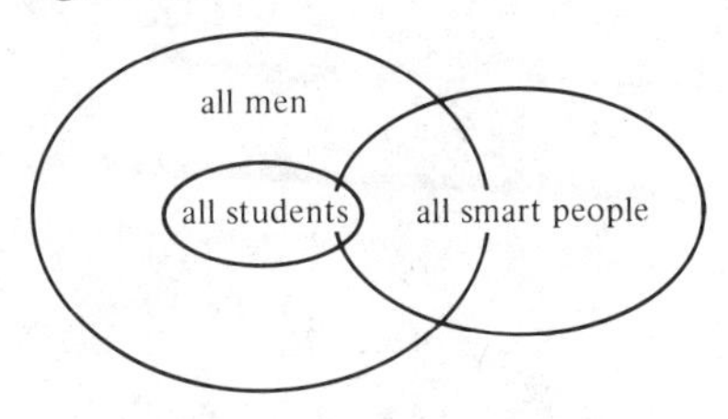

2 If $A = \{1, 2, 3\}$, list all the subsets of A including $\varnothing$ and A.

3 For each of the following pairs of statements, draw a Venn diagram and then derive a logical conclusion from the diagram:

(a) All students are tall.
All girls are students.

(b) No girls are smart.
All teachers are girls.

(c) All athletes are men.
No men are smart.

(d) All players are fast.
Bill is a player.

(e) No men play ball.
All students are men.

4 Draw a logical conclusion from each of the following sets of statements:

(a) Some players are fast.
All players are tall.

(b) All students are smart.
Some girls are students.

(c) All girls are students.
All students are smart.
All smart people read books.

(d) All men eat bread.
No fat people are fast.
Everyone who eats bread is fat.

5 Is it possible to draw a logical conclusion from the following two statements? Draw a Venn diagram to illustrate the situation.

All communists are socialists.
Fred is a socialist.

/review test 1/

NAME ______________________

DATE ______________________

1 Perform the following computations:

(a) $7 + 4$ ________ (b) $9 - 3$ ________ (c) $6 + 9$ ________

(d) 5×8 ________ (e) $12 \div 4$ ________ (f) $15 - 7$ ________

(g) 7×9 ________ (h) $22 - 15$ ________ (i) $28 \div 4$ ________

(j) $17 - 8$

2 Write out the words for the following numbers:

(a) 6,482 ______________________

(b) 914,812 ______________________

(c) 32,459,691 ______________________

(d) 4,008 ______________________

(e) 285,168 ______________________

(f) 506,003 ______________________

(g) 90,023 ______________________

(h) 1,000,101 ______________________

3 Write the numerals for the following numbers:

(a) Eighty-two million, six hundred and forty ______________________

(b) Three thousand, nine hundred and four ______________________

(c) Six hundred and three thousand, and five ______________________

(d) Three thousand and eighty-five ______________________

4 Add:

(a)
```
  582
  793
  514
  693
 +217
 ----
```

(b)
```
 4328
  159
 2645
 5014
 +928
 ----
```

5 Subtract:

(a)
```
  91432
 -36281
 ------
```

(b)
```
  5162
 -2849
 -----
```

6 Multiply:

(a) $\begin{array}{r} 9143 \\ \times 618 \\ \hline \end{array}$

(b) $\begin{array}{r} 14032 \\ \times 893 \\ \hline \end{array}$

7 Divide:

(a) $65\overline{)84639}$

(b) $739\overline{)407392}$

8 Find the results:

(a) $15(8 + 7)$ ________________

(b) $(18 - 7)(6 + 4)$________________

(c) $(15 \times 7)(3 + 13)$____________

(d) $(7 \times 9) - (4 \times 8)$ ____________

(e) $54 \div (12 - 3)$ ______________

(f) $(28 - 7) \div 3$ ________________

(g) $(13 + 12) \div (7 - 2)$ ________

(h) $(12 - 3)(2 + 5)$________________

9 If 84 books are to be divided equally among 7 people, how many will each person get?

10 If 79 books are to be divided equally among 8 people, how many will each person get and how many will be left over?

11 If you have \$452 and you buy a sofa for \$146 and a TV set for \$120, will you have enough money left over to buy a table for \$200?

12 If your car will go 14 miles on 1 gal of gasoline, how many miles can you travel on 12 gal of gasoline?

13 If your car will go 14 miles on 1 gal of gasoline, how much gasoline will you use for a trip of 294 miles?

14 Which property of our number system does each of the following illustrate?

(a) $5 \times 6 = 6 \times 5$ ____________

(b) $9(3 + 8) = (9 \times 3) + (9 \times 8)$ ____________

(c) $(4 + 6) + 2 = 4 + (6 + 2)$ ____________

(d) $(2 \times 3) \times 8 = 2 \times (3 \times 8)$ ____________

(e) $(2 \times 5) + (2 \times 7) = 2(5 + 7)$ ____________

(f) $6 + 5 = 5 + 6$ ____________

15 Factor these numbers into primes:

(a) 8196 ____________

(b) 3648 ____________

(c) 4300 ____________

(d) 720 ____________

(e) 6482 ____________

(f) 5165 ____________

16 Find the greatest common divisor of the following pairs of numbers:

(a) (18, 24)________ (b) (9, 21) ________ (c) (15, 35)________

(d) (51, 26)________ (e) (456, 218) ________ (f) (648, 158)________

17 (a) Change 1242 from base 10 to base 6. ____________

(b) Change 863 from base 10 to base 8. ____________

(c) Change 129 from base 10 to base 2. ____________

(d) Change 563 from base 7 to base 10. ____________

(e) Change 110111 from base 2 to base 10. ____________

(f) Change 3213 from base 4 to base 10. ____________

18 Add in base 5

$$\begin{array}{r} 214 \\ 342 \\ 403 \\ +142 \\ \hline \end{array}$$

19 Multiply in base 6

$$\begin{array}{r} 542 \\ \times 214 \\ \hline \end{array}$$

20 Subtract in base 8

$$\begin{array}{r} 51626 \\ -37143 \\ \hline \end{array}$$

21 Multiply in base 2

$$\begin{array}{r} 101101 \\ \times 11011 \\ \hline \end{array}$$

22 Draw a logical conclusion from each of the following pairs of statements:

(a) All cats are black.
Felix is a cat.

(b) No cats are fast.
Felix is a cat.

(c) All men are smart.
All students are men.

(d) All students are men.
No men are short.

(e) Some men are stupid.
All men are athletes.

(f) All rabbits are fast.
No pretty animals are fast.

23 State the number of elements in each of the following sets:

(a) The set of fingers on your right hand ____________

(b) The set of students in your classroom ____________

(c) The set of chairs in your classroom ____________

(d) The set of elephants in your classroom ____________

(e) The set of teachers in your classroom ____________

(f) The set of legs on your chair ____________

(g) The set of United States senators ____________

24 If $A = \{1, 2, 3, 4, 5, 6, 7, 8\}$, $B = \{1, 3, 5, 7\}$, $C = \{2, 4, 6, 8\}$, and $D = \{2, 3, 4, 5, 9\}$, describe the following sets:

(a) $A \cup D$ ____________ (b) $A \cap D$ ____________

(c) $A \cap C$ ____________ (d) $B \cup C$ ____________

(e) $B \cap D$ ____________ (f) $B \cap C$ ____________

25 (a) Name a set that has 8 elements. ____________

(b) Name a set that has 20 elements. ____________

(c) Name a set that has 144 elements. ____________

pretest 2

NAME ______________________

DATE ______________________

A Perform the following computations:

(1) $5 - (-3)$ ______________ (2) $-8 + 7$ ______________

(3) $-8 - 7$ ______________ (4) $(-4) \times (-2)$ ______________

(5) $(-18) \div 3$ ______________ (6) $(-18) \div (-3)$ ______________

(7) $(-5) \times (-4)$ ______________ (8) $(-6) \times 4$ ______________

B Reduce to lowest terms:

(1) $\frac{24}{36}$ ______________ (2) $\frac{14}{35}$ ______________ (3) $\frac{54}{36}$ ______________

C Compute:

(1) $5\frac{2}{5} + 1\frac{2}{3}$ ______________ (2) $6\frac{3}{8} - 2\frac{5}{6}$ ______________

(3) $2\frac{1}{4} \times 3\frac{1}{7}$ ______________ (4) $8\frac{2}{5} \div 3\frac{2}{9}$ ______________

(5) $4\frac{2}{3}(1\frac{1}{8} + 3\frac{2}{3})$ ______________ (6) $5\frac{1}{2} - (4\frac{2}{3} \div 1\frac{4}{5})$ ______________

(7) $\frac{8}{9} \times \frac{15}{6} \times \frac{15}{21}$ ______________ (8) $(1\frac{1}{2} \times 1\frac{1}{3}) \div 1\frac{1}{4}$ ______________

D Are you more in debt if you receive 4 bills for $6 each or 5 bills for $5 each?

E (1) Change to decimals:

(a) $4\frac{3}{5}$ ______________ (b) $7\frac{1}{12}$ ______________

(2) Change to fractions:

(a) 9.64 ______________ (b) 8.141414. . . ______________

F Add $1.87 + 29.6 + 3.002$.

G Subtract $19.6 - 4.65$.

H Multiply 11.964×3.07.

I Divide $14.92 \div 9.3$.

J What is the ratio of 25 seconds to 2 hours?

K (1) Change to percent:

(a) 6.42 ____________ (b) $\frac{3}{8}$ ____________

(2) Change to decimals:

(a) 5.72% ____________ (b) 1.5% ____________

L (1) If you can get a 15 percent discount on a TV set costing $180, how much would you have to pay for it?

(2) Twenty-four is what percent of 560?

chapter two

REAL NUMBERS

2.1/integers

We saw in Chapter One that whenever we add two natural numbers, we always get a natural number for an answer. We describe this by saying that the natural numbers are *closed* under addition.

Let us check now whether the natural numbers are closed under subtraction; that is, let us see whether we always get a natural number for an answer when we subtract one natural number from another. We consider first the problem

$8 - 3$. This problem asks what you have left when you take 3 objects away from a set of 8 objects, or, another way of looking at it is to ask: What number must be added to 3 in order to get 8? We can write this problem in the form $\_ + 3 = 8$. No matter which way we look at this problem we see that the answer is always 5, which is a natural number.

Now consider the problem $3 - 8$. This problem asks what you have left when you take 8 objects away from a set of 3 objects, and we know that this cannot be done. If we try to do this problem by asking what natural number must be added to 8 in order to get 3, we know that there is no such number. No matter how we look at it, there is no answer to the problem $3 - 8$. In other words, the natural numbers are not closed under subtraction. When we try to subtract one natural number from another, we cannot always be sure that we will get a natural number for an answer. We know from experience that (1) the problem $m - n$ will have a natural number for an answer when m is greater than n (we will use the notation $m > n$ to indicate that m is greater than n) and that (2) it will not have a natural number for an answer when m is less than n (we will use the notation $m < n$ to indicate that m is less than n). When $m = n$, we know that $m - n = 0$.

We are aware that there are many situations in real life that call for the subtraction of numbers to get the solutions of certain problems, and it would be desirable not to have to worry each time that we have to subtract whether the answer will exist. In other words, we would like to arrange things so that the operation of subtraction will be closed. This may be done by making up some new numbers to serve as answers for problems of the type $m - n$ when $m < n$. We will therefore enlarge our system of natural numbers so that the problem $m - n$ will always have an answer, regardless of which number is larger.

Since $m + 0 = m$, we know that $m - n = 0$ when $m = n$, so we first agree to include zero in our enlarged system of numbers. Next, since $0 < m$ for any natural number m, we know that $0 - m$ is not a natural number, so we will have to make up a new number to serve as the answer for the problem $0 - m$ for each natural number m. Since the addition of zero does not change a number, and it is therefore usually correct to omit a zero when it is added, we will agree to write $-m$ as the answer to the problem $0 - m$ for each natural number m, and we will refer to $-m$ as the *negative* of the natural number m.

Note that since $0 - m = -m$, we have $m + (-m) = (-m) + m = 0$; that is, $-m$ is the number that must be added to m to give a result of 0. This new number $-m$ is also called the *inverse* of the number m under addition. Thus the inverse under addition is the same as the negative of a number. We can also say that m is the inverse of $-m$ under addition, since m is the number that must be added to $-m$ to give an answer of 0.

We see that each natural number, m, has an additive inverse, $-m$ (which is not a natural number, because we cannot take the union of two nonempty sets and get the empty set as an answer). Each of these new numbers $-m$ has the natural number m as its additive inverse, and 0 is its own additive inverse since $0 + 0 = 0$; every number in our enlarged system of numbers, therefore, has an additive inverse. No number can have more than one inverse, because if $m + (-m) = 0$ and $m + n = 0$, then we have $m + (-m) = m + n$,[1] and if we subtract m from both sides of this equation, we get $-m = n$.[2]

We may now enlarge the list of properties of our number system to include the following property:

> For each number m, there is a unique number $-m$, called the inverse of m, such that $m + (-m) = (-m) + m = 0$.

It turns out that once we have added 0 and the negative, $-m$, of each natural number m to our system of natural numbers, we do not have to add any more numbers; the enlarged system of numbers will now be closed under subtraction. To show this, we mention first that in enlarging our system of natural numbers, we do not want to spoil any of the properties that hold for the natural numbers. In other words, we want to construct this new system so that the commutative, associative, and distributive properties are still valid. If we agree to this, we can prove Theorem 2.1.

theorem 2.1 *If m and n are natural numbers, then $m - n = m + (-n)$.*

proof Assume that there is an answer to the problem $m - n$ and let p stand for this answer so that $m - n = p$, which means that $m = p + n$. Then $m + (-n) = (p + n) + (-n) = p + n + (-n) = p + 0 = p$. Since $m - n = p$ and $m + (-n) = p$, this proves that $m - n = m + (-n)$.

Theorem 2.1 tells us that if the properties of the natural numbers are to hold in our new system of numbers, then subtraction must be the same as addition of the inverse.

With this fact in mind we can now show that our enlarged system is closed under subtraction. If m and n are natural numbers, we know that $m - n$ is a natural number, when $m > n$, and that $m - n = 0$, when $m = n$. It remains only to show that $m - n$ is in our enlarged system of numbers when $m < n$. If $m < n$, then $n > m$ and $n - m$ is a natural number p. Note that $n - m$ must be the inverse of $m - n$, since $(n - m) + (m - n) = n + (-m) + m + (-n) = n + (-n) + m + (-m) = 0$. This means that $m - n = -(n - m) = -p$, which is already in our enlarged system of numbers because p is a natural number. Thus our new system contains the answers to the problems of the form $m - n$ as well as to all problems of the form $m + n$ when m and n are natural numbers.

This enlarged system of numbers, which consists of the natural numbers, zero, and the newly constructed numbers, $-m$, one for each natural number m, is known

[1]There is a basic rule in mathematics that says that things equal to the same thing are equal to each other.

[2]There is another basic rule in mathematics that says that if the same thing (or equal things) is added to or subtracted from equals, the results are equal.

as the system of *integers*. The following section will show that this system is closed under the operations of addition, subtraction, and multiplication, and it will show how to carry out these operations by giving a set of rules for each operation.

examples

1 What is the additive inverse of the integer 7?

► The number that must be added to 7 in order to give an answer of 0 is -7. Thus -7 is the additive inverse of 7.

2 What is the additive inverse of the integer -4?

► The number that must be added to -4 in order to give an answer of 0 is 4. Thus the additive inverse of -4 is 4.

3 Fill in the blank space:

$$9 - \_\_ = -6 + 8$$

► Since $-6 + 8 = 2$, we are asking: What number subtracted from 9 gives 2 as an answer? We know that $9 - 7 = 2$, so our answer is 7.

exercises 2.1

1 Fill in the blanks:

(a) $\_\_ + 6 = 0$ (b) $8 + \_\_ = 0$ (c) $-2 + \_\_ = 0$
(d) $25 + \_\_ = 0$ (e) $\_\_ + (-14) = 0$ (f) $0 + \_\_ = 0$
(g) $m + \_\_ = 0$ (h) $\_\_ + (-m) = 0$

2 What are the additive inverses of the following integers?

(a) 7 (b) -3 (c) 2
(d) -25 (e) 0 (f) -12

3 Fill in the blanks:

(a) $12 - 8 = 12 + \_\_$ (b) $-3 - 8 = -3 + \_\_$
(c) $7 - \_\_ = 7 + (-2)$ (d) $4 - \_\_ = 4 + (-10)$
(e) $-5 - \_\_ = -5 + (-4)$

4 Fill in the blanks:

(a) $0 - 10 = \_\_$ (b) $\_\_ - 6 = -6$
(c) $\_\_ - 2 = 7 + (-2)$ (d) $9 - \_\_ = 7 - 2$
(e) $-6 - 4 = -6 + \_\_$ (f) $7 - \_\_ = 7$

5 Write the additive inverse of each of the following:

(a) $12 - 5$ (b) $2 + 8$ (c) $7 - 3$
(d) $4 - 9$ (e) $-3 + 5$ (f) $-6 + 4$
(g) $-2 - 9$ (h) $-8 - 5$

2.2/operations on integers

The system of integers constructed in Section 2.1 consists of three mutually disjoint subsets: the natural numbers, which will be referred to hereafter as the *positive integers*; the number 0; and the numbers $-m$ (one for each positive integer, m), which will hereafter be referred to as the *negative integers*. We want to define the operations of addition, subtraction, and multiplication on the integers so that the properties of the natural numbers will be preserved.

We consider first the operation of addition. Since 0 is to be the identity for addition, we know that $m + 0 = 0 + m = m$ for any integer m. When both numbers are positive, we follow the rule for the addition of natural numbers given in Chapter One.

This leaves two cases to be considered, the case where one integer is positive and the other is negative and the case where both integers are negative.

Consider the situation where one number, to be denoted by m, is positive and the other, to be denoted by $-n$, is negative. Since the commutative law is to hold, it does not matter which we write first. If $m > n$, then we already know how to find $m + (-n)$, since it is equal to $m - n$ and we already know how to do this. If $n > m$, then we have seen in Section 2.1 that $m + (-n) = m - n = -(n - m)$, and we know how to find $n - m$ when $n > m$.

When both numbers, which will be denoted by $-m$ and $-n$, are negative, we see that $(-m) + (-n) + m + n = (-m) + m + (-n) + n = 0 + 0 = 0$, so $(-m) + (-n)$ is the additive inverse of $m + n$ and is therefore equal to $-(m + n)$.

Theorem 2.1 tells us how to subtract one integer from another. Since $m - n = m + (-n)$ for any two positive integers m and n, we extend this same rule to all the integers by stipulating that to subtract one integer from another we change the sign[1] of the integer being subtracted and then we add.

The rules[2] for the addition and subtraction of integers may now be summarized.

ADDITION

Case 1: *Both integers have the same sign.*

Rule: Add the integers as you would for natural numbers and use the sign they both have.

[1]By changing the sign of an integer we mean that if it is positive, we change it to negative; if it is negative, we change it to positive; and if it is zero, we leave it unchanged. We usually think of a positive integer as having a + sign in front of it, a negative integer as having a − sign in front of it, and 0 as having no sign.

[2]When we talk about adding or subtracting the integers in these rules, the sign of the integer should be ignored and the integer taken as a positive integer or a natural number. This dropping of the sign is called taking the *absolute value* of the integer. The absolute value of an integer m is always positive and is denoted by $|m|$.

Case 2: *The integers have different signs.*

Rule: Subtract the smaller from the larger as you would for natural numbers and use the sign of the larger one. (If the integers are the same except for their signs, the answer is 0.)

Case 3: *One of the integers is* 0.

Rule: The sum is the other integer.

SUBTRACTION

Rule: Change the sign of the integer being subtracted and then add the resulting integers.

These rules are now illustrated with several examples.

EXAMPLES OF ADDITION

Case 1 (a) $3 + 5 = 8$ (b) $(-2) + (-7) = -9$
(c) $(-6) + (-5) = -11$

Case 2 (a) $5 + (-3) = 5 - 3 = 2$ (b) $(-5) + (3) = -(5 - 3) = -2$
(c) $(-4) + 7 = 7 - 4 = 3$ (d) $4 + (-7) = -(7 - 4) = -3$

Case 3 (a) $0 + 6 = 6$ (b) $(-8) + 0 = -8$
(c) $0 + (-2) = -2$

The process of the addition of integers may be seen more easily by making use of the number line illustrated in Figure 2.1. This number line does not end in either direction; it continues endlessly to the right and to the left. When we want to add two integers, we consider a positive integer as a trip to the right and a negative integer as a trip to the left. The sum of the two integers is the result of starting at 0 and taking the two corresponding trips. For example, to add $(-5) + 3$, we start at 0, go 5 units to the left, and then go 3 units to the right, and we wind up at -2. This tells us that $(-5) + 3 = -2$. To add $(-4) + (-2)$, we start at 0 and go 4 units to the left and then 2 more units to the left, and we wind up at -6. This means that $(-4) + (-2) = -6$. To add $6 + (-3)$, start at 0 and go 6 units to the right and then 3 units to the left and wind up at 3, so that $6 + (-3) = 3$. Try checking the examples of addition given above by using the number line.

Before we look at some examples of subtraction, we should emphasize that the minus sign, $-$, serves two purposes—one to indicate the process of subtraction and the second to indicate a negative number. For example, in the expression

figure 2.1

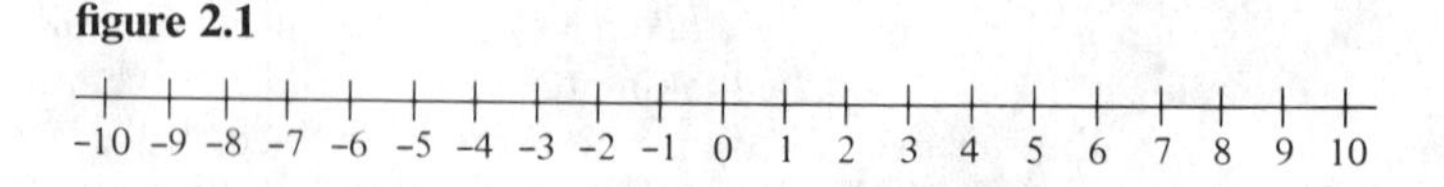

$-4 - 3$, the first minus sign indicates a negative 4 and the second indicates subtraction. In the expression $-4 - (-3)$, the first minus sign indicates a negative 4, the second indicates subtraction, and the third indicates a negative 3. When the reader sees a minus sign in a computational problem, he should be clear whether it indicates a negative number or the operation of subtraction.

EXAMPLES OF SUBTRACTION

(a) $8 - 5 = 8 + (-5) = 3$
(b) $4 - 9 = 4 + (-9) = -(9 - 4) = -5$
(c) $-2 - 6 = -2 + (-6) = -8$
(d) $-2 - (-6) = -2 + 6 = 6 - 2 = 4$
(e) $0 - 7 = 0 + (-7) = -7$
(f) $0 - (-7) = 0 + 7 = 7$
(g) $-4 - 0 = -4 + 0 = -4$
(h) $8 - (-3) = 8 + 3 = 11$

Note in example (d) that subtracting a negative number is equivalent to adding a positive number. This is an example of the well-known saying that two negatives make a positive, which will come up again in our discussion of the multiplication of integers.

We can use the number line to give a very simple set of rules for the addition and subtraction of integers. Any addition or subtraction problem consists of numbers, parentheses, plus signs, and minus signs, some of which indicate subtraction and some of which indicate negative integers. If we agree to ignore parentheses, the following set of rules will give the answer to any problems in addition or subtraction of integers:

1. Start at 0 facing to the right.
2. When you see a plus sign, ignore it.
3. When you see a minus sign, turn around.
4. When you see a number, count off that many spaces and when you finish, face right.
5. The point where you finish indicates the answer.

This procedure will now be applied to the problem $-5 - (-7)$. We start at 0 facing right (Figure 2.2a). We see a minus sign first, so we turn around and face left (Figure 2.2b). Then we see a 5, so we count off 5 spaces and then face right (Figure 2.2c). We then see two minus signs in succession, so we turn around twice, which leaves us facing right again (Figure 2.2c). Then we see a 7, so we count off 7 spaces and end up at 2, which is the answer to the problem (Figure 2.2d).

As an exercise, try to explain why this procedure works. Remember the meaning of a plus sign and the two possible meanings of a minus sign.

In order to derive the rules for multiplication of integers it will be necessary to prove three theorems. We will assume that the properties of the natural numbers

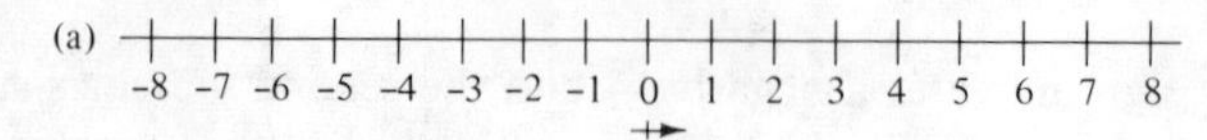

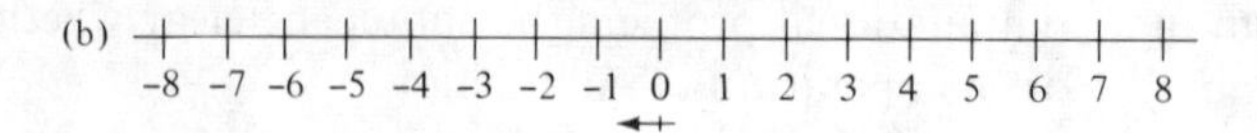

(c)
-8 -7 -6 -5 -4 -3 -2 -1 0 1 2 3 4 5 6 7 8

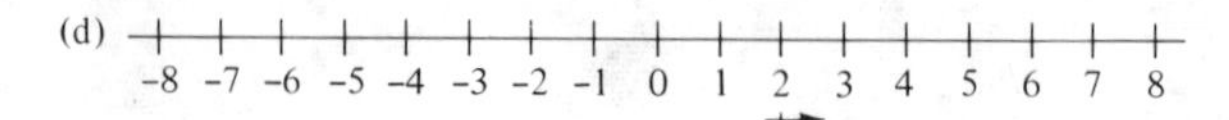

figure 2.2

are to hold for the integers, and the theorems will give us the rules for multiplying integers.

theorem 2.2 *If a is an integer, then* $a \cdot 0 = 0 \cdot a = 0$.[1]

proof $a \cdot 0 = a(0 + 0) = a \cdot 0 + a \cdot 0$ by the distributive property. Since multiplication is closed, $a \cdot 0$ is an integer and must therefore have an inverse $-(a \cdot 0)$. Adding this to both sides of the equation $a \cdot 0 = a \cdot 0 + a \cdot 0$, we get $-(a \cdot 0) + a \cdot 0 = -(a \cdot 0) + a \cdot 0 + a \cdot 0$ or $0 = 0 + a \cdot 0$, which reduces to $a \cdot 0 = 0$. By the commutative property, we have also $0 \cdot a = 0$.

theorem 2.3 *If a and b are integers, then* $a(-b) = (-a)b = -(ab)$.

proof We have $a[b + (-b)] = a \cdot 0 = 0$ and $a[b + (-b)] = ab + a(-b)$, so, since things equal to the same thing are equal to each other, we have $ab + a(-b) = 0$. This says that $a(-b)$ is the inverse of ab or that $a(-b) = -(ab)$. By a similar argument $(-a)b = -(ab)$.

theorem 2.4 *If a and b are integers, then* $(-a)(-b) = ab$.

proof We have $(-a)[b + (-b)] = (-a) \cdot 0 = 0$ and $(-a)[b + (-b)] = (-a)(b) + (-a)(-b)$, so $(-a)(b) + (-a)(-b) = 0$. Since $(-a)(b) = -(ab)$, this becomes $-(ab) + (-a)(-b) = 0$. This says that $(-a)(-b)$ is the inverse of $-(ab)$. Since ab is also the inverse of $-(ab)$ and the inverse of an integer is unique, we have $(-a)(-b) = ab$.

[1] $a \cdot 0$ is another way of writing $a \times 0$.

As a specific example of this last theorem, consider the following problem:

$$(-2)[(7) + (-3)] = -8$$
$$(-2)(7) + (-2)(-3) = -8$$
$$-14 + (-2)(-3) = -8$$

Since $-14 + 6 = -8$, we must have $(-2)(-3) = 6$, if we wish the distributive property to be true.

These three theorems give us the rules[1] for multiplying integers, which may be summarized as follows:

MULTIPLICATION

Case 1: One or both of the integers is zero.

Rule: Their product is zero.

Case 2: The two integers have different signs.

Rule: Multiply the integers as you would for natural numbers and put a minus sign in front of the answer. (In other words, a positive integer times a negative integer yields a negative integer.)

Case 3: Both integers have the same sign.

Rule: Multiply the integers as you would for natural numbers and take the positive integer obtained as the answer. (When the integers are both negative, this is another example of two negatives making a positive.)

EXAMPLES OF MULTIPLICATION

(a) $5 \times 4 = 20$ (b) $(-6) \cdot (3) = -18$ (c) $(-7) \cdot (-5) = 35$
(d) $6 \times (-4) = -24$ (e) $8 \times 0 = 0$ (f) $0 \times 0 = 0$

If we return to our original definition of multiplication of natural numbers in terms of sets, we may obtain a converse to Theorem 2.2. According to that definition, the direct product of two nonempty sets must be nonempty because each set has at least one element in it; therefore, that product set must have at least the pair consisting of these two elements in it. This means that the product of two nonzero natural numbers must be different from zero, and our rules for multiplication imply that this may be extended to the integers by Theorem 2.5.

theorem 2.5 *If a and b are integers and $ab = 0$, then at least one of a or b must be* 0.

This theorem allows us to add one more property to our list of properties of the integers:

[1]Here again, as in the rules for addition, when you multiply the integers, forget about their signs and use their absolute values.

The Cancelation Law: If a and b are integers, c a nonzero integer, and $ac = bc$, then $a = b$.

This follows because if we subtract bc from both sides of the equation $ac = bc$, we get $ac - bc = 0$ or $(a - b)c = 0$. Since $c \neq 0$, $a - b$ must be 0 by Theorem 2.5, and this means that $a = b$.

Going back now to our definition for division, we recall that $12 \div 4$ means: What number must be multiplied by 4 to give 12? We see that the answer is 3. Similarly, $12 \div (-4)$ means: What number must be multiplied by -4 to give 12? We see that the answer is -3. Applying this same definition, we see that $(-12) \div 4 = -3$ and $(-12) \div (-4) = 3$. These examples show us that to get the rules for the division of integers we can go back to cases 2 and 3 of the rules for multiplication and substitute the word "divide" for the word "multiply." Remember that we never divide by 0.

EXAMPLES OF DIVISION

(a) $8 \div 2 = 4$
(b) $(-15) \div 3 = -5$
(c) $(-28) \div (-4) = 7$
(d) $14 \div (-7) = -2$
(e) $0 \div 6 = 0$
(f) $0 \div (-8) = 0$

exercises 2.2

1 Compute the following:

(a) $2 - (-8)$
(b) $-4 - 7$
(c) $-3 - (-4)$
(d) $5 - 8$
(e) $5 + (-4 + 7)$
(f) $(8 - 3) - 10$
(g) $-7 + 6 - 2$
(h) $(3 - 8) + 2$
(i) $3 - (8 + 2)$
(j) $[3 - (-8)] + 2$
(k) $3 - [-8 + 2]$
(l) $(-5 + 4) - (2 - 3)$

2 Compute the following:

(a) $7 \times (-4)$
(b) $(-3) \times (-8)$
(c) $(-4) \times 6$
(d) $(-3)[-4 + (-2)]$
(e) $(-3)(-4 - 2)$
(f) $(-3)(-4 + 4)$
(g) $(-2)[3 + (-5)]$
(h) $[5 \times (-4)] + 7$
(i) $5 \times [-4 + 7]$
(j) $-6[-5 - (-8)]$
(k) $[2 - [(-5) \times (-3)]]$
(l) $[2 - (-5)] \times (-3)$

3 Compute the following:

(a) $(-24) \div 6$
(b) $(-18) \div (-3)$
(c) $32 \div (-4)$
(d) $[(-10) \div 5] + (-7)$
(e) $(-4)[(-15) \div (-3)]$
(f) $(-3) + [(-8) \div (-4)]$

4 If you travel 12 miles east and then 7 miles west, how far are you from where you started and in which direction?

5 If you travel 15 miles north and then 26 miles south, how far are you from where you started and in which direction?

6 If you receive a check for \$36 and a bill for \$53, how far are you ahead or behind?

7 If you receive 8 bills for \$9 each, how far are you ahead or behind?

8 If you are notified that 3 bills for \$6 each have been canceled, how far ahead or behind does this put you?

9 If you receive 4 bills for \$11 each and 9 checks for \$5 each, how far ahead or behind does this put you?

10 If you lose \$8, then find \$13, and then lose \$7 all in one day, how far ahead or behind are you for that day?

11 Prove that $-(a - b) = b - a$ by showing that $(a - b) + (b - a) = 0$.

12 Prove that $a(b - c) = ab - ac$ by writing $a(b - c) = a[b + (-c)]$ and using the distributive law.

13 Prove the following extension of the distributive law: If a, b, c, and d are integers, then $(a + b)(c + d) = ac + ad + bc + bd$. [*Hint:* Consider $(a + b)$ as one integer and write $(a + b)(c + d) = (a + b)c + (a + b)d$.]

2.3 / order properties

When the natural numbers were defined, they were arranged in the order 1, 2, 3, . . . , and we said that m was less than n if it came before n in this ordering and that m was greater than n if it came after n. Now that we have constructed the system of integers, we can extend this ordering to the integers.

We have seen that the set of integers is the union of three mutually disjoint subsets: the positive integers, the negative integers, and zero. In other words, given any integer, it must be positive, negative, or zero. Since the subtraction of integers is closed, if we are given two integers m and n, their difference $m - n$ must be positive, or negative, or zero. This allows us to order the integers in the following way:

> **definition 2.1** Given two integers m and n, we say that m is greater than n if $m - n$ is positive, we say that m is less than n if $m - n$ is negative, and we say that m is equal to n if $m - n = 0$.

This order relation is seen to satisfy the following two properties:

Order Property 1 (*The Law of Trichotomy*) Given two integers m and n, exactly one of the following is true: (a) $m > n$, (b) $m < n$, or (c) $m = n$.

Order Property 2 If m, n, p, and q are integers, $m > p$, and $n > q$, then $(m + n) > (p + q)$.

This last fact follows because $m - p$ and $n - q$ are positive and therefore so is $(m - p) + (n - q) = m + n - p - q = (m + n) - (p + q)$.

We see that to find out which of two integers m and n is larger, we take the difference $m - n$. If this is positive, then $m > n$; if this is negative, then $n > m$; if this is zero, then $m = n$. This same rule will be applied later to fractions to determine which of two fractions is larger.

Several theorems dealing with order properties of the integers now follow from our rules for operations on the integers.

theorem 2.6 (*The Transitive Property*) *If $a > b$ and $b > c$, then $a > c$.*

proof $a > b$ means $a - b$ is positive, and $b > c$ means $b - c$ is positive. The sum $(a - b) + (b - c) = a - c$ must therefore be positive, and this means that $a > c$.

theorem 2.7 *If a and b are integers, c a positive integer, and $a > b$, then $ac > bc$.*

proof Since $a > b$, $a - b$ is positive. Then $(a - b)c = ac - bc$ is positive because the product of two positive integers is positive. This means that $ac > bc$.

theorem 2.8 *If a and b are integers, c a positive integer, and $ac > bc$, then $a > b$.*

proof Since $ac > bc$, we have $ac - bc$ is positive, and therefore $a - b$ must be positive, for if $a - b$ were negative, then $(a - b)c = ac - bc$ would be negative, and if $a - b$ were 0, then $(a - b)c = ac - bc$ would be 0. This implies that $a > b$.

theorem 2.9 *If a and b are integers, c a negative integer, and $a > b$, then $ac < bc$.*

proof $a - b$ is positive and c is negative, so $(a - b)c = ac - bc$ is negative. This means that $-(ac - bc) = bc - ac$ is positive or $bc > ac$, which is the same as $ac < bc$.

theorem 2.10 *If m and n are integers and $m > n$, then $(-m) < (-n)$.*

proof We have that $m - n$ is positive, and so $-(m - n) = n - m$ is negative. We also have that $(-m) - (-n) = -m + n = n - m$, which was seen to be negative. This means that $(-m) < (-n)$.

Another way of looking at the order properties of the integers is in terms of the number line shown in Figure 2.1.

Given two integers m and n, if we say that $m > n$ when m is to the right of n on the number line, $n > m$ when m is to the left of n on the number line, and $m = n$ when m and n are the same point, this ordering of the integers will agree with that given in our definition.

exercises 2.3

1 For each of the following pairs of integers, pick the larger integer:
(a) (5, 7) (b) (−5, 7) (c) (−5, −7) (d) (5, −7)

2 In each of the following sets of integers, pick the largest and the smallest integers:
(a) (7, 3, 12, 2, 0, 5, 19) (b) (−7, −3, −12, −2, 0, −5, −19)
(c) (1, −2, 3, −4, 5, −6, 7, −8) (d) (−3, −7, 9, −2, 8, −6)
(e) (12, 15, −4, 17, −19, 0) (f) (−2, 14, −9, 2, 7, −1)

3 For each part of problem 2, draw a number line, mark the location of each integer on the number line, and then check to see whether the one that you picked as the largest is farthest to the right and the one that you picked as the smallest is farthest to the left.

4 Which of the following is largest and which is smallest?
(a) $8 - (-4)$ (b) $-2 - 7$ (c) -9×5
(d) $3 \times (-12)$ (e) $-8 + 42$

5 Which is better, getting 6 checks for $12 each or getting 11 checks for $7 each?

6 Which is worse, getting 8 bills for $10 each or getting 9 bills for $9 each?

7 Which is worse, getting 4 checks for $8 each or getting 7 checks for $5 each?

8 Which is better, getting 9 bills for $6 each or getting 8 bills for $7 each?

9 If $a > b$, use the definition to prove that $(a + c) > (b + c)$ for any integer c.

10 If $m > -n$, prove that $n > -m$.

11 Explain how to find the product $(-4) \times 3$ using the number line.

12 If m is any integer and n is a positive integer, explain how you would find the product mn using the number line.

2.4 / rational numbers

In Section 2.1 we enlarged the system of natural numbers to obtain the system of integers in order that the operation of subtraction be closed. With respect to the integers the operations of addition, subtraction, and multiplication are closed, but the operation of division is not closed. In this section we shall attempt to remedy this situation by enlarging the system of integers to a larger system called the *rational numbers*, in which the operation of division will be closed except for the case in which one tries to divide by zero. We know that if we try to work the problem $1 \div 2$, there is no integer that could be the answer because there is no integer that can be multiplied by 2 to give 1 as the answer. We will remedy this situation by making up a new number that will be the answer to the problem $1 \div 2$ and that we will write $\frac{1}{2}$.

In the same way we will make up new numbers that will stand for the quotients obtained by dividing the multiplicative identity 1 by each of the positive integers to obtain the numbers $\frac{1}{1}, \frac{1}{2}, \frac{1}{3}, \frac{1}{4}, \ldots$, where $\frac{1}{2}$ means $1 \div 2$, $\frac{1}{3}$ means $1 \div 3$, etc. With the exception of the first one, these numbers are all new ones.

In a similar manner, we form the quotients $\frac{1}{-1}, \frac{1}{-2}, \frac{1}{-3}, \frac{1}{-4}, \ldots$, all of which will be new except for the first one. The quotient $1/m$, formed by dividing 1 by the integer m, will be the multiplicative inverse[1] of m, since $1 \div m = 1/m$ implies that $(1/m) \cdot m = m \cdot (1/m) = 1$. For the same reason, the inverse of $1/m$ is m.

In this enlarged system of numbers all the integers except 0 now have inverses. Unfortunately, we are prohibited from dividing by 0, for if we attempted to form a number $m/0$, where m is any nonzero number, we would have $m \div 0 = m/0$, and therefore $(m/0) \cdot 0 = m$, which would be a contradiction of Theorem 2.2. This proves one of the basic rules of mathematics: *You are never allowed to divide by zero.*

This enlarged system consisting of all the integers and the numbers, $\ldots, \frac{1}{-3}, \frac{1}{-2}, \frac{1}{-1}, \frac{1}{1}, \frac{1}{2}, \frac{1}{3}, \ldots$, is still not closed under division; in fact, it is not even closed now with respect to multiplication. For example, the product of 2 and $\frac{1}{3}$ is not one of the numbers in our enlarged system. To remedy this situation, we now proceed to form all the products $m \times (1/n)$ where m and n are any two integers with $n \neq 0$, and we agree to write the product $m \times (1/n)$ in the form m/n. Since any integer m divided by 1 gives m again, we see that any integer m may be written in the form $m/1$, and so this enlarged set of numbers of the form m/n where m and n are integers and $n \neq 0$ contains the set of integers as a subset.

definition 2.2 The set consisting of all numbers of the form m/n where m and n are integers and $n \neq 0$ is called *the rational numbers*, and the symbols m/n used to denote these rational numbers are called *fractions*. The top (or left) number of the fraction is referred to as the *numerator* and the bottom (or right) number is referred to as the *denominator*.

The set of rational numbers has one striking difference from the set of integers. When two integers are represented by two different numerals, then these two

[1]The multiplicative inverse of a number is defined similarly to the additive inverse. The multiplicative inverse of a number m is that number n such that $mn = nm = 1$, the identity for multiplication. The multiplicative inverse of the number m is often denoted by m^{-1}.

integers are different; that is, they are not equal. In the case of the rational numbers, however, we know, for example, that $12 \div 3 = 8 \div 2$, and therefore $\frac{12}{3} = \frac{8}{2}$; that is, two entirely different fractions turn out to be equal. To avoid utter confusion, we must establish a criterion for telling when two fractions are equal.

Notice in the example above that $12 \times 2 = 8 \times 3$. Examination of other equal pairs of fractions discloses that this cross multiplication always yields equal results, and we adopt this as our *rule for the equality of fractions*:

$$\frac{a}{b} = \frac{c}{d} \quad \text{if and only if} \quad ad = bc$$

The fractions $\frac{5}{8}$ and $\frac{8}{13}$ are not equal, for example, since 5×13 is not equal to 8×8. The fractions $\frac{7}{5}$ and $\frac{21}{15}$, however, are equal, since $7 \times 15 = 5 \times 21$.

This rule for the equality of fractions immediately implies a *cancelation rule for fractions*. Notice that if a, b, and c are integers with $b \neq 0$ and $c \neq 0$, then

$$\frac{ac}{bc} = \frac{a}{b}$$

by the rule of the equality of fractions. The cancelation rule tells us that if we multiply or divide the top and bottom of any fraction by any nonzero integer, the resulting fraction is equal to the original one.

The cancelation law tells us, for example, that the fraction $\frac{48}{112}$ may be reduced to the fraction $\frac{3}{7}$ by dividing the top and bottom by 16; $\frac{3}{7}$ is a much easier fraction to work with than $\frac{48}{112}$, so, as we shall see later, the cancelation rule is a useful tool for simplifying computation.

When a fraction cannot be reduced by dividing the top and bottom by some integer, that is, when the top and bottom are relatively prime, we say that the fraction is *reduced to lowest terms*. When we divide the numerator and the denominator of a fraction by the greatest common divisor of the numerator and the denominator, the fraction that we get will be reduced to lowest terms because we have already divided out the largest number possible; there is nothing else that can be canceled from the numerator and the denominator. If we did not know that 16 divided both 48 and 112, we could reduce $\frac{48}{112}$ by dividing top and bottom by 2 to get $\frac{24}{56}$, then again by 2 to get $\frac{12}{28}$ and then by 4 to get $\frac{3}{7}$, which is the same answer that we got before, although it took us a little longer.

Our objective in enlarging the system of integers to the rational numbers was to obtain a system that was closed under division (except for division by 0) and that satisfied all the properties satisfied by the integers. To determine whether we have been successful in doing this, we must define addition, subtraction, multiplication, and division of fractions.

We want to preserve the original definition of the multiplication of integers, so we must have

$$\frac{m}{1} \times \frac{n}{1} = \frac{mn}{1}$$

for any integers m and n.

This suggests that our rule for the multiplication of fractions must be

$$\boxed{\frac{m}{n} \times \frac{p}{q} = \frac{mp}{nq}}$$

and this is the rule that we adopt for the multiplication of fractions. The rule tells us that the product of two fractions is a fraction, the numerator of which is the product of the two numerators and the denominator of which is the product of the two denominators.

It is easy to check that $m/m = 1$ is the identity for multiplication and that the multiplicative inverse of m/n is n/m when $m \neq 0$ and $n \neq 0$, since $(m/n) \times (n/m) = mn/mn$, which is equal to $1/1 = 1$ by the cancelation rule. (The fraction n/m is also called the *reciprocal* of m/n.) It is also easy to check that the multiplication of fractions is closed and satisfies both the associative and commutative properties.

To show that the division of fractions is closed we extend our definition of division to apply to fractions and prove Theorem 2.11.

theorem 2.11 *If m/n and p/q are fractions with $p \neq 0$, then $(m/n) \div (p/q) = (m/n) \cdot (q/p)$; that is, to divide by a fraction, we multiply by its reciprocal.*

proof By our definition of division, the quotient $(m/n) \div (p/q)$ must be a fraction that when multiplied by p/q gives m/n, and $(m/n) \cdot (q/p) = mq/np$ times p/q gives $mqp/npq = m/n$.

Theorem 2.11 tells us that the division of fractions is closed (except for division by 0), because it is actually multiplication by the inverse.

We must now determine the rule for the addition of fractions in order that the properties satisfied by the integers be preserved for the rational numbers. It turns out that we have no choice in this matter; if we wish to preserve these properties, there is only one way in which the addition of fractions may be defined. We must have

$$\begin{aligned}\frac{a}{b} + \frac{c}{d} &= \frac{ad}{bd} + \frac{bc}{bd} \\ &= \frac{1}{bd} \cdot \frac{ad}{1} + \frac{1}{bd} \cdot \frac{bc}{1} \\ &= \frac{1}{bd}\left(\frac{ad}{1} + \frac{bc}{1}\right) \\ &= \frac{1}{bd}\left(\frac{ad + bc}{1}\right) \\ &= \frac{ad + bc}{bd}\end{aligned}$$

Our rule for the addition of fractions is therefore given by

$$\boxed{\frac{a}{b} + \frac{c}{d} = \frac{ad + bc}{bd}}$$

This rule tells us that to add two fractions we multiply the numerator of each one by the denominator of the other and take the sum of these two products for the numerator of the answer. For the denominator we take the product of the two denominators.

If you have difficulty in following the derivation of this formula, put any four integers in place of a, b, c, and d (except that neither b nor d can be 0) and work the derivation through with these specific fractions. Then substitute four other integers and follow it through again. See if you can give the reason for each step. The use of the letters a, b, c, and d merely means that this derivation will work no matter what two fractions we start with. As an example, let us add $\frac{7}{8} + \frac{9}{5}$. We have

$$\frac{7}{8} + \frac{9}{5} = \frac{(7 \times 5) + (8 \times 9)}{(8 \times 5)} = \frac{35 + 72}{40} = \frac{107}{40}$$

As another example, let us add $\frac{5}{7} + \frac{8}{7}$. We have

$$\frac{5}{7} + \frac{8}{7} = \frac{(5 \times 7) + (7 \times 8)}{7 \times 7} = \frac{35 + 56}{49} = \frac{91}{49}$$

which can be reduced to $\frac{13}{7}$ by dividing the numerator and the denominator by 7. We see that $\frac{5}{7} + \frac{8}{7} = \frac{13}{7}$. Similarly,

$$\frac{3}{10} + \frac{14}{10} = \frac{(3 \times 10) + (10 \times 14)}{10 \times 10} = \frac{30 + 140}{100} = \frac{170}{100} = \frac{17}{10}$$

We see that $\frac{3}{10} + \frac{14}{10} = \frac{17}{10}$. The last two examples indicate to us that *to add two fractions with the same denominator, we add the numerators and put the sum over this denominator.*

It is easy to see that the addition of fractions is closed and satisfies the commutative property. To show that it satisfies the associative property it must be shown that

$$\left(\frac{a}{b} + \frac{c}{d}\right) + \frac{e}{f} = \frac{a}{b} + \left(\frac{c}{d} + \frac{e}{f}\right)$$

(which may be done as an exercise).

The identity for addition is $0/c = 0/1 = 0$, since

$$\frac{a}{b} + \frac{0}{c} = \frac{ac + 0}{bc} = \frac{ac}{bc} = \frac{a}{b}$$

and the inverse $-(a/b)$ of the fraction a/b is the fraction $(-a)/b$, since

$$\frac{a}{b} + \frac{-a}{b} = \frac{ab + (-ab)}{bb} = \frac{0}{bb} = 0$$

Applying the rule the equality of fractions, we have

$$-\frac{a}{b} = \frac{-a}{b} = \frac{a}{-b}$$

so that if a fraction has exactly one minus sign, it does not matter whether we put it on top, on the bottom, or out in front of the fraction; we also have

$$\frac{-a}{-b} = \frac{a}{b}$$

so if a fraction has a minus sign on the top and the bottom we may cancel them both out.

Applying Theorem 2.1 to fractions we have

$$\frac{a}{b} - \frac{c}{d} = \frac{a}{b} + \frac{-c}{d} = \frac{ad + (-bd)}{bd}$$

so the rule for the subtraction of fractions may be written

$$\boxed{\frac{a}{b} - \frac{c}{d} = \frac{ad - bc}{bd}}$$

To subtract $\frac{5}{8}$ from $\frac{1}{3}$, for example, we have

$$\frac{1}{3} - \frac{5}{8} = \frac{(1 \times 8) - (3 \times 5)}{3 \times 8} = \frac{8 - 15}{24} = \frac{-7}{24} = -\frac{7}{24}$$

The rules for the addition, subtraction, multiplication, and division of fractions may now be summarized:

$$\frac{a}{b} + \frac{c}{d} = \frac{ad + bc}{bd}$$

$$\frac{a}{b} - \frac{c}{d} = \frac{ad - bc}{bd}$$

$$\frac{a}{b} \cdot \frac{c}{d} = \frac{ac}{bd}$$

$$\frac{a}{b} \div \frac{c}{d} = \frac{a}{b} \cdot \frac{d}{c} = \frac{ad}{bc}$$

It should be emphasized that to add, subtract, multiply, and divide fractions, you must know how to add, subtract, multiply, and, where possible, to divide integers.

Although our objective in deriving the system of rational numbers in the form of fractions was to obtain a system that was closed with respect to the four basic operations (except for division by zero), there is a very practical aspect to fractions. Just as the natural numbers were used to count the number of objects in a set, the fractions may be used to measure portions of objects less than a whole object and a portion of an object left over.

As an example, the fraction $\frac{5}{8}$ may be used to denote that portion of an object obtained if we divide it into 8 equal parts and then take 5 of these parts (see Figure 2.3). Each of the 8 equal parts constitutes $\frac{1}{8}$ of the object, and 5 of these parts constitutes $5 \times \frac{1}{8}$ or $\frac{5}{8}$ of the object.

If you have 11 candy bars that are to be divided equally among 3 people, you can give each person 3 whole candy bars and you will then have 2 candy bars left over. If each of these are then divided into 3 equal parts, each of which will be $\frac{1}{3}$ candy bar, this will yield $\frac{6}{3}$, so each person may now be given an additional $\frac{2}{3}$, and thus each person winds up with $3\frac{2}{3}$ candy bars. Expressed as an arithmetic problem, this may be written

$$11 \div 3 = \frac{11}{3} = 3 + \frac{2}{3} = 3\frac{2}{3}$$

figure 2.3

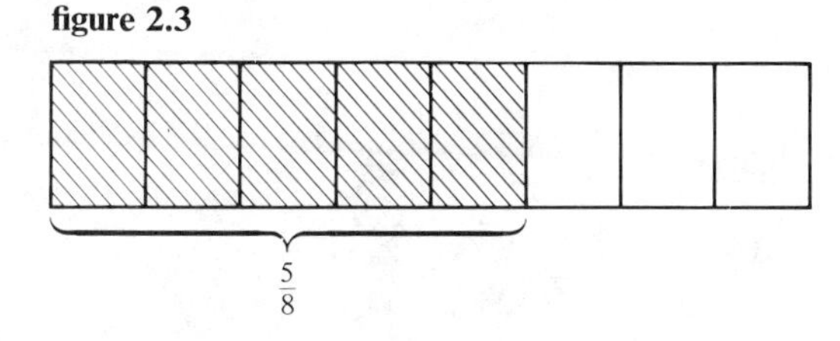

This suggests that certain fractions in which the numerator is equal to or greater than the denominator (usually referred to as *improper fractions*) may be written as a whole number plus a fraction (usually referred to as a *mixed number*). As a working rule, to change an improper fraction to a mixed number, divide the numerator by the denominator. Take the quotient as the whole number and the remainder over the denominator of the original fraction as the fraction. For example, to change $\frac{27}{4}$ to a mixed number, we divide 27 by 4 and get a quotient of 6 and a remainder of 3. This means that $\frac{27}{4} = 6\frac{3}{4}$. As another example, to change $\frac{218}{23}$ to a mixed number, we compute

$$\begin{array}{r|l} & 9 \\ \hline 23 & 218 \\ & 207 \\ \hline & 11 \end{array}$$

so that $\frac{218}{23} = 9\frac{11}{23}$. For practice change these improper fractions to mixed numbers: $\frac{9}{4}, -\frac{17}{3}, \frac{8}{8}, \frac{7}{5}, -\frac{29}{6}, \frac{32}{4}, \frac{68}{3}, -\frac{19}{7}$.

To change a mixed number to a fraction we must remember that a mixed number such as $5\frac{2}{3}$ means $\frac{5}{1} + \frac{2}{3}$, which is equal to $[(5 \times 3) + (2 \times 1)]/3$, which may be written $[(5 \times 3) + 2]/3$. The rule for changing a mixed number to a fraction may therefore be given as follows: Multiply the whole number by the denominator of the fraction, add the numerator, and then put all this over the denominator. For example, to change $9\frac{11}{23}$ to a fraction,

$$9\frac{11}{23} = \frac{(9 \times 23) + 11}{23} = \frac{207 + 11}{23} = \frac{218}{23}$$

For practice try changing these mixed numbers to improper fractions: $3\frac{5}{8}$, $6\frac{2}{9}$, $-9\frac{2}{3}$, $-7\frac{1}{4}$, $1\frac{1}{8}$, $-12\frac{2}{9}$.

These last two rules permit us to change mixed numbers to fractions and fractions to mixed numbers whenever we find it convenient to do so.

/ examples /

1 $5\frac{2}{3} + 7\frac{3}{8}$

► $$5\frac{2}{3} + 7\frac{3}{8} = \frac{17}{3} + \frac{59}{8} = \frac{(17 \times 8) + (3 \times 59)}{24}$$

$$= \frac{136 + 177}{24} = \frac{313}{24} = 13\frac{1}{24}$$

2 $4\frac{2}{5} - 2\frac{7}{8}$

$$\blacktriangleright\ 4\frac{2}{5} - 2\frac{7}{8} = \frac{22}{5} - \frac{23}{8} = \frac{(22 \times 8) - (5 \times 23)}{5 \times 8}$$

$$= \frac{176 - 115}{40} = \frac{61}{40} = 1\frac{21}{40}$$

3 $2\frac{7}{9} \times 5\frac{2}{5}$

$$\blacktriangleright\ 2\frac{7}{9} \times 5\frac{2}{5} = \frac{25}{9} \times \frac{27}{5} = \frac{675}{45} = 15$$

4 $8\frac{1}{3} \div \left(-3\frac{5}{6}\right)$

$$\blacktriangleright\ 8\frac{1}{3} \div \left(-3\frac{5}{6}\right) = \frac{25}{3} \div \left(-\frac{23}{6}\right) = \frac{25}{3} \div \frac{-23}{6} = \frac{25}{3} \times \frac{6}{-23}$$

$$= \frac{150}{-69} = -2\frac{12}{69} = -2\frac{4}{23}$$

In the last problem, although it is not incorrect not to reduce the fractional part of the answer to lowest terms, it is, nonetheless, a good habit to reduce all fractions to lowest terms because this will make any succeeding computation much simpler.

/ example /

5 $4\frac{3}{5}\left(2\frac{5}{6} + 4\frac{1}{9}\right)$

$$\blacktriangleright\ 2\frac{5}{6} + 4\frac{1}{9} = \frac{17}{6} + \frac{37}{9} = \frac{153 + 222}{54} = \frac{375}{54} = \frac{125}{18}$$

$$4\frac{3}{5}\left(\frac{125}{18}\right) = \frac{23}{5} \times \frac{125}{18} = \frac{2875}{90} = 31\frac{85}{90} = 31\frac{17}{18}$$

Try this last problem without reducing $\frac{375}{54}$ to $\frac{125}{18}$ and notice that the computation is harder.

The addition and subtraction of mixed numbers may also be accomplished somewhat more simply by working with the whole numbers and fractions separately. In Example 1 we can add the 5 and the 7 to obtain 12, and then we can add $\frac{2}{3} + \frac{3}{8}$ to obtain

$$\frac{16 + 9}{24} = \frac{25}{24} = 1\frac{1}{24}$$

and then we can add 12 and $1\frac{1}{24}$ to obtain $13\frac{1}{24}$, which is the answer to the problem.

/example/

6 $9\frac{6}{7} + 4\frac{3}{5}$

► $9 + 4 = 13$

$$\frac{6}{7} + \frac{3}{5} = \frac{30 + 21}{35} = \frac{51}{35} = 1\,\frac{16}{35}$$

$$13 + 1\,\frac{16}{35} = 14\,\frac{16}{35}$$

Try the last example as well as the following one the other way as a check.

/example/

7 $9\frac{6}{7} - 4\frac{3}{5}$

► $9 - 4 = 5$

$$\frac{6}{7} - \frac{3}{5} = \frac{30 - 21}{35} = \frac{9}{35}$$

$$5 + \frac{9}{35} = 5\,\frac{9}{35}$$

The method by which the rule for adding fractions was derived,

$$\frac{a}{b} + \frac{c}{d} = \frac{ad}{bd} + \frac{bc}{bd} = \frac{ad + bc}{bd}$$

consisted of multiplying the numerator and the denominator of each fraction by the denominator of the other so that both fractions had the same denominator (the product of the two original denominators), and then we were able to add like quantities. It turns out in many cases that we can find a smaller common denominator than the product of the two denominators. For example, when the denominators are 6 and 9, we can use 18 as a common denominator; we do not have to go to 54. In order to make the computation as simple as possible, we would use the *lowest common denominator* (abbreviated LCD). *This lowest common denominator is the least common multiple* of the two denominators.

There is nothing wrong with doing a problem without worrying about the lowest common denominator; however, if we are willing to go to the trouble of finding the lowest common denominator, we may save ourselves some work in the long run.

/example/

8 $5\frac{1}{6} + 8\frac{5}{9}$

► *Without finding the LCD:*

$$5 + 8 = 13$$

$$\frac{1}{6} + \frac{5}{9} = \frac{9 + 30}{54} = \frac{39}{54} = \frac{13}{18}$$

$$13 + \frac{13}{18} = 13\frac{13}{18}$$

Finding the LCD: When doing a problem this way, it is usually more convenient to write the numbers one below the other. The LCD of 6 and 9 is 18; therefore,

$$\begin{array}{rcr} 5\frac{1}{6} & = & 5\frac{3}{18} \\ + 8\frac{5}{9} & = & 8\frac{10}{18} \\ \hline & & 13\frac{13}{18} \end{array}$$

Notice that the numerator and denominator of each fraction were multiplied by that number that would give 18 as the denominator.

We could likewise add three or more rational numbers by finding the LCD of all the fractions.

/example/

9

$$\begin{array}{r} 4\frac{2}{3} \\ 7\frac{5}{8} \\ + 5\frac{1}{6} \\ \hline \end{array}$$

► The LCD of 3, 8, and 6 is 24. We therefore write

$$4\frac{2}{3} = 4\frac{16}{24}$$

$$7\frac{5}{8} = 7\frac{15}{24}$$

$$+\,5\frac{1}{6} = 5\frac{4}{24}$$

$$16\frac{35}{24} = 17\frac{11}{24}$$

We could also do this problem without finding the LCD by adding the first two numbers and then adding the third. We have

$$4\frac{2}{3} + 7\frac{5}{8} = 11 + \frac{2}{3} + \frac{5}{8} = 11 + \frac{16 + 15}{24} = 11\frac{31}{24} = 12\frac{7}{24}$$

Then

$$12\frac{7}{24} + 5\frac{1}{6} = 17 + \frac{7}{24} + \frac{1}{6} = 17 + \frac{42 + 24}{244} = 17\frac{66}{144} = 17\frac{11}{24}$$

which is the same answer that we got before.

/example/

10 $7\frac{4}{15} - 2\frac{5}{6}$

► *Without finding the LCD:*

$$7\frac{4}{15} - 2\frac{5}{6} = \frac{109}{15} - \frac{17}{6} = \frac{654 - 255}{90} = \frac{399}{90}$$

$$= 4\frac{39}{90} = 4\frac{13}{30}$$

Finding the LCD:
The LCD of 15 and 6 is 30.

$$7\frac{4}{15} = 7\frac{8}{30}$$

$$-2\frac{5}{6} = -2\frac{25}{30}$$

Since we cannot take $\frac{25}{30}$ from $\frac{8}{30}$, we "borrow" $1 = \frac{30}{30}$ from 7 or we "rearrange" the number $7\frac{8}{30}$ to obtain $7\frac{8}{30} = 6\frac{38}{30}$. This gives

$$\begin{array}{rcr} 7\dfrac{8}{30} & = & 6\dfrac{38}{30} \\[2ex] -2\dfrac{25}{30} & = & -2\dfrac{25}{30} \\ \hline & & 4\dfrac{13}{30} \end{array}$$

If we worked this problem by keeping the whole numbers and fractions separate and without finding the LCD, we would have

$$7 - 2 = 5$$

$$\frac{4}{15} - \frac{5}{6} = \frac{24 - 75}{90} = \frac{-51}{90}$$

$$5 + \frac{-51}{90} = \frac{5}{1} - \frac{51}{90} = \frac{450 - 51}{90} = \frac{399}{90} = 4\frac{39}{90} = 4\frac{13}{30}$$

example

11 $5\frac{2}{3} + 7\frac{3}{5}$

► *Without finding the LCD:*

$$5 + 7 = 12$$

$$\frac{2}{3} + \frac{3}{5} = \frac{10 + 9}{15} = \frac{19}{15} = 1\frac{4}{15}$$

$$12 + 1\frac{4}{15} = 13\frac{4}{15}$$

Finding the LCD:

$$\begin{array}{rcr} 5\dfrac{2}{3} & = & 5\dfrac{10}{15} \\[2ex] +7\dfrac{3}{5} & = & 7\dfrac{9}{15} \\ \hline & & 12\dfrac{19}{15} = 13\dfrac{4}{15} \end{array}$$

The last example illustrates the fact that when the lowest common denominator is the product of the two denominators (this occurs when the two denominators are relatively prime), then the two methods are the same. Remember that in any case either method is correct. Finding the lowest common denominator may save some time; not finding the lowest common denominator may take a little longer, but if you reduce the final answer to lowest terms, you will get the same answer as you would by finding the lowest common denominator.

It is sometimes possible to simplify the computation of the product of two fractions. Consider, for example, the problem

$$\frac{15}{8} \times \frac{3}{10} = \frac{45}{30} = \frac{9}{16}$$

This may be written in the form

$$\frac{5 \times 3}{8} \times \frac{3}{5 \times 2} = \frac{5 \times 3 \times 3}{8 \times 5 \times 2}$$

We may then cancel the 5 from the numerator and the denominator to obtain

$$\frac{3 \times 3}{8 \times 2} = \frac{9}{16}$$

The point here is that we do not have to wait until after we multiply to cancel the 5's; the 5 in the numerator of the first fraction will carry across to the numerator of the answer, and the 5 in the denominator of the second fraction will carry across to the denominator of the answer. Therefore, instead of waiting until after we multiply, we may cancel the 5's before we multiply. The problem would then be done as follows:

$$\frac{{}^{3}\cancel{15}}{8} \times \frac{3}{\cancel{10}_{2}} = \frac{9}{16}$$

Thus *in multiplying fractions, one may cancel a common factor from the numerator of one fraction and from the denominator of another.* This rule holds even when we are multiplying more than two fractions together.

examples

12 $6\frac{2}{3} \times 3\frac{4}{15}$

► $6\frac{2}{3} \times 3\frac{4}{15} = \frac{{}^{4}\cancel{20}}{3} \times \frac{49}{\cancel{15}_{3}} = \frac{196}{9} = 21\frac{7}{9}$

13 $1\frac{7}{8} \times \frac{4}{7} \times \frac{9}{10}$

$$\blacktriangleright \quad \frac{^{3}\not{15}}{_{2}\not{8}} \times \frac{^{1}\not{4}}{7} \times \frac{9}{\not{10}_{2}} = \frac{3 \times 1 \times 9}{2 \times 7 \times 2} = \frac{27}{28}$$

Notice in the example above that when the 4 on top was canceled, this left a 1 on top, not a 0. This example illustrates the fact that the rule of cancelation for the multiplication of fractions can save us a lot of computation.

examples

14 $2\frac{1}{3} \times 1\frac{5}{8} \times 4\frac{1}{14}$

$$\blacktriangleright \quad 2\frac{1}{3} \times 1\frac{5}{8} \times 4\frac{1}{14} = \frac{^{1}\not{7}}{_{1}\not{3}} \times \frac{13}{8} \times \frac{\not{57}^{19}}{\not{14}_{2}} = \frac{1 \times 13 \times 19}{1 \times 8 \times 2} = \frac{247}{16} = 15\frac{7}{16}$$

15 $4\frac{5}{6} \div 3\frac{1}{9}$

$$\blacktriangleright \quad 4\frac{5}{6} \div 3\frac{1}{9} = \frac{29}{6} \div \frac{28}{9} = \frac{29}{_{2}\not{6}} \times \frac{\not{9}^{3}}{28} = \frac{29 \times 3}{2 \times 28} = \frac{87}{56} = 1\frac{31}{56}$$

16 Find the GCD and LCM of 84 and 126.

► By the method described in Chapter One, we find that the GCD (84,126) = 42. Now the LCM (84,126) = (84 × 126)/42. It saves a lot of work here to cancel before we multiply, so we have

$$\text{LCM } (84{,}126) = \frac{^{2}\not{84} \times 126}{\not{42}_{1}} = \frac{252}{1} = 252$$

Whenever you find the least common multiple of two numbers by the method described in Chapter One, you will save a lot of time if you cancel before you multiply.

The ordering of integers may be extended to the fractions by saying that a fraction m/n is positive if m and n are either both positive or both negative, m/n is negative if m and n have different signs, and m/n is zero if $m = 0$. We then say that m/n is greater than p/q, denoted again by $(m/n) > (p/q)$ if $(m/n) - (p/q)$ is positive, we say that m/n is less than p/q, denoted by $(m/n) < (p/q)$ if $(m/n) - (p/q)$ is negative, and $(m/n) = (p/q)$ if $(m/n) - (p/q) = 0$.

If you were asked, for example, which is larger, $\frac{35}{126}$ or $\frac{59}{231}$, you would subtract $\frac{35}{126} - \frac{59}{231}$ and get

$$\frac{(35 \times 231) - (59 \times 126)}{126 \times 231} = \frac{8085 - 7434}{29106} = \frac{651}{29106}$$

which is positive, so you could say that $\frac{35}{126}$ is larger than $\frac{59}{231}$.

The fractions, like the integers, may be represented by points on the number line. The fraction $\frac{10}{6}$, for example, corresponds to the point obtained by breaking the interval between 0 and 1 (this is usually referred to as the *unit interval*) into 6 equal parts and then counting off 10 of these parts to the right, starting from 0 (see Figure 2.4). Notice that if we had used $\frac{5}{3}$, we would have ended up with the same point.

In general, the fraction m/n corresponds to the point on the number line obtained by dividing the unit interval into n equal parts and then counting off m of these parts, to the right if m/n is positive, and to the left if m/n is negative, starting from 0.

It turns out, just as for the integers, that $m/n > p/q$ if m/n is to the right of p/q on the number line, $m/n < p/q$, if m/n is to the left of p/q, and $m/n = p/q$ if m/n and p/q are represented by the same point.

definition 2.3 Points on the number line that correspond to rational numbers are called *rational points.*

The rational numbers satisfy the *density property*, which may be stated as follows:

Between any two distinct rational numbers, there is another rational number.

This is easy to see, for if the two numbers are a/b and c/d, then the number

$$\frac{1}{2}\left(\frac{a}{b} + \frac{c}{d}\right) = \frac{ad + bc}{2bd}$$

is a rational number that is located exactly halfway between a/b and c/d.

We can now find another rational number between a/b and $(ad + bc)/2bd$, and if we were to continue indefinitely in this manner we would see that *between*

figure 2.4

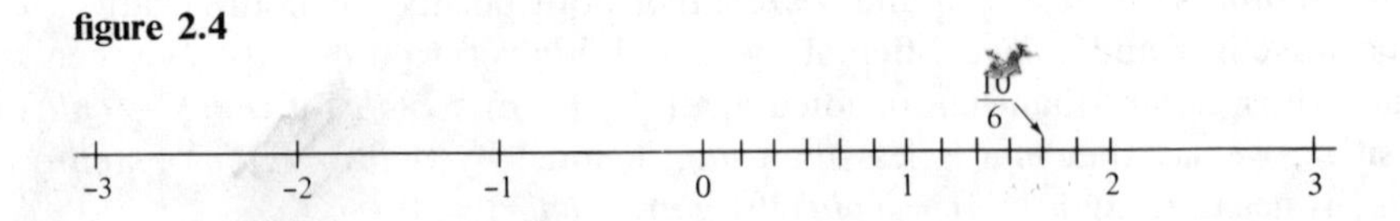

any two distinct rational numbers, no matter how close together, there are an infinite number of rational numbers. This would lead us to believe that the set of all rational points on the number line should completely fill up the line—there should be no holes left. It turns out that this is not so; in spite of the density property of the rational numbers, there are a great many holes left in the number line, and we shall show in Section 2.5 how we can fill these in by extending the system of rational numbers to the system of real numbers.

We conclude this section by summarizing the properties of the rational numbers (m, n, and p stand for rational numbers):

1. Closure for addition:

$$m + n = p$$

2. Closure for multiplication:

$$m \times n = p$$

3. Commutativity for addition:

$$m + n = n + m$$

4. Commutativity for multiplication:

$$m \times n = n \times m$$

5. Associativity for addition:

$$(m + n) + p = m + (n + p)$$

6. Associativity for multiplication:

$$(m \times n) \times p = m \times (n \times p)$$

7. Distributivity:

$$m(n + p) = mn + mp$$

8. Identity for addition:

$$0 + m = m + 0 = m$$

9. Identity for multiplication:

$$1 \times m = m \times 1 = m$$

10. Inverse for addition:

$$m + (-m) = (-m) + m = 0$$

11. Cancelation for multiplication:

$$\text{If } m \times p = n \times p \quad \text{and} \quad p \neq 0, \quad \text{then} \quad m = n$$

12. Cancelation for addition:

$$\text{If } m + p = n + p, \quad \text{then} \quad m = n$$

13. Inverse for multiplication:

$$m \times m^{-1} = m^{-1} \times m = 1$$

14. Density property:
For each m and n with $m < n$, there exists a p such that $m < p < n$

exercises 2.4

1 Reduce to lowest terms:

(a) $\frac{56}{84}$ (b) $\frac{72}{144}$ (c) $\frac{25}{80}$ (d) $\frac{156}{812}$ (e) $\frac{36}{60}$

2 Compute:

(a) $\frac{5}{7} + \frac{3}{8}$ (b) $\frac{2}{3} - \frac{9}{10}$ (c) $\frac{-5}{8} \times \frac{2}{-3}$

(d) $6\frac{3}{8} + 7\frac{1}{4}$ (e) $\frac{5}{6}\left(\frac{7}{8} - \frac{3}{4}\right)$ (f) $5\frac{1}{3}\left(1\frac{3}{7} - 2\frac{5}{8}\right)$

(g) $2\frac{1}{3} \div 6\frac{3}{5}$ (h) $5\frac{1}{6} - 3\frac{4}{9}$

3 Compute:

(a) $4\frac{3}{5} + 2\frac{1}{8} - 1\frac{1}{3}$ (b) $\left(7\frac{1}{8} \times 1\frac{5}{6}\right) - 3\frac{2}{5}$

(c) $6\frac{3}{5} - \left(2\frac{1}{3} \times 1\frac{3}{4}\right)$ (d) $\frac{3}{7} - \frac{1}{21} + \frac{5}{6}$

(e) $\dfrac{\frac{5}{8} - \frac{3}{5}}{\frac{3}{4} - \frac{7}{9}}$ (f) $\dfrac{5\frac{1}{8} + 2\frac{3}{7}}{4\frac{2}{3} - 3\frac{1}{6}}$

(remember that a/b means $a \div b$)

4 For each of the following pairs of fractions, determine which is larger:

(a) $\left(\frac{5}{8}, \frac{3}{5}\right)$ (b) $\left(\frac{56}{97}, \frac{138}{245}\right)$ (c) $\left(\frac{1968}{3159}, \frac{4796}{7106}\right)$

5 Which of the following four fractions is the largest?

(a) $\frac{4}{3}$ (b) $\frac{8712}{6302}$ (c) $\frac{7119}{5029}$ (d) $\frac{116}{87}$

6 Multiply:

(a) $\frac{12}{5} \times \frac{15}{7} \times \frac{21}{8}$ (b) $\frac{9}{5} \times \frac{25}{14} \times \frac{21}{12}$ (c) $\frac{39}{10} \times \frac{9}{26} \times \frac{15}{6}$

7 Find the GCD and LCM of 198 and 330.

8 Prove that the addition of fractions is commutative.

9 Prove that the distributive law holds for fractions.

10 Prove that if two fractions have the same denominator, we subtract by subtracting the numerators and putting the difference over this denominator.

11 Which fractions do these points on the number line represent?

(a)

(b)

7 8 9

(c)

-7 -6 -5

(d)

-2 -1 0

12 For each of the following pairs of rational numbers, find the point located halfway between them.

(a) $\left(\frac{2}{3}, \frac{7}{8}\right)$ (b) $\left(\frac{9}{2}, -\frac{4}{5}\right)$ (c) $\left(-2\frac{1}{3}, -5\frac{1}{8}\right)$

(d) $\left(-3\frac{1}{5}, 6\frac{1}{8}\right)$ (e) $\left(-1\frac{1}{3}, 0\right)$ (f) $\left(\frac{1}{9}, \frac{1}{10}\right)$

2.5/decimals and real numbers

In this section we are going to fill in the holes that were left in the real-number line after all the rational points were taken care of. Our decimal system of representation of the natural numbers by numerals is a place-value system in which the first place on the right represents singles, the next place to the left represents tens, the next place hundreds, etc. As we move from right to left, the value of each place is multiplied by 10, and as we move from left to right, the value of each place is divided by 10 until we reach the singles place. If we were to keep going past the singles place and keep on dividing by 10 as we move along, the next place would represent tenths, the one after that hundredths, etc. We agree to put a decimal point after the units place to separate the whole-number part from the fractional part of the number. Thus a number that is written

$$86.315$$

represents 8 tens, 6 singles, 3 tenths, 1 hundredth, and 5 thousandths, or, since 3 tenths is equal to 300 thousandths and 1 hundredth is equal to 10 thousandths, this number represents 86 and 315 thousandths.

In Section 2.4 we used fractions to represent the rational numbers. Rational numbers may also be represented by decimals. Thus any rational number may be represented by a fraction or by a decimal, and we shall show how to change a fraction into the equivalent decimal and vice versa.

Remember that a fraction a/b means $a \div b$. If we wish to change a fraction such as $\frac{109}{43}$ to a decimal we merely divide 109 by 43 using long division, but we write 109 in the form 109.000 . . . (the addition of any number of 0's at the end of a number to the right of the decimal point does not change the number). The first part of the computation gives us

$$\begin{array}{r|l} & 2. \\ \hline 43 & 109.000\ldots \\ & \underline{86} \\ & 23 \end{array}$$

From our process of long division we know that we get a quotient (in this case 2) and a remainder (in this case 23), which must be less than 43; in other words, the remainder in this long-division process must always be one of the numbers 0, 1, 2, . . . , 42. This means that if we continue the long division past the decimal point, one of two things must happen; at some point we get a remainder of zero, in which case the process is completed and we get a *terminating* decimal for our answer, or we get a remainder that we had before, in which case the sequence of digits in the quotient repeats itself from that previous point on and we get a nonterminating *repeating*[1] decimal for our answer.

[1]Some people refer to a terminating decimal as a repeating decimal also, since we could say that it repeats zeros from some point on.

The same two possibilities occur when any integer m is divided by any integer n, which gives Theorem 2.12.

theorem 2.12 *When a rational number is represented by a decimal, it is represented either by a terminating or a repeating decimal.*

examples

1 Change $4\frac{3}{8}$ to a decimal.

$$4\frac{3}{8} = \frac{35}{8}$$

$$\begin{array}{r} 4.375 \\ 8\overline{)35.000} \\ \underline{32} \\ 30 \\ \underline{24} \\ 60 \\ \underline{56} \\ 40 \\ \underline{40} \\ 0 \end{array}$$

2 Change $2\frac{5}{6}$ to a decimal.

$$2\frac{5}{6} = \frac{17}{6}$$

$$\begin{array}{r} 2.83 \\ 6\overline{)17.000} \\ \underline{12} \\ 50 \\ \underline{48} \\ 20 \\ \underline{18} \\ 2 \end{array}$$

Since the remainder 2 is a repetition of the previous remainder, the quotient repeats from this point on; that is, $2\frac{5}{6} = 2.83333\ldots$.

3 Change $5\frac{3}{11}$ to a decimal.

$$► \quad 5\frac{3}{11} = \frac{58}{11}$$

```
      5.27
  11|58.000
     55
      30
      22
       80
       77
        3
```

Since the remainder 3 is a repetition of a previous remainder, the quotient is a repeating decimal; that is, $5\frac{3}{11} = 5.272727\ldots$.

We will show that the converse of Theorem 2.12 is true also by showing how to convert a terminating decimal or a repeating decimal to a fraction or a mixed number.

For a terminating decimal, the whole-number part (the part to the left of the decimal point) is left unchanged, and the fractional part (the part to the right of the decimal) is changed to a fraction by writing the digits to the right of the decimal as the numerator and (as indicated by our place-value system) by putting a 1 followed by as many 0's as there are digits on top for the denominator.

examples

4 $89.1437 = 89\frac{1437}{10000}$

5 $153.00274 = 153\frac{00274}{100000} = 153\frac{274}{100000}$

6 $0.00730 = \frac{00730}{100000} = \frac{730}{100000} = \frac{73}{10000}$

The last example could have been simplified somewhat by eliminating the last 0 on the right, because adding or removing 0's at the end of a number to the right of the decimal point does not change it.

Consider now a repeating decimal such as 17.493493493 .... Again the 17 is left unchanged. Let the symbol x stand for the repeating-decimal part of the number, so that

$$x = 0.493493493\ldots$$

Multiplying both sides of this equation by 1000 yields

$$1000x = 493.493493493\ldots$$

Subtracting the equation for x from the equation for $1000x$ yields

$$999x = 493$$

and, finally, dividing both sides by 999 yields $x = \frac{493}{999}$.

The same procedure may be applied to any repeating decimal. To change a repeating decimal to a fraction or a mixed number, therefore, leave the whole-number part as it is and change the repeating-decimal part to a fraction by writing one sequence of repeating digits as the numerator and putting as many 9's in the denominator as there are digits in the numerator.

examples

7 $18.17041704\ldots = 18\,\frac{1704}{9999}$

8 $283.070707\ldots = 283\,\frac{07}{99} = 283\,\frac{7}{99}$

9 $4.090090090\ldots = 4\,\frac{090}{999} = 4\,\frac{90}{999} = 4\,\frac{10}{111}$

For a repeating decimal that does not start repeating right after the decimal point, see problem 12 of Exercises 2.5.

We must consider now how to add, subtract, multiply, or divide rational numbers when they are written as decimals. Of course, we could change the decimals to fractions, perform the indicated operation on the fractions, and change the answer back to a decimal; it is easier, however, to work directly with the decimals.

In adding or subtracting decimals, we must remember to keep track of the singles, tens, hundreds, etc., and also of the tenths, hundredths, thousandths, etc. We can be sure of doing this if we always line up the decimal points below each other.

/ examples /

10 Add $51.27 + 1.89 + 0.074$.

► We line up the decimal points below each other and then add in the ordinary way:

$$\begin{array}{r} 51.27 \\ 1.89 \\ +0.074 \\ \hline 53.234 \end{array}$$

11 Add $182.75 + 96.842 + 137$.

► The number 137 has no decimal point, but we agree that when a number has no decimal point, it means that the decimal point is at the end of the number; therefore, we can write 137. for 137 so as to line up the decimals points, or if we are careful, we can leave out the decimal point for 137 and write

$$\begin{array}{l} 182.75 \\ \ \ 96.842 \\ +137 \\ \hline 416.592 \end{array}$$

12 Subtract $149.62 - 15.7391$.

► We add two 0's to the right of the decimal point in 149.62 so that each number will have four decimal places. This will permit us to subtract:

$$\begin{array}{r} 149.6200 \\ -15.7391 \\ \hline 133.8809 \end{array}$$

If we take a number such as 43.87 and multiply it by 10, we get 438.7. Notice that moving the decimal point one place to the right is the same as multiplying by 10. Similarly, moving the decimal point two places to the right is the same as multiplying by 100, moving it three places to the right is the same as multiplying by 1000, etc. Since multiplication and division are inverse operations, this means that moving the decimal point one place to the left is the same as dividing by 10, moving it two places to the left is the same as dividing by 100, etc.

If we want to multiply two numbers like 15.742 and 98.39, let us multiply 15.742 by 1000 by moving the decimal point three places to the right and let us multiply 98.39 by 100 by moving the decimal point two places to the right. We now have 15742 and 9839. We multiply these two numbers in the ordinary way.

```
   15742
  ×9839
  141678
  47226
 125936
141678
154885538
```

Since we multiplied one number by 1000 and the other by 100, we must make up for this by dividing the answer by 1000 and then by 100. We can do this all at once by moving the decimal point five places to the left in the answer. Thus we have 15.472 × 98.39 = 1548.85538.

From this example we see that the rule for multiplying decimals is as follows:

> Ignore the decimal points and multiply the numbers in the ordinary way. Then add the number of places (starting from the right) to the decimal point in the two numbers and count off this many places (starting from the right) to locate the decimal point in the answer.

examples

13 Multiply 1.873 × 19.5.

► Since 1.873 has three decimals places and 19.5 has one decimal place, we multiply the numbers and then point off four places in the answer:

```
   1.873
  ×19.5
    9365
  16857
  1873
 36.5235
```

14 Multiply 4.92 × 0.0075.

► The first number has two decimal places and the second has four decimal places, so we multiply and then point off six decimal places in the answer:

```
     4.92
  ×0.0075
     2460
    3444
 0.036900
```

Notice that we had to add on a 0 on the left so that we could point off six places.

When we do a division problem such as 186.957 ÷ 7.64, we must remember that this is the same as 186.957/7.64. The long-division process is designed for division by whole numbers, so we move the decimal point two places to the right in the denominator, and we must then make up for this by moving the decimal point two places to the right in the numerator. What we have done is to multiply the denominator and numerator by 100. We see that 186.957 ÷ 7.64 is the same as 18695.7 ÷ 764. We carry out this long division and get

```
          24.4
    764|18695.7
        1528
         3415
         3056
           3597
           3056
            541
```

Notice that the decimal point is moved straight up into the quotient.

If we stop here, we have 24.4 as the answer to one decimal place. It is not exact because we had a remainder left over. If we wanted to carry it out further, we could add several 0's and have

```
          24.4708
    764|18695.70000
        1528
         3415
         3056
           3597
           3056
            5410
            5348
               620
               000
               6200
               6112
```

We have now carried the answer out to four decimal places.

examples

15 Divide 15.7 by 1.965.

► We have to move the decimal point three places to the right in both numbers, and in order to do this we must add two 0's on the right to 15.7:

```
                7.9898
1.965.|15.700.0000
       13 755
        1 9450
        1 7685
          17650
          15720
           19300
           17685
            16150
            15720
              430
```

We have added four more 0's to the right of 15700 so that we could carry out the answer to four decimal places.

16 Divide 21.18 by 0.017; carry out the answer to three decimal places.

```
►           1 245.882
0.017.|21.180.000
       17
        41
        34
         78
         68
         100
          85
          150
          136
           140
           136
              40
              34
               6
```

17 Divide 0.019 by 4.8; carry out the answer to five decimal places.

```
►        0.00395
4.8.|0.0.19000
        144
         460
         432
          280
          240
           40
```

The statement that every rational number, when expressed in decimal form, must either be a terminating or repeating decimal leads one to ask: What about the nonterminating, nonrepeating decimals? The fact that there are such decimals leads one to believe that there are numbers that are not rational numbers. This turns out to be the case, and these *irrational* numbers fill in the holes left in the number line after all the rational numbers have been accounted for.

There are numbers such as $\sqrt{2}$ (read "the square root of 2" and defined as that number which when multiplied by itself gives 2 as an answer) that are irrational. This means that there is no rational number that can be multiplied by itself to give 2 as an answer. The number $\frac{10}{7}$ comes close but does not quite make it. Other irrational numbers are $\sqrt{3}$, $\sqrt{5}$, $\sqrt{6}$, $\sqrt{7}$, $\sqrt{8}$, $\sqrt{10}$, $\sqrt{11}$, etc.

We have seen that a decimal represents a rational number if and only if it is either terminating or repeating. This means that a nonterminating, nonrepeating decimal must be an irrational number and vice versa, so we have the following classification of decimals:

$$\left.\begin{array}{r}\text{terminating decimals}\\ \text{repeating decimals}\end{array}\right\} \quad \text{rational numbers}$$

$$\text{nonterminating, nonrepeating decimals}\} \quad \text{irrational numbers}$$

definition 2.4 The set of all decimals (both rational and irrational numbers) constitutes the *real numbers.*

Since nonterminating, nonrepeating decimals can never be written in their entirety, we shall not be concerned here with addition, subtraction, multiplication, or division of such decimals, although the system of real numbers is closed with respect to addition, subtraction, multiplication, and division (except for division by zero), and it satisfies all the properties of the rationals enumerated at the end of Section 2.4. The four arithmetic operations on the real numbers may be carried out geometrically by making use of the number line. We shall show in Section 3.3 ways of handling irrational numbers that may be written in compact form, such as $\sqrt{2}$.

To show that the real numbers actually do fill up the entire number line, we shall show that any point on the number line may be represented by a decimal. Take a point such as the one indicated by x in Figure 2.5. We will assume that the point is to the right of 0. (If it is to the left of 0, it is a negative number, and a procedure similar to one that we shall describe—counting from right to left instead

figure 2.5

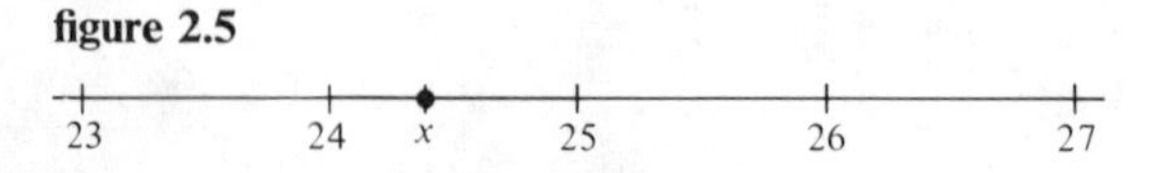

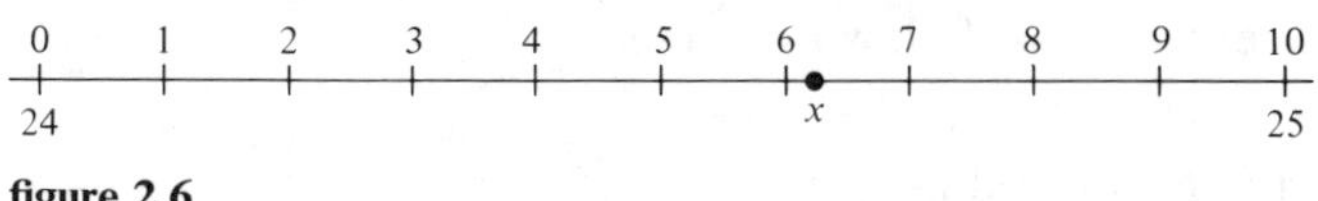

figure 2.6

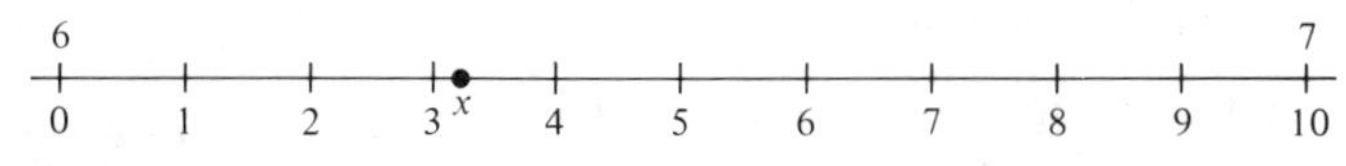

figure 2.7

of from left to right and putting a minus sign in front of the decimal obtained—will yield the answer.)

For the whole-number part of the decimal, take the whole number immediately to the left of x. In Figure 2.5 we take 24 for the whole-number part of our decimal. (If x falls on a whole-number division, then we are done, because any whole number may be considered as a decimal by putting a decimal point at the end of it.) Then divide the interval in which x lies into 10 equal parts.

For the first decimal place take the digit between 0 and 9 immediately to the left of x. (In Figure 2.6 we take 6.) If x falls on one of the division points, we use this digit in the first decimal place and we are done. If x does not fall on a division point, divide the interval in which it is now contained into 10 equal parts and put the digit immediately to the left of x in the second decimal place. (In Figure 2.7, we use 3, so the decimal part now reads 0.63, and our number now reads 24.63.) Continuing in this way we get the decimal that is represented by the point x. If at some point x falls on one of the division points, we have a terminating decimal, and the process is complete. If x never falls on a division point, then we have a nonterminating decimal (which may be a repeating one or not), and we can write out as many places of the infinite decimal as we wish.

Since every point on the number line corresponds to a decimal as derived by the process above, we see that the real numbers entirely fill up the number line, which for this reason is often referred to as the *real-number line*. The property of filling up the real-number line is referred to as the *completeness property* of the real numbers.

The system of real numbers is about as far as we need go in deriving a number system suitable for our needs, since it satisfies properties 1 through 14 listed in Section 2.4 as well as the completeness property, and it suffices for most practical applications in mathematics and in related areas.

exercises 2.5

1 Express the following numbers in words:

(a) 8.12 (b) 173.008 (c) 19.1431 (d) 80.9

2 Write the numerals for the following numbers:

(a) Twenty-six and thirty-two hundredths
(b) Forty and four hundredths
(c) Eight and three hundred and two thousandths
(d) Sixteen and four ten thousandths

3 Which of the following are rational numbers?

(a) $4\frac{3}{8}$ (b) 5.172 (c) 1.646464 . . . (d) $\sqrt{2}$

(e) $\sqrt{3}$ (f) $\sqrt{4}$ (g) -10

4 Change to decimals:

(a) $4\frac{3}{8}$ (b) $2\frac{4}{7}$ (c) $26\frac{1}{3}$ (d) $9\frac{3}{5}$ (e) $3\frac{1}{12}$ (f) $\frac{1}{7}$

5 Change to fractions:

(a) 4.173 (b) 9.2714 (c) 3.151515 . . . (d) 26.2222 . . . (e) 50.005

6 Add:

(a) $6.192 + 58.7 + 0.0025 + 183.154$
(b) $26.9 + 0.0004 + 17 + 2.9$
(c) $0.0005 + 0.003 + 0.01 + 0.8$

7 Subtract:

(a) $19.257 - 8.3$ (b) $14 - 0.742$ (c) $1.8 - 0.935$

8 Multiply:

(a) $\begin{array}{r} 58.624 \\ \times 4.93 \\ \hline \end{array}$ (b) $\begin{array}{r} 159.7 \\ \times 0.0024 \\ \hline \end{array}$ (c) $\begin{array}{r} 0.053 \\ \times 0.129 \\ \hline \end{array}$

9 Divide:

(a) $12\overline{)787.924}$ (b) $18.7\overline{)39.215}$ (c) $0.015\overline{)581.37}$

10 Which decimals do the following points represent?

(a) 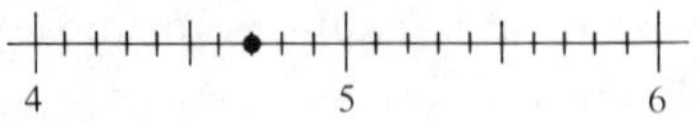 4 5 6

(b) 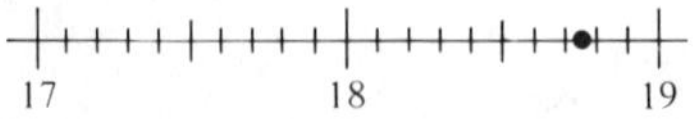 17 18 19

(c) 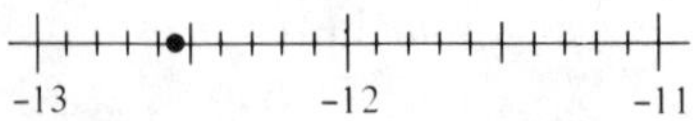 -13 -12 -11

11 (a) Multiply 8.7×4.625.
(b) Change the numbers in part (a) to fractions, multiply them, and then change the product back to a decimal to check your answer in part (a).

12 Describe a method for changing a repeating decimal to a fraction when the decimal does not start repeating right after the decimal point.

2.6 / ratio and percent

We have seen that $6 \div 5$ means the same thing as $\frac{6}{5}$. In other words, 6 divided by 5 is the same as the fraction $\frac{6}{5}$. Another way that this is sometimes written is $6:5$, and when it is written this way, it is called the *ratio* of 6 to 5. The ratio of 6 to 5 means $6 \div 5$ or $\frac{6}{5}$. If you were asked, "What is the ratio of 25 to 40?" you should answer, "The ratio of 25 to 40 is $\frac{25}{40}$ or $\frac{5}{8}$."

examples

1 What is the ratio of 2 in. to 5 ft?
► Since 5 ft is the same as 60 in., the ratio of 2 in. to 5 ft is $2:60$, which may be written $\frac{2}{60}$ or $\frac{1}{30}$, so the ratio 2 in. to 5 ft is $\frac{1}{30}$.

2 What is the ratio of $\frac{2}{3}$ to $\frac{7}{8}$?

$$► \frac{2}{3} : \frac{7}{8} = \frac{2}{3} \div \frac{7}{8} = \frac{2}{3} \cdot \frac{8}{7} = \frac{16}{21}$$

3 If John got a grade of 78 on an exam and Bill got a grade of 96, what is the ratio of their grades?
► The ratio of their grades is $78:96$ or $\frac{78}{96}$, which may be reduced to $\frac{13}{16}$.

4 What is the ratio of \$5.00 to \$0.12?
► Since \$5.00 is the same as 500 cents, the ratio is $500:12$, which is the same as $\frac{500}{12}$ or $\frac{125}{3}$.

5 Are the numbers 6 and 8 in the same ratio as the numbers 15 and 20?
► Since $\frac{6}{8} = \frac{3}{4}$ and $\frac{15}{20} = \frac{3}{4}$, these numbers are in the same ratio.

6 What is the ratio of a side of a square to the perimeter of the square?
► Since the perimeter of a square is the sum of the four equal sides, the ratio of the side to the perimeter is $1:4$ or $\frac{1}{4}$.

Another way of expressing ratios is by percent. *Percent means hundredths.* Twenty-five percent means twenty-five hundredths, which may be written 0.25 or $\frac{1}{4}$. Similarly, 12 percent means 0.12, 5 percent means 0.05, etc. We see that to *change a percent to a decimal we move the decimal point two places to the left.*

In the opposite direction, since 0.12 means 12 hundredths or 12 percent, we see that *to change a decimal to percent we move the decimal point two places to the right.*

examples

7 Change the following to decimals[1]:

(a) 18% (b) 12.5% (c) 350% (d) 2% (e) 0.1% (f) 0.06%

► (a) 18% = 0.18 (b) 12.5% = 0.125 (c) 350% = 3.5
(d) 2% = 0.02 (e) 0.1% = 0.001 (f) 0.06% = 0.0006

8 Change the following to percent:

(a) 0.15 (b) 1.87 (c) $\frac{3}{4}$ (d) 0.006 (e) $1\frac{5}{8}$ (f) $\frac{1}{1000}$

► (a) 0.15 = 15% (b) 1.87 = 187%

(c) $\frac{3}{4} = 0.75 = 75\%$ (d) 0.006 = 0.6%

(e) $1\frac{5}{8} = 1.625 = 162.5\%$ (f) $\frac{1}{1000} = 0.001 = 0.1\%$

9 Change the following to fractions or mixed numbers:

(a) 18% (b) 12.5% (c) 125% (d) 0.2% (e) 5% (f) 350%

► (a) $18\% = 0.18 = \frac{18}{100} = \frac{9}{50}$ (b) $12.5\% = 0.125 = \frac{125}{1000} = \frac{1}{8}$

(c) $125\% = 1.25 = 1\frac{25}{100} = 1\frac{1}{4}$ (d) $0.2\% = 0.002 = \frac{2}{1000} = \frac{1}{500}$

(e) $5\% = 0.05 = \frac{5}{100} = \frac{1}{20}$ (f) $350\% = 3.5 = 3\frac{1}{2}$

When we ask How much is 12% of 250? we mean How much is 12% times 250? or How much is 0.12 × 250? In other words, *to take a certain percent of a number, we change the percent to a decimal, and then we multiply.* This tells us that 12% of 250 is the same as 250 × 0.12, which is 30.

examples

10 How much is 5% of 700?

► 5% of 700 = 700 × 0.05 = 35

[1]The symbol % means "percent."

11 How much is 40% of 800?

► 40% of 800 = 800 × 0.40 = 320

12 If you get a 10% discount on a radio costing \$35.00, how much must you pay for it?

► First we figure 10% of \$35.00. This is \$35.00 × 0.10 = \$3.50. We then subtract this from \$35.00:

$$\begin{array}{r} \$35.00 \\ -3.50 \\ \hline \$31.50 \end{array}$$

The radio would thus cost you \$31.50.

13 If you are paying \$140.00 rent per month, and your landlord tells you that he is increasing your rent by 6%, how much will you have to pay?

► First, 6% of \$140.00 = \$140.00 × 0.06 = \$8.40. Adding this to \$140.00, you will have to pay \$140.00 + \$8.40 = \$148.40.

14 If your income tax comes to \$568.00 and you have to pay a 5% surcharge on this, how much is the surcharge?

► The surcharge is 5% of \$568.00 or \$568.00 × 0.05, which is equal to \$28.40.

15 If you have to pay \$0.80 tax on a purchase of \$16.00, what percent tax are you paying?

► The ratio of \$0.80 to \$16.00 (which is equal to 1600 cents) is $\frac{80}{1600}$ or $\frac{1}{20}$. We change this to a decimal and then to percent. $\frac{1}{20} = 0.05 = 5\%$, so your tax rate is 5%.

16 If you have to pay \$23 interest on a loan of \$400, what percent interest are you paying?

► The ratio of \$23 to \$400 is $\frac{23}{400}$. Changing $\frac{23}{400}$ to a decimal yields 0.0575, and then changing this to percent gives 5.75%. Thus you are paying 5.75% or $5\frac{3}{4}\%$ interest on the loan.

17 If you make a loan of \$800, and you must pay 6% of this amount in interest each year for 5 years, how much interest will you pay?

► Each year you will pay 800 × 0.06 = \$48 in interest. Thus, for 5 years you will pay \$48 × 5 or \$240 in interest on the loan.

exercises 2.6

1 Express the following ratios as fractions:

(a) 5 : 9 (b) 6 : 9 (c) 12 : 10
(d) 4 : 2 (e) 250 : 10 (f) 17 : 68

2 Express the following ratios as decimals:

(a) 12 : 9 (b) 2 : 8 (c) 7 : 4
(d) 26 : 2 (e) 240 : 6 (f) 5 : 1000

3 Express the following ratios as percent:

(a) 24 : 10 (b) 4 : 150 (c) 7 : 2000

(d) 8 : 152 (e) 12 : 5 (f) 12 : 500

4 What is the ratio of 3 hours to 5 days?

5 What is the ratio of 8 yd to 3 in.?

6 What is the ratio of 5 lb to 6 oz?

7 Change the following to percent:

(a) 4.9 (b) 0.05 (c) 62.5

(d) 0.0003 (e) $\frac{5}{2}$ (f) $\frac{5}{8}$

8 Change the following to decimals:

(a) 22% (b) 1.8% (c) 1525%

(d) 0.1% (e) $\frac{1}{2}$ % (f) $\frac{2}{3}$ %

9 How much is 15% of 380?

10 How much is 120% of 600?

11 If the price of a ticket originally costing $2.80 is raised by 5%, what is the new price of the ticket?

12 If you can get a 30% discount on a record marked $4.20, how much will it cost you?

13 If you deposit $250 in a bank and you get 4% interest on this deposit each year, how much will you have at the end of the year?

14 If you pay $0.12 tax on a purchase of $3.00, what percent tax are you paying?

15 If you receive $13.40 interest on a deposit of $320.00, what percent interest are you receiving?

16 What percent of 120 is 18?

17 What percent of 300 is 125?

18 What percent of 35 is 275?

review test 2

NAME ______________

DATE ______________

1 What are the additive inverses of the following integers?

(a) -7 ________ (b) 4 ________ (c) 0 ________ (d) -16 ________

2 Perform the following computations:

(a) $-7 - (-3)$ ______________ (b) $4 \times (-6)$ ______________

(c) $(-21) \div (-3)$ ______________ (d) $-8 + (-2)$ ______________

(e) $7 + (-4)$ ______________ (f) $-3 - 8$ ______________

(g) $(-20) \div 5$ ______________ (h) $(-6) \times (-9)$ ______________

3 Perform the following computations:

(a) $(-5)[4 - (-6)]$ ______________

(b) $18 \div (7 - 9)$ ______________

(c) $[6 \times (-3)] \div (-9)$ ______________

(d) $(-5 - 3) \times (-2)$ ______________

(e) $[(-4) \times 3] - 6$ ______________

(f) $(-8) + [(-3) - (-2)]$ ______________

(g) $[(-8) + (-3)] - (-2)$ ______________

(h) $[8 - (-3)] \times (-3)$ ______________

4 If you receive 6 bills for \$3 each and 5 checks for \$4 each, how far are you ahead or behind?

5 If you travel 6 miles north, then 9 miles south, then 5 miles north, how far are you from where you started and in which direction?

6 If you add \$10 to your bank account, then withdraw \$8, and then withdraw \$6, how much has your account increased or decreased?

7 In each of the following sets of integers, pick the largest and the smallest integers:

(a) $(-3, 6, 0, -4, 9)$ ______________________

(b) $(-17, 22, 39, -21, 6)$ ______________________

(c) $(17, -28, 5, -3, 1)$ ______________________

(d) $(-4, -7, -12, 0, -5)$ ______________________

(e) $(8, 19, 0, 26, 5)$ ______________________

8 Which would you rather have, three bills for \$9 each, seven bills for \$4 each, or five bills for \$6 each?

9 Reduce to lowest terms:

(a) $\frac{28}{35}$ ______________ (b) $\frac{15}{25}$ ______________ (c) $\frac{4}{9}$ ______________

(d) $\frac{18}{24}$ ______________ (e) $\frac{26}{39}$ ______________

10 Compute:

(a) $\frac{3}{5} - \frac{7}{8}$ ______________ (b) $4\frac{2}{3} + 5\frac{1}{9}$ ______________

(c) $\frac{8}{5} \div 2\frac{1}{3}$ ______________ (d) $1\frac{7}{8} \times 3\frac{2}{5}$ ______________

(e) $4\frac{1}{8} - 2\frac{2}{7}$ ______________ (f) $3\frac{1}{2} \div 5\frac{1}{4}$ ______________

11 Compute:

(a) $5\frac{1}{2}\left(4\frac{2}{3} - 3\frac{1}{5}\right)$

(b) $\left(1\frac{1}{5} \times 3\frac{2}{3}\right) - 2\frac{1}{4}$

(c) $\dfrac{\dfrac{1}{2}+\dfrac{1}{3}}{\dfrac{1}{4}-\dfrac{1}{5}}$

(d) $1\dfrac{1}{2}+\left(2\dfrac{1}{9}\div 1\dfrac{1}{4}\right)$

(e) $\dfrac{2}{5}\times\dfrac{21}{16}\times\dfrac{10}{7}$

(f) $\left(\dfrac{5}{8}\times\dfrac{14}{15}\right)\div\dfrac{35}{24}$

12 Find the GCD and LCM of 78 and 126.

13 Change to decimals:

(a) $3\dfrac{5}{8}$ ______________________ (b) $4\dfrac{1}{6}$ ______________________

14 Change to fractions:

(a) 2.936 ______________________ (b) 7.323232 . . . ______________________

15 Add $3.74 + 19.006 + 38.1$.

16 Subtract $14.1 - 6.32$.

17 Multiply 14.029×1.65.

18 Divide $19.28 \div 2.043$.

19 What is the ratio of 2 in. to 3 yd?

20 Change to percent:

(a) 1.06 ______________ (b) $\frac{3}{5}$ ______________

21 Change to decimals:

(a) 2.5% ______________ (b) $\frac{1}{10}$ % ______________

22 If you can get a 12% discount on an item costing $55, how much would you have to pay for it?

23 If your rent is \$120 per month and it is increased by 8%, how much rent will you have to pay?

24 If you pay \$0.08 tax on a purchase of \$2.40, what percent tax are you paying?

25 What percent of 800 is 64?

chapter three

EXPONENTS AND LOGARITHMS

3.1/properties of exponents

The rule that we derived for the multiplication of natural numbers in terms of product sets suggests that multiplication is really repeated addition. If we wanted to take the sum $3 + 3 + 3 + 3 + 3$, we would merely write 5×3. In the same way exponents are merely a way of writing repeated multiplication in a compact form. If we wished to express the product $3 \times 3 \times 3 \times 3 \times 3$, we would write 3^5. The small 5 written up above the line is called an *exponent* and tells us how many 3's we are to multiply. The expression 4^3, for example, would mean $4 \times 4 \times 4$, and $4^3 = 64$. Similarly, 5^4 would stand for $5 \times 5 \times 5 \times 5$ or 625. We could express the definition in a more general form by writing

$$x^n = \overbrace{xxx \cdots x}^{n \text{ factors}}$$

where x is any number and n is a natural number. The expression x^n is read "x raised to the nth power" or, more simply, "x to the nth." The exponent thus tells us to what power we are to raise x. For some of the smaller powers we have special names: x^2 is read "x squared," and x^3 is read "x cubed." The meaning of exponents is expressed formally by Definition 3.1.

definition 3.1 The expression x^n stands for the product of n x's. The number n is called the *exponent* in this case.

By using exponents we can write certain expressions in much more compact form than would otherwise be possible. The expression $2x^3y^2z^6$, for example, stands for $2 \cdot xxxyyzzzzzz$ and is certainly much more convenient to use than the latter expression. This use of the symbols x, y, and z to stand for any numbers will be discussed in more detail in Chapter Four. For the present, we wish to derive a set of rules for doing computations with exponents.

Suppose that we wish to multiply the quantity x^m by the quantity x^n, where x is any number and m and n are natural numbers. x^m stands for the product of m x's and x^n stands for the product of n x's, so we have

$$x^m \cdot x^n = \overbrace{xxx \cdots x}^{m \text{ factors}} \cdot \overbrace{xxx \cdots x}^{n \text{ factors}} = \overbrace{xxx \cdots x}^{m + n \text{ factors}}$$

This merely says that when we take the product of m x's and multiply this by the product of n x's, we have the product of $(m + n)$ x's, and this gives us our first property of exponents:

Property I:

$$x^m x^n = x^{m+n}$$

Many people refer to this property by saying that when we multiply, we add exponents.

Next we get the analogous property for division. Suppose that we want to divide x^m by x^n where $m > n$. This may be written

$$\frac{x^m}{x^n} = \frac{\overbrace{xxx \cdots x}^{m \text{ factors}}}{\underbrace{xxx \cdots x}_{n \text{ factors}}}$$

Each of the n x's on the bottom may be canceled with one of the m x's on the top. This will cancel n of the m x's on the top and will leave the product of $(m - n)$ x's on the top which may be written x^{m-n}. For example $2^5 = 32$ and $2^3 = 8$. When we divide 2^5 by 2^3 we get 2^{5-3} or 2^2, which is 4. This checks out since $32 \div 8 = 4$. Our second property may be written as follows:

Property II:

$$\frac{x^m}{x^n} = x^{m-n}$$

This property may be described by saying that when we divide, we subtract exponents. For the case where $n > m$, we would have some extra x's left over on the bottom, so the second property would read

Property II′:

$$\frac{x^m}{x^n} = \frac{1}{x^{n-m}}$$

It should be emphasized for this property as well as for the others in this section that the symbol x always stands for the same number wherever it appears in any formula; that is, Property I tells us that $3^4 \cdot 3^7 = 3^{11}$, but it would not be correct to write $3^4 \cdot 2^7 = 6^{11}$.

Suppose that we consider next $(2 \cdot 4)^3 = 8^3 = 512$. This is the same as $2^3 \cdot 4^3 = 8 \times 64 = 512$. Similarly, for any numbers x and y and any natural number m, we have

$$(xy)^m = \overbrace{(xy)(xy)(xy)\cdots(xy)}^{m \text{ products}}$$

$$= (\overbrace{xxx\cdots x}^{m \text{ factors}})(\overbrace{yyy\cdots y}^{m \text{ factors}}) = x^m y^m$$

This is our next property:

Property III:

$$(xy)^m = x^m y^m$$

It may be stated in words by saying that to raise a product to a power, we raise each factor to that power and then multiply the results.

For our next property, consider the expression $(3^4)^2$. This is equal to $(81)^2$, which equals 6561. This is the same as $3^{4 \times 2} = 3^8$. In a similar manner, we have for any number x and any two natural numbers m and n

$$(x^m)^n = \overbrace{x^m x^m x^m \cdots x^m}^{n \text{ factors}}$$

$$= x^{\overbrace{m+m+m+m+\cdots+m}^{n \text{ terms}}} = x^{mn}$$

Therefore,

Property IV:

$$(x^m)^n = x^{mn}$$

Stated in words, Property IV says that when we raise a number to one power and then raise the result to another power, we have the original number raised to the product of the two powers.

Properties III and IV may be combined into a single property as follows:

by Property III

$$(x^m y^n)^p = (x^m)^p (y^n)^p$$

by Property IV

$$= x^{mp} y^{np}$$

so

$$(x^m y^n)^p = x^{mp} y^{np}$$

The rule for handling a quotient raised to a power will complete our set of properties for exponents for this section. Consider the expression $(8/4)^3 = 2^3 = 8$. This is the same as $8^3/4^3 = 512/64 = 8$. In general, we would have

$$\left(\frac{x}{y}\right)^n = \overbrace{\frac{x}{y} \cdot \frac{x}{y} \cdot \frac{x}{y} \cdots \frac{x}{y}}^{n \text{ factors}} = \frac{\overbrace{xxx \cdots x}^{n \text{ factors}}}{\underbrace{yyy \cdots y}_{n \text{ factors}}} = \frac{x^n}{y^n}$$

Therefore,

Property V:

$$\left(\frac{x}{y}\right)^n = \frac{x^n}{y^n}$$

In words, we say that to raise a quotient to a power, we raise the numerator to that power and the denominator to that power and then divide.

The last three properties may be combined into one by writing

$$\left(\frac{x^m y^n}{z^p}\right)^q = \frac{x^{mq} y^{nq}}{z^{pq}}$$

examples

1 $3^4 = 3 \cdot 3 \cdot 3 \cdot 3 = 81$

2 $(-8)^2 = (-8)(-8) = 64$

3 $-(8)^2 = -(8 \times 8) = -64$

4 $(-2)^3 = (-2)(-2)(-2) = -8$

5 $-2^3 = -(2 \cdot 2 \cdot 2) = -8$

6 $\left(\frac{1}{4}\right)^3 = \frac{1}{4} \cdot \frac{1}{4} \cdot \frac{1}{4} = \frac{1}{64}$

7 $\frac{3^2 \cdot 2^3}{5^2} = \frac{9 \cdot 8}{25} = \frac{72}{25}$

8 $\dfrac{8^3}{4^5} = \dfrac{8 \cdot 8 \cdot 8}{4 \cdot 4 \cdot 4 \cdot 4 \cdot 4} = \dfrac{{}^2\cancel{8} \cdot {}^2\cancel{8} \cdot {}^2\cancel{8}}{\cancel{4} \cdot \cancel{4} \cdot \cancel{4} \cdot 4 \cdot 4} = \dfrac{2 \cdot 2 \cdot 2}{4 \cdot 4} = \dfrac{8}{16} = \dfrac{1}{2}$

9 $\dfrac{9^3}{3^2} = \dfrac{9 \cdot 9 \cdot 9}{3 \cdot 3} = \dfrac{{}^3\cancel{9} \cdot {}^3\cancel{9} \cdot 9}{\cancel{3} \cdot \cancel{3}} = 3 \cdot 3 \cdot 9 = 81$

10 $\dfrac{(-4)^2}{-(6)^3} = \dfrac{(-4)(-4)}{-(6 \cdot 6 \cdot 6)} = \dfrac{16}{-216} = -\dfrac{4}{54}$

11 $(2x^3y^2z^4)^5 = 2^5(x^3)^5(y^2)^5(z^4)^5 = 32x^{15}y^{10}z^{20}$

12 $\dfrac{15x^2y^5z^3}{3x^5y^2z} = \dfrac{5y^3z^2}{x^3}$

13 $(2xy^3z^2)^4(3x^2yz^3)^2 = 16x^4y^{12}z^8 \cdot 9x^4y^2z^6 = 144x^8y^{14}z^{14}$

14 $\dfrac{(3x^2y^3z)^5}{(2xy^6z^2)^4} = \dfrac{243x^{10}y^{15}z^5}{16x^4y^{24}z^8} = \dfrac{243x^6}{16y^9z^3}$

15 $\dfrac{(2x^3y^2z^4)^2(5x^4yz^3)^4}{(4xy^3z^2)^3} = \dfrac{4x^6y^4z^8 \cdot 625x^{16}y^4z^{12}}{64x^3y^9z^6}$

$$= \dfrac{625x^{22}y^8z^{20}}{16x^3y^9z^6} = \dfrac{625x^{19}z^{14}}{16y}$$

Notice in the last example that we have applied our rule for the reduction of fractions and have canceled 4 from the numerator and denominator.

The rules for working with exponents are summarized below:

1. $x^m x^n = x^{m+n}$
2. $\dfrac{x^m}{x^n} = x^{m-n}$
3. $(xy)^m = x^m y^m$
4. $(x^m)^n = x^{mn}$

(3 and 4 combined: $(x^m y^n)^p = x^{mp} y^{np}$)

5. $\left(\dfrac{x}{y}\right)^m = \dfrac{x^m}{y^m}$

Remember that thus far we have discussed only the case in which the exponents are natural numbers. In Section 3.2 we shall investigate the case in which the exponents are negative integers, zero, or fractions.

exercises 3.1

1 Evaluate:

(a) 4^3 (b) 8^2 (c) $(\frac{1}{2})^4$
(d) 9^3 (e) $(1.8)^2$ (f) $(1.524)^3$

2 Simplify:

(a) $\frac{2^3 \cdot 3^2}{5^3}$ (b) $\frac{5^2 \cdot 3^3}{2^3}$ (c) $\frac{7^3}{2^4}$ (d) $\frac{6^2}{3^4}$

3 Simplify the following expressions:

(a) $\frac{(3x^2y^4z^2)^4(2x^4y^2z^4)^5}{(2x^3y^2z^5)^4}$ (b) $\frac{(2xy^2z^4)^3(4x^2yz^3)^4}{(3x^2yz^2)^2}$

(c) $\frac{(5x^2yz)(2x^4y^2z^4)^5}{(3xyz^2)^3}$ (d) $\frac{(x^3y^2z)^4(2x^3yz^5)^3}{(3xy^2z)^4}$

4 Evaluate:

(a) $(-6)^4$ (b) $(-2)^3(3)^2$ (c) $(-5)^2$

(d) $-(5^2)$ (e) $(-2)^2(-3)^3$

3.2 / zero, negative, and fractional exponents

The definition that x^n means the product of n x's would not make any sense if n were 0, a negative integer, or a fraction. In keeping with our practice of enlarging number systems to obtain more useful systems that retain all the properties of the original system, we shall attach a meaning to the expression x^n in these cases in such a way that properties I–V for computation with exponents will remain valid for the enlarged system. As was the case previously, we shall see that we have no choice in these definitions; if we wish the properties to remain valid, we must define these expressions in one and only one way.

First, to determine what x^0 must stand for, we note that for any positive integer n we must have, by property II,

$$\frac{x^n}{x^n} = x^0$$

We know, however, that any number (except 0) divided by itself yields an answer of 1; therefore, by the fundamental axiom of mathematics that states that things equal to the same thing are equal to each other, we must have

$$x^0 = 1$$

Next, let us consider negative exponents by looking at a specific example. Consider the expression x^2/x^5. If property II is to remain valid, we must have $x^2/x^5 = x^{-3}$. We know, however, that $x^2/x^5 = 1/x^3$, so we must define $x^{-3} = 1/x^3$.

In general, we have

$$\frac{x^0}{x^m} = \frac{1}{x^m} \quad \text{and} \quad \frac{x^0}{x^m} = x^{0-m} = x^{-m}$$

by property II, so we may write

$$x^{-m} = \frac{1}{x^m}$$

for any positive integer m.

Note that if m were a negative integer, say $m = -n$ for some positive integer n, then $x^{-m} = x^n$ and

$$\frac{1}{x^m} = \frac{1}{x^{-n}} = 1 \div x^{-n} = 1 \div \frac{1}{x^n} = 1 \cdot x^n = x^n$$

so $x^{-m} = 1/x^m$. Also, when $m = 0$, we have $x^{-m} = x^0 = 1$, and $1/x^m = 1/1 = 1$, so $x^{-m} = 1/x^m$, when m is any integer—positive, negative, or zero. Thus we have Definition 3.2.

definition 3.2 For any integer m, $x^{-m} = 1/x^m$.

It is easy to check that properties I–V for exponents remain valid for negative exponents. For example,

$$x^{-5}x^3 = \frac{1}{x^5} \cdot \frac{x^3}{1} = \frac{1}{x^2} = x^{-5+3} = x^{-2}$$

To illustrate property II,

$$\frac{x^2}{x^{-4}} = x^2 \div \frac{1}{x^4} = x^2 \cdot \frac{x^4}{1} = x^6 \quad \text{and} \quad \frac{x^2}{x^{-4}} = x^{2-(-4)} = x^6$$

Note that $1/x^{-4}$ is the same as $x^4/1$; that is, a negative exponent on the bottom is the same as a positive exponent on top.

It should be apparent that you must remember your rules for computation with integers to perform computations involving exponents.

As an example of property III, we have

$$(xy)^{-4} = \frac{1}{(xy)^4} \quad \text{and} \quad (xy)^{-4} = x^{-4}y^{-4} = \frac{1}{x^4} \cdot \frac{1}{y^4} = \frac{1}{(xy)^4}$$

To illustrate property IV

$$(x^3)^{-2} = \frac{1}{(x^3)^2} = \frac{1}{x^6} = x^{-6} \quad \text{and} \quad (x^3)^{-2} = x^{(3)(-2)} = x^{-6}$$

For property V, note that

$$\left(\frac{x}{y}\right)^{-3} = \frac{1}{(x/y)^3} = 1 \div \left(\frac{x}{y}\right)^3 = 1 \div \frac{x^3}{y^3} = 1 \cdot \frac{y^3}{x^3} = \frac{y^3}{x^3}$$

and

$$\left(\frac{x}{y}\right)^{-3} = \frac{x^{-3}}{y^{-3}} = x^{-3} \div y^{-3} = \frac{1}{x^3} \div \frac{1}{y^3} = \frac{1}{x^3} \cdot \frac{y^3}{1} = \frac{y^3}{x^3}$$

We have illustrated these properties by use of specific examples, but it should be clear that they will hold when the exponents are any integers.

examples

1 Simplify $(3x^{-2}y^4z^{-3})^{-2}$.

$$(3x^{-2}y^4z^{-3})^{-2} = (3)^{-2}x^4y^{-8}z^6 = \frac{x^4z^6}{9y^8}$$

2 Simplify $(2x^3y^{-1}z)^3(4xy^{-3}z^2)^{-2}$.

$$\begin{aligned}(2x^3y^{-1}z)^3(4xy^{-3}z^2)^{-2} &= 2^3x^9y^{-3}z^34^{-2}x^{-2}y^6z^{-4} \\ &= 8 \cdot 4^{-2}x^7y^3z^{-1} \\ &= \frac{8x^7y^3}{16z} = \frac{x^7y^3}{2z}\end{aligned}$$

3 Simplify $(3x^2y^{-1}z^{-4})^2/(2x^{-3}y^3z^{-2})^{-3}$.

$$\frac{(3x^2y^{-1}z^{-4})^2}{(2x^{-3}y^3z^{-2})^{-3}} = \frac{9x^4y^{-2}z^{-8}}{2^{-3}x^9y^{-9}z^6} = 9 \cdot 8x^{-5}y^7z^{-14} = \frac{72y^7}{x^5z^{14}}$$

To determine what meaning should be attached to fractional exponents, let us consider the expression $x^{1/2}$. According to property I, we must have $x^{1/2} \cdot x^{1/2} = x^1 = x$, so $x^{1/2}$ must be defined as that number which when multiplied by itself gives x as an answer. We have a special name for this number and another way of denoting it.

definition 3.3 The number a, for which $a^2 = x$, is called the *square root* of x and is denoted by $\sqrt{x}$ or by $x^{1/2}$. Similarly, the number b, for which $b^n = x$, is called the *n*th *root* of x and is denoted by $\sqrt[n]{x}$ or by $x^{1/n}$. The symbol $\sqrt{}$ used to denote these roots is called a *radical.*

examples

4 $\sqrt{36} = 6$, since $6^2 = 36$

5 $\sqrt[3]{8} = 2$, since $2^3 = 8$

6 $\sqrt[5]{32} = 2$, since $2^5 = 32$

7 $\sqrt{144} = 12$, since $12^2 = 144$

8 $\sqrt[4]{81} = 3$, since $3^4 = 81$

9 $\sqrt[4]{16} = 2$, since $2^4 = 16$

According to the definition, we have defined $x^{1/n}$ as that number b such that $b^n = x$, and this satisfies property III since $(x^{1/n})^n = x^{(1/n)(n)} = x^1 = x$.

It should be pointed out here that a number may have more than one square root; for example, both 5 and -5 qualify as square roots of 25. To prevent any confusion, we shall agree that $\sqrt{25}$ will stand for the positive square root 5; if we wish to talk about the negative square root, we will write $-\sqrt{25}$.

definition 3.4 The expression $\sqrt{x}$ denotes the positive square root of x. The negative square root is denoted by $-\sqrt{x}$. Similarly, $\sqrt[n]{x}$ denotes the positive real number whose nth power is x.

At the end of this section we will describe a method for computing the square root of a number written in decimal form.

We have determined the meaning that we must attach to a fractional exponent when the numerator is 1. Property IV tells us that $x^{m/n}$ is equal to $(x^{1/n})^m = (x^m)^{1/n}$, so $x^{m/n}$ must mean the mth power of the nth root of x or the nth root of the mth power of x. Both of these are the same. For example, $16^{3/4} = (16^{1/4})^3 = (\sqrt[4]{16})^3 = 2^3 = 8$ and $16^{3/4} = (16^3)^{1/4} = \sqrt[4]{16^3} = \sqrt[4]{4096} = 8$ (since $8^4 = 4096$).

definition 3.5 For any integers m and n ($n \neq 0$), $x^{m/n} = (x^{1/n})^m = (x^m)^{1/n}$.

The example above illustrates that it is usually easier to take the nth root first and then to raise this to the mth power rather than to take the mth power first and then take the nth root of this.

examples

10 Evaluate $8^{2/3}$.

$$8^{2/3} = (8^{1/3})^2 = 2^2 = 4$$

11 Evaluate $64^{4/3}$.

$$64^{4/3} = (64^{1/3})^4 = 4^4 = 256$$

12 Evaluate $27^{-2/3}$.

$$27^{-2/3} = \frac{1}{27^{2/3}} = \frac{1}{(27^{1/3})^2} = \frac{1}{3^2} = \frac{1}{9}$$

As a working rule, when we have a negative exponent, we may ignore the minus until we are finished with the computation and then merely put 1 over the answer we obtained without the minus sign.

examples

13 Evaluate $64^{2/3}$.

$$64^{2/3} = (64^{1/3})^2 = 4^2 = 16$$

14 Evaluate $64^{4/6}$.

$$64^{4/6} = (64^{1/6})^4 = 2^4 = 16$$

Notice that $64^{2/3}$ leads to the same result as $64^{4/6}$, since $\frac{2}{3} = \frac{4}{6}$.

15 Evaluate $27^{5/3}$.

$$27^{5/3} = (27^{1/3})^5 = 3^5 = 243$$

16 Evaluate $9^{1.5}$.

$$9^{1.5} = 9^{3/2} = (9^{1/2})^3 = 3^3 = 27$$

17 Evaluate $16^{2.75}$.

$$16^{2.75} = 16^{2\,3/4} = 16^{11/4} = (16^{1/4})^{11} = 2^{11} = 2048$$

18 Simplify $(9x^{1/3}y^{-1/2}z^{3/4})^{1/2}$.

$$(9x^{1/3}y^{-1/2}z^{3/4})^{1/2} = 9^{1/2}x^{1/6}y^{-1/4}z^{3/8} = \frac{3x^{1/6}z^{3/8}}{y^{1/4}}$$

19 Simplify $(16x^{-1}y^{2/3}z^{1/2})^{-1/4}(4x^2y^{1/3}z^{-2})^{1/2}$.

$$(16x^{-1}y^{2/3}z^{1/2})^{-1/4}(4x^2y^{1/3}z^{-2})^{1/2}$$

$$= (16^{-1/4}x^{1/4}y^{-1/6}z^{-1/8})(4^{1/2}x^1y^{1/6}z^{-1})$$

$$= \frac{4^{1/2}}{16^{1/4}}x^{5/4}y^0z^{-9/8} = \frac{2}{2}\cdot x^{5/4}\cdot 1\cdot z^{-9/8} = \frac{x^{5/4}}{z^{9/8}}$$

20 Simplify $[(2^{-4}\cdot 8^{2/3})/16^{-3/4}]^{-3}$.

$$\left(\frac{2^{-4}\cdot 8^{2/3}}{16^{-3/4}}\right)^{-3} = \left(\frac{\frac{1}{16}\cdot 4}{\frac{1}{8}}\right)^{-3} = \left(\frac{1}{4}\cdot 8\right)^{-3} = 2^{-3} = \frac{1}{8}$$

21 Simplify $(16/9)^{1/2}$.

$$\left(\frac{16}{9}\right)^{1/2} = \frac{16^{1/2}}{9^{1/2}} = \frac{4}{3}$$

The last example tells us that $\sqrt{m/n} = \sqrt{m}/\sqrt{n}$; that is, to take the square root of a fraction, we take the square root of the top and put it over the square root of the bottom.

Exponents also provide a way for writing very large or very small numbers in a compact form. Remember that to multiply by 10, we move the decimal point one place to the right; to multiply by 100, we move it two places to the right; and, in general, to multiply by 10^n (which is a 1 followed by n 0's), we move the decimal point n places to the right. To divide a number by 10^n (which is the same as multiplying by 10^{-n}), we move the decimal point n places to the left.

Suppose that we are dealing with astronomical quantities and have to write the number 342,000,000,000,000. This is the same as 3.42×10^{14}, which is much simpler to write. Conversely, if we want to know what 3.42×10^{14} stands for, we merely move the decimal point 14 places to the right, adding as many zeros as necessary.

If we are dealing with subatomic particles and want to write a number such as 0.0000000024, we could write 2.4×10^{-9}. Conversely, to write 2.4×10^{-9} in the long form, we would move the decimal point 9 places to the left, adding as many zeros before the 2 as necessary. This compact form of writing very large and very small numbers is known as *scientific notation.*

examples

22 Write the number 2,870,000,000,000 in scientific notation.

► If we move the decimal (from the end of the number where it is considered located) 12 places to the left, we have 2.87. This is equivalent to division by 10^{12}, so we must multiply by 10^{12} to compensate. Therefore, 2,870,000,000,000 is the same as 2.87×10^{12}.

23 Write 0.000000024 in scientific notation.

► $0.000000024 = 2.4 \times 10^{-8}$

We will use a specific example to show how to take the square root of a number written in decimal form. Consider the problem of finding $\sqrt{54832.7136}$. The first thing we do is to start at the decimal point and break the number up into groups of two digits each to the left and to the right as follows:

$$\sqrt{5'48'32.'71'36}$$

The last groups to the left and right may have one or two digits. We then take the first group on the left (in this case simply 5) and determine the largest square root in 5, which is 2 (since $2^2 = 4$ and $3^2 = 9$), and we put this number directly over this first group:

$$\begin{array}{l} \;\;2 \\ \sqrt{5'48'32.'71'36} \end{array}$$

We then square the 2, put the result, 4, under the 5, and then subtract:

$$\begin{array}{l} \;\;2 \\ \sqrt{5'48'32.'71'36} \\ \;\;4 \\ \hline \;\;1 \end{array}$$

We then bring down the next group of two digits:

$$\begin{array}{l} \quad 2 \\ \sqrt{5'48'32.'71'36} \\ \quad 4 \\ \hline \quad 1\;48 \end{array}$$

We then form a trial divisor by doubling the 2 on top to get 4 and leaving a blank space next to the 4 for a digit yet to be determined; the trial divisor is therefore forty-something.

$$\begin{array}{r|l} & \quad 2 \\ & \sqrt{5'48'32.'71'36} \\ & \quad 4 \\ \hline 4\_ & 1\;48 \end{array}$$

Forty goes into 148 about 3 times; so we put a 3 on top above the group 48 and also in the blank space, and we multiply 3×43, put the result, 129, under 148 and subtract:

$$\begin{array}{r|l} & \quad 2\;\;3 \\ & \sqrt{5'48'32.'71'36} \\ & \quad 4 \\ \hline 43 & 1\;48 \\ & 1\;29 \\ \hline & \quad\;\; 19 \end{array}$$

If the product had been larger than 143, we would have erased the two 3's, used 2's instead, and proceeded as before. If the remainder upon subtraction had been greater than 43, we would have gone up to 4's instead of 3's.

We now bring down the next group of two digits, double the 23 on top to get 46, and leave a blank space to form a new trial divisor, four hundred and sixty-something, and proceed as before:

$$\begin{array}{r|l} & \quad 2\;\;3 \\ & \sqrt{5'48'32.'71'36} \\ & \quad 4 \\ \hline 4\underline{3} & 1\;48 \\ & 1\;29 \\ \hline 46\_ & \quad\;\; 19\;32 \end{array}$$

Four hundred and sixty goes into 1932 about four times, so the next number that we put on top (over the group 32) and in the blank space is 4, and we finish the problem by repeating the above sequence of steps until we have carried the answer out to as many decimal places as we wish. We can always add as many zeros to the right of the decimal point as we wish if it is desired to carry the answer to more than two decimal places:

```
              2  3  4.  1  6
            √5'48'32.'71'36'
              4
       43 | 1 48
            1 29
      464 |   19 32
              18 56
     4681 |      76 71
                 46 81
    46826 |      29 90 36
                 28 09 56
                  1 80 80
```

The decimal point is located in the answer by moving it straight upward. To check our answer we multiply 234.16 by itself and then add 1.8080 to see that we get back the original number 54832.7136.

examples

24 Find $\sqrt{1000}$ to one decimal place.

```
               3  1.  6  2
             √10'00.'00'00
               9
        61 |  1 00
                61
       626 |    39 00
                37 56
      6322 |     1 44 00
                 1 26 44
                    17 56
```

Rounding the answer off to one decimal place yields $\sqrt{1000} = 31.6$.

25 Find $\sqrt{2}$ to two decimal places.

►

```
            1. 4  1  4
          √2./00/00/00
            1
      24 | 1 00
             96
     281 |    4 00
              2 81
    2824 |    1 19 00
              1 12 96
                 6 04
```

Rounding 1.414 off to two decimal places yields $\sqrt{2} = 1.41$ as our answer.

exercises 3.2

1 Evaluate:

(a) 2^{-3} (b) 3^{-2} (c) $27^{2/3}$ (d) $27^{-2/3}$ (e) $-27^{2/3}$

(f) $\dfrac{1}{3^2}$ (g) $\dfrac{1}{3^{-2}}$ (h) $16^{1/2}$ (i) $16^{2/4}$ (j) $16^{-1/2}$

2 Simplify:

(a) $(3xy^{-3}z)^{-2}(2x^{-3}yz^2)^2$ (b) $\dfrac{(2x^{-1}y^{-3}z)^2}{(3xy^{-3}z)^{-2}}$

(c) $\dfrac{(3x^2y^{-1}z^3)^2(2x^4y^{-1}z^{-2})^{-3}}{(2x^{-3}y^2z^{-3})^4}$ (d) $\dfrac{(2xy^{-3}z^{-2})^{-4}(3xy^{-1}z^{-2})^2}{(2x^4y^{-2}z^{-3})^{-3}}$

3 Evaluate:

(a) $16^{1.5}$ (b) $16^{0.75}$ (c) $32^{1.4}$
(d) $32^{-1.4}$ (e) $81^{1.25}$ (f) $81^{-0.75}$

4 Simplify:

(a) $\dfrac{2^{-3}}{5^2}$ (b) $\dfrac{3^{-2}\cdot 2^3}{4^{1.5}}$ (c) $(x^{2/3}y^{1/2})^2$ (d) $\dfrac{(4x^{1/2}y^{3/5})^{1/2}}{(2x^{3/2}y^{1/3})^2}$

5 Compute the square root of the following numbers to two decimal places:

(a) 10 (b) 2 (c) 1.65 (d) 18412.9 (e) 61.935 (f) 218.15

3.3/operations on irrational expressions

In Chapter Two we saw that the system of rational numbers was closed with respect to addition, subtraction, multiplication, and division (except by zero); therefore, given two rational numbers, we know how to find their sum, difference, product, and quotient (when the divisor is not zero), whether they are written as fractions or decimals. We can even perform these operations when one is written as a fraction and the other as a decimal by changing the decimal to a fraction or vice versa.

In the case of irrational numbers, however, it is a different story. Suppose that we are asked to find the sum of $\sqrt{2}$ and $\sqrt{5}$. If we were asked to find the sum of 2 and 5, we would say that the answer is 7 and would mean by this that 7 is another way of writing $2 + 5$. In the case of $\sqrt{2} + \sqrt{5}$, however, there is not a single symbol that is used to stand for $\sqrt{2} + \sqrt{5}$, and all that we could write by way of an answer would be $\sqrt{2} + \sqrt{5}$. If we are willing to settle for an approximate answer, we can evaluate $\sqrt{2}$ and $\sqrt{5}$ as decimals, rounded off to say three decimal places, and obtain

$$\begin{aligned} \sqrt{2} &\approx 1.414 \\ \sqrt{5} &\approx 2.236 \\ \hline \sqrt{2} + \sqrt{5} &\approx 3.650 \end{aligned}$$

The symbol $\approx$ means "is approximately equal to." We cannot use an $=$ sign here, because when we round off decimals, we are making an approximation.

It is true that, in general, when we wish to add or subtract two expressions involving radicals, all that we can do is rewrite the original sum or difference; there is no simpler form in which to write it. In some cases, however, it is possible to reduce expressions involving radicals to a simpler form.

To do this, we go back to the distributive law, $m(n + p) = mn + mp$ for any real numbers m, n, and p. Written backward it reads $mn + mp = m(n + p)$, and when written this way it is usually referred to as a *factoring rule.* In words, this rule says that when we have a sum of two terms and each term has the same factor, then this sum may be written as the product of this common factor by the sum of the other two factors. Making this simplication is referred to as *factoring out a common factor.*

We apply this factoring procedure by considering a sum such as $\sqrt{27} + \sqrt{48}$. We know that $\sqrt{27}$ may be written $(27)^{1/2} = (3 \cdot 9)^{1/2}$, which, according to property III for exponents, is equal to $3^{1/2} \cdot 9^{1/2} = \sqrt{3} \cdot \sqrt{9} = \sqrt{3} \cdot 3$. Similarly, $\sqrt{48} = \sqrt{3 \cdot 16} = \sqrt{3} \cdot \sqrt{16} = \sqrt{3} \cdot 4$. We therefore have $\sqrt{27} + \sqrt{48} = \sqrt{3} \cdot 3 + \sqrt{3} \cdot 4$ and factoring out the common factor $\sqrt{3}$ yields $\sqrt{3}(3 + 4) = \sqrt{3} \cdot 7$, which is usually written $7\sqrt{3}$.[1]

[1]We do not really need the distributive law here if we are willing to agree that $\sqrt{3} \cdot 3 + \sqrt{3} \cdot 4 = 3\sqrt{3} + 4\sqrt{3}$ is the same as $7\sqrt{3}$. The distributive law written backward as a factoring rule guarantees that this is so, in case we have any doubts about it.

examples

1 Simplify $\sqrt{50} + \sqrt{32}$.

► $\sqrt{50} + \sqrt{32} = \sqrt{25 \cdot 2} + \sqrt{16 \cdot 2} = \sqrt{25} \cdot \sqrt{2} + \sqrt{16} \cdot \sqrt{2} = 5\sqrt{2} + 4\sqrt{2} = 9\sqrt{2}$

2 Simplify $\sqrt{20} + \sqrt{45} - \sqrt{500}$.

► $\sqrt{20} + \sqrt{45} - \sqrt{500} = \sqrt{4 \cdot 5} + \sqrt{9 \cdot 5} - \sqrt{100 \cdot 5} = \sqrt{4}\,\sqrt{5} + \sqrt{9}\,\sqrt{5} - \sqrt{100}\,\sqrt{5} = 2\sqrt{5} + 3\sqrt{5} - 10\sqrt{5} = -5\sqrt{5}$

3 Simplify $\sqrt{1/2} + \sqrt{2/9}$.

$$► \sqrt{\frac{1}{2}} + \sqrt{\frac{2}{9}} = \sqrt{\frac{2}{4}} + \sqrt{\frac{2}{9}} = \sqrt{2 \cdot \frac{1}{4}} + \sqrt{2 \cdot \frac{1}{9}}$$

$$= \sqrt{2} \cdot \sqrt{\frac{1}{4}} + \sqrt{2} \cdot \sqrt{\frac{1}{9}} = \sqrt{2} \cdot \frac{1}{2} + \sqrt{2} \cdot \frac{1}{3}$$

$$= \sqrt{2}\left(\frac{1}{2} + \frac{1}{3}\right) = \sqrt{2}\left(\frac{5}{6}\right) = \frac{5}{6}\sqrt{2}$$

which may also be written $(5\sqrt{2})/6$.

4 Simplify $\sqrt{1/3} + \sqrt{12}$.

$$► \sqrt{\frac{1}{3}} + \sqrt{12} = \sqrt{\frac{3}{9}} + \sqrt{3 \cdot 4} = \sqrt{3} \cdot \sqrt{\frac{1}{9}} + \sqrt{3} \cdot \sqrt{4}$$

$$= \sqrt{3} \cdot \frac{1}{3} + \sqrt{3} \cdot 2 = \sqrt{3} \cdot \left(\frac{7}{3}\right) = \frac{7\sqrt{3}}{3}$$

Property III for exponents tells us that $\sqrt{a}\,\sqrt{b} = \sqrt{ab}$; that is, the product of two square roots is the square root of the product.

5 Simplify $\sqrt{18} \cdot \sqrt{2}$.

► $\sqrt{18} \cdot \sqrt{2} = \sqrt{36} = 6$

6 Simplify $\sqrt{3} \cdot \sqrt{24} \cdot \sqrt{2}$.

► $\sqrt{3} \cdot \sqrt{24} \cdot \sqrt{2} = \sqrt{3 \cdot 24 \cdot 2} = \sqrt{144} = 12$

7 Simplify $\sqrt{24}/\sqrt{96}$.

By property V for exponents

$$\frac{\sqrt{24}}{\sqrt{96}} = \sqrt{\frac{24}{96}} = \sqrt{\frac{1}{4}} = \frac{1}{2}$$

Alternatively,

$$\frac{\sqrt{24}}{\sqrt{96}} = \frac{\sqrt{4 \cdot 6}}{\sqrt{16 \cdot 6}} = \frac{\sqrt{4} \cdot \sqrt{6}}{\sqrt{16} \cdot \sqrt{6}} = \frac{\sqrt{4}}{\sqrt{16}} = \frac{2}{4} = \frac{1}{2}$$

8 Simplify $(\sqrt{27} + \sqrt{75})/\sqrt{108}$.

$$\blacktriangleright \quad \frac{\sqrt{27} + \sqrt{75}}{\sqrt{108}} = \frac{\sqrt{9 \cdot 3} + \sqrt{25 \cdot 3}}{\sqrt{36 \cdot 3}} = \frac{\sqrt{9} \cdot \sqrt{3} + \sqrt{25} \cdot \sqrt{3}}{\sqrt{36} \cdot \sqrt{3}}$$

$$= \frac{3\sqrt{3} + 5\sqrt{3}}{6\sqrt{3}} = \frac{8\sqrt{3}}{6\sqrt{3}} = \frac{8}{6} = \frac{4}{3}$$

It should be emphasized that these examples have been specially prepared to permit simplification. If you make up an expression involving radicals arbitrarily or if you encounter one in a practical problem, chances are that you will not be able to simplify it but will probably have to leave it as is.

Another method for simplifying expressions involving radicals is the process known as *rationalizing the denominator*. Suppose that we wish to express the number $2/\sqrt{5}$ as a decimal carried out to three decimal places. We write $\sqrt{5} \approx 2.236$ and then we have to perform the long-division computation:

$$\begin{array}{r|r}
 & 0.8944 \\
\hline
2.236 & 2.000\;0000 \\
 & 1\;788\;8 \\
\hline
 & 211\;20 \\
 & 201\;24 \\
\hline
 & 9\;960 \\
 & 8\;944 \\
\hline
 & 1\;0160 \\
 & 8944 \\
\hline
 & 1216
\end{array}$$

Rounding the answer off to three decimal places yields $2/\sqrt{5} = 0.894$.

We can avoid the tedious long-division computation if we multiply the numerator and denominator of $2/\sqrt{5}$ by $\sqrt{5}$ to obtain

$$\frac{2}{\sqrt{5}} \cdot \frac{\sqrt{5}}{\sqrt{5}} = \frac{2\sqrt{5}}{5}$$

Then $2\sqrt{5} = 2(2.236) = 4.472$ and dividing this by 5 yields

$$\frac{2\sqrt{5}}{5} = \frac{4.472}{5} = 0.894$$

as before, but with much less work. This process is called rationalizing the denominator because we have converted a fraction with an irrational denominator into one with a rational denominator.

examples

9 Evaluate $5/\sqrt{7}$ to three decimal places by rationalizing the denominator.

$$\blacktriangleright \quad \frac{5}{\sqrt{7}} = \frac{5\sqrt{7}}{7} = \frac{5(2.6457)}{7} = \frac{13.2285}{7} = 1.8897$$

which rounds off to 1.890.

We carried $\sqrt{7}$ out to four decimal places even though the problem called for an answer carried out to three decimal places, because when 5 is multiplied by the fourth decimal place, this could have an effect on the third decimal place in the final answer.

10 Evaluate $4\sqrt{8}/\sqrt{3}$ to three decimal places.

$$\blacktriangleright \quad \frac{4\sqrt{8}}{\sqrt{3}} = \frac{4\sqrt{8}\sqrt{3}}{\sqrt{3}\sqrt{3}} = \frac{4\sqrt{24}}{3} = \frac{4\sqrt{4}\sqrt{6}}{3} = \frac{4 \cdot 2\sqrt{6}}{3}$$

$$= \frac{8\sqrt{6}}{3} \approx \frac{8(2.4498)}{3} = \frac{19.5984}{3} = 6.5328$$

which rounds off to 6.533.

Consider next the problem of rationalizing the denominator of the fraction $5/(\sqrt{7} + \sqrt{3})$. Notice that if we multiply $\sqrt{7} + \sqrt{3}$ by $\sqrt{7} - \sqrt{3}$ we have, by the distributive law,

$$\begin{aligned}(\sqrt{7} + \sqrt{3})(\sqrt{7} - \sqrt{3}) &= (\sqrt{7} + \sqrt{3})\sqrt{7} - (\sqrt{7} + \sqrt{3})\sqrt{3} \\ &= \sqrt{7}\sqrt{7} + \sqrt{3}\sqrt{7} - \sqrt{7}\sqrt{3} - \sqrt{3}\sqrt{3} \\ &= 7 - 3 \\ &= 4\end{aligned}$$

Thus if we multiply the numerator and denominator of $5/(\sqrt{7} + \sqrt{3})$ by $\sqrt{7} - \sqrt{3}$, we obtain

$$\frac{5}{\sqrt{7} + \sqrt{3}} \cdot \frac{\sqrt{7} - \sqrt{3}}{\sqrt{7} - \sqrt{3}} = \frac{5(\sqrt{7} - \sqrt{3})}{4}$$

$$\approx \frac{5(2.6457 - 1.7321)}{4} = \frac{5(0.9136)}{4} = \frac{4.5680}{4} = 1.1420$$

which rounds off to 1.142.

As a working rule, we may say that when the denominator is of the form $\sqrt{m} + \sqrt{n}$, it may be rationalized by multiplying the numerator and the denominator by $\sqrt{m} - \sqrt{n}$, and when the denominator is of the form $\sqrt{m} - \sqrt{n}$, it may be rationalized by multiplying the numerator and the denominator by $\sqrt{m} + \sqrt{n}$.

examples

11 Evaluate $6/(\sqrt{8} - \sqrt{3})$.

$$\blacktriangleright\ \frac{6}{\sqrt{8} - \sqrt{3}} = \frac{6}{\sqrt{8} - \sqrt{3}} \cdot \frac{\sqrt{8} + \sqrt{3}}{\sqrt{8} + \sqrt{3}} = \frac{6(\sqrt{8} + \sqrt{3})}{8 - 3}$$

$$\approx \frac{6(2.8284 + 1.7321)}{5} = \frac{6(4.5605)}{5} = \frac{27.3630}{5} = 5.4726$$

which rounds off to 5.473.

12 Evaluate $4\sqrt{3}/(\sqrt{2} + \sqrt{5})$.

$$\blacktriangleright\ \frac{4\sqrt{3}}{\sqrt{2} + \sqrt{5}} = \frac{4\sqrt{3}}{\sqrt{2} + \sqrt{5}} \cdot \frac{\sqrt{2} - \sqrt{5}}{\sqrt{2} - \sqrt{5}} = \frac{4\sqrt{3}(\sqrt{2} - \sqrt{5})}{2 - 5}$$

$$\approx \frac{4(1.7321)(1.4142 - 2.2361)}{-3} = \frac{4(1.7321)(-0.8219)}{-3}$$

$$= \frac{(6.9284)(-0.8219)}{-3} = \frac{-5.69445196}{-3} = 1.89815065$$

which rounds off to 1.898.

13 Evaluate $5\sqrt{2} + (1/\sqrt{2})$.

$$\blacktriangleright\ 5\sqrt{2} + \frac{1}{\sqrt{2}} = \frac{5\sqrt{2}}{1} + \frac{1}{\sqrt{2}} = \frac{5\sqrt{2}\sqrt{2} + 1}{\sqrt{2}}$$

$$= \frac{5 \cdot 2 + 1}{\sqrt{2}} = \frac{11}{\sqrt{2}} = \frac{11\sqrt{2}}{2}$$

$$= \frac{11(1.414)}{2} = \frac{15.554}{2} = 7.777$$

14 Evaluate $(1/\sqrt{5}) + 6$.

$$\blacktriangleright\ \frac{1}{\sqrt{5}} + 6 = \frac{1}{\sqrt{5}} + \frac{6}{1} = \frac{1 + 6\sqrt{5}}{\sqrt{5}} = \frac{(1 + 6\sqrt{5})\sqrt{5}}{\sqrt{5}\sqrt{5}}$$

$$= \frac{\sqrt{5} + 6 \cdot 5}{5} = \frac{\sqrt{5} + 30}{5} = \frac{2.236 + 30}{5}$$

$$= \frac{32.236}{5} = 6.447$$

exercises 3.3

1 Simplify:

(a) $\sqrt{48} + \sqrt{27} + \sqrt{75}$ (b) $\sqrt{28} + \sqrt{112} + \sqrt{63}$

(c) $\sqrt{20} + \sqrt{80} - \sqrt{45}$ (d) $\sqrt{\frac{5}{9}} + \sqrt{20}$

(e) $\sqrt{\frac{3}{4}} + \sqrt{\frac{3}{2}}$

2 Simplify:

(a) $\sqrt{32} \cdot \sqrt{2}$ (b) $\sqrt{2} \cdot \sqrt{8}$ (c) $\sqrt{20} \cdot \sqrt{5}$

(d) $\sqrt{6} \cdot \sqrt{3} \cdot \sqrt{3}$ (e) $\frac{\sqrt{2}}{\sqrt{18}}$ (f) $\frac{\sqrt{72}}{\sqrt{2}}$

3 Simplify:

(a) $\frac{\sqrt{45} + \sqrt{20}}{\sqrt{80}}$ (b) $\frac{\sqrt{112}}{\sqrt{28} + \sqrt{63}}$

4 Simplify by rationalizing the denominator:

(a) $\frac{4}{\sqrt{5}}$ (b) $\frac{2}{\sqrt{3}}$ (c) $\frac{3}{\sqrt{8}}$ (d) $\frac{2}{\sqrt{8} - \sqrt{2}}$

(e) $\frac{5}{\sqrt{3} + \sqrt{7}}$ (f) $\frac{1}{\sqrt{2} + \sqrt{3}}$ (g) $\frac{3}{\sqrt{7} - \sqrt{2}}$

3.4 / computations with logarithms

Exponents are not only useful for writing certain mathematical expressions in compact form; they are also helpful in shortening various arithmetic computations. To illustrate this we will use the following list of the first 20 powers of 2:

$2^1 = 2$ $2^6 = 64$

$2^2 = 4$ $2^7 = 128$

$2^3 = 8$ $2^8 = 256$

$2^4 = 16$ $2^9 = 516$

$2^5 = 32$ $2^{10} = 1024$

$2^{11} = 2048$ $\quad$ $2^{16} = 65{,}536$

$2^{12} = 4096$ $\quad$ $2^{17} = 131{,}072$

$2^{13} = 8192$ $\quad$ $2^{18} = 262{,}144$

$2^{14} = 16{,}384$ $\quad$ $2^{19} = 524{,}288$

$2^{15} = 32{,}768$ $\quad$ $2^{20} = 1{,}048{,}576$

Suppose that we wish to compute the product 256×1024. Since $256 = 2^8$ and $1024 = 2^{10}$, $256 \times 1024 = 2^{18} = 262{,}144$. Next let us divide 1,048,576 by 4096. We know that $1{,}048{,}576 = 2^{20}$ and $4096 = 2^{12}$, so $1{,}048{,}576 \div 4096 = 2^8 = 256$. If we wish to find 64^3, we note that $64 = 2^6$, so $64^3 = (2^6)^3 = 2^{18} = 262{,}144$.

This works very nicely as long as we are dealing with numbers that are powers of 2. If we have numbers that are not powers of 2, we can get a very rough estimate of the answer by estimating the powers of 2. For example, consider 418×650. If we estimate 418 to be about $2^{8.7}$ and 650 to be about $2^{9.2}$, we get $418 \times 650 \approx 2^{17.9}$, which we might estimate as about 250,000. Actual computation shows the answer to be 271,700, so our estimate by this crude method is quite far off.

Let us now try the same approach using 10 instead of 2. The first few powers of 10 starting with 10^0 are as follows:

$10^0 = 1$ $\quad$ $10^5 = 100{,}000$

$10^1 = 10$ $\quad$ $10^6 = 1{,}000{,}000$

$10^2 = 100$ $\quad$ $10^7 = 10{,}000{,}000$

$10^3 = 1000$ $\quad$ $10^8 = 100{,}000{,}000$

$10^4 = 10{,}000$ $\quad$ $10^9 = 1{,}000{,}000{,}000$

Problems such as 100×1000 are very easy, of course. The above list along with the rules for exponents merely verifies the rules of computation for multiplying by 10, 100, 1000, etc.

Going back to the problem 418×650, we might estimate $418 \approx 10^{2.4}$ and $650 \approx 10^{2.7}$ to obtain $418 \times 650 \approx 10^{5.1}$, which we might estimate at about 180,000. This estimate is even further off than the one that we obtained using 2. It is obvious that unless we can improve substantially our estimates of the fractional parts of the exponents, this method is useless.

Fortunately, this difficulty can be overcome by the use of a table of exponents or a table of *logarithms*, as these exponents are called when they are used in this way. A table of logarithms such as that given in Table 3.1 actually gives only the fractional part of the logarithm; the whole-number part is very easy to determine when using 10 as a base by looking up the powers of 10 or by counting the number of digits to the left of the decimal point and subtracting 1.

table 3.1 logarithms

N	0	1	2	3	4	5	6	7	8	9
10	0000	0043	0086	0128	0170	0212	0253	0294	0334	0374
11	0414	0453	0492	0531	0569	0607	0645	0682	0719	0755
12	0792	0828	0864	0899	0934	0969	1004	1038	1072	1106
13	1139	1173	1206	1239	1271	1303	1335	1367	1399	1430
14	1461	1492	1523	1553	1584	1614	1644	1673	1703	1732
15	1761	1790	1818	1847	1875	1903	1931	1959	1987	2014
16	2041	2068	2095	2122	2148	2175	2201	2227	2253	2279
17	2304	2330	2355	2380	2405	2430	2455	2480	2504	2529
18	2553	2577	2601	2625	2648	2672	2695	2718	2742	2765
19	2788	2810	2833	2856	2878	2900	2923	2945	2967	2989
20	3010	3032	3054	3075	3096	3118	3139	3160	3181	3201
21	3222	3243	3263	3284	3304	3324	3345	3365	3385	3404
22	3424	3444	3464	3483	3502	3522	3541	3560	3579	3598
23	3617	3636	3655	3674	3692	3711	3729	3747	3766	3784
24	3802	3820	3838	3856	3874	3892	3909	3927	3945	3962
25	3979	3997	4014	4031	4048	4065	4082	4099	4116	4133
26	4150	4166	4183	4200	4216	4232	4249	4265	4281	4298
27	4314	4330	4346	4362	4378	4393	4409	4425	4440	4456
28	4472	4487	4502	4518	4533	4548	4564	4579	4594	4609
29	4624	4639	4654	4669	4683	4698	4713	4728	4742	4757
30	4771	4786	4800	4814	4829	4843	4857	4871	4886	4900
31	4914	4928	4942	4955	4969	4983	4997	5011	5024	5038
32	5051	5065	5079	5092	5105	5119	5132	5145	5159	5172
33	5185	5198	5211	5224	5237	5250	5263	5276	5289	5302
34	5315	5328	5340	5353	5366	5378	5391	5403	5416	5428
35	5441	5453	5465	5478	5490	5502	5514	5527	5539	5551
36	5563	5575	5587	5599	5611	5623	5635	5647	5658	5670
37	5682	5694	5705	5717	5729	5740	5752	5763	5775	5786
38	5798	5809	5821	5832	5843	5855	5866	5877	5888	5899
39	5911	5922	5933	5944	5955	5966	5977	5988	5999	6010
40	6021	6031	6042	6053	6064	6075	6085	6096	6107	6117
41	6128	6138	6149	6160	6170	6180	6191	6201	6212	6222
42	6232	6243	6253	6263	6274	6284	6294	6304	6314	6325
43	6335	6345	6355	6365	6375	6385	6395	6405	6415	6425
44	6435	6444	6454	6464	6474	6484	6493	6503	6513	6522
45	6532	6542	6551	6561	6571	6580	6590	6599	6609	6618
46	6628	6637	6646	6656	6665	6675	6684	6693	6702	6712
47	6721	6730	6739	6749	6758	6767	6776	6785	6794	6803
48	6812	6821	6830	6839	6848	6857	6866	6875	6884	6893
49	6902	6911	6920	6928	6937	6946	6955	6964	6972	6981
50	6990	6998	7007	7016	7024	7033	7042	7050	7059	7067
51	7076	7084	7093	7101	7110	7118	7126	7135	7143	7152
52	7160	7168	7177	7185	7193	7202	7210	7218	7226	7235
53	7243	7251	7259	7267	7275	7284	7292	7300	7308	7316
54	7324	7332	7340	7348	7356	7364	7372	7380	7388	7396
N	0	1	2	3	4	5	6	7	8	9

table 3.1 (*continued*)

N	0	1	2	3	4	5	6	7	8	9
55	7404	7412	7419	7427	7435	7443	7451	7459	7466	7474
56	7482	7490	7497	7505	7513	7520	7528	7536	7543	7551
57	7559	7566	7574	7582	7589	7597	7604	7612	7619	7627
58	7634	7642	7649	7657	7664	7672	7679	7686	7694	7701
59	7709	7716	7723	7731	7738	7745	7752	7760	7767	7774
60	7782	7789	7796	7803	7810	7818	7825	7832	7839	7846
61	7853	7860	7868	7875	7882	7889	7896	7903	7910	7917
62	7924	7931	7938	7945	7952	7959	7966	7973	7980	7987
63	7993	8000	8007	8014	8021	8028	8035	8041	8048	8055
64	8062	8069	8075	8082	8089	8096	8102	8109	8116	8122
65	8129	8136	8142	8149	8156	8162	8169	8176	8182	8189
66	8195	8202	8209	8215	8222	8228	8235	8241	8248	8254
67	8261	8267	8274	8280	8287	8293	8299	8306	8312	8319
68	8325	8331	8338	8344	8351	8357	8363	8370	8376	8382
69	8388	8395	8401	8407	8414	8420	8426	8432	8439	8445
70	8451	8457	8463	8470	8476	8482	8488	8494	8500	8506
71	8513	8519	8525	8531	8537	8543	8549	8555	8561	8567
72	8573	8579	8585	8591	8597	8603	8609	8615	8621	8627
73	8633	8639	8645	8651	8657	8663	8669	8675	8681	8686
74	8692	8698	8704	8710	8716	8722	8727	8733	8739	8745
75	8751	8756	8762	8768	8774	8779	8785	8791	8797	8802
76	8808	8814	8820	8825	8831	8837	8842	8848	8854	8859
77	8865	8871	8876	8882	8887	8893	8899	8904	8910	8915
78	8921	8927	8932	8938	8943	8949	8954	8960	8965	8971
79	8976	8982	8987	8993	8998	9004	9009	9015	9020	9025
80	9031	9036	9042	9047	9053	9058	9063	9069	9074	9079
81	9085	9090	9096	9101	9106	9112	9117	9122	9128	9133
82	9138	9143	9149	9154	9159	9165	9170	9175	9180	9186
83	9191	9196	9201	9206	9212	9217	9222	9227	9232	9238
84	9243	9248	9253	9258	9263	9269	9274	9279	9284	9289
85	9294	9299	9304	9309	9315	9320	9325	9330	9335	9340
86	9345	9350	9355	9360	9365	9370	9375	9380	9385	9390
87	9395	9400	9405	9410	9415	9420	9425	9430	9435	9440
88	9445	9450	9455	9460	9465	9469	9474	9479	9484	9489
89	9494	9499	9504	9509	9513	9518	9523	9528	9533	9538
90	9542	9547	9552	9557	9562	9566	9571	9576	9581	9586
91	9590	9595	9600	9605	9609	9614	9619	9624	9628	9633
92	9638	9643	9647	9652	9657	9661	9666	9671	9675	9680
93	9685	9689	9694	9699	9703	9708	9713	9717	9722	9727
94	9731	9736	9741	9745	9750	9754	9759	9763	9768	9773
95	9777	9782	9786	9791	9795	9800	9805	9809	9814	9818
96	9823	9827	9832	9836	9841	9845	9850	9854	9859	9863
97	9868	9872	9877	9881	9886	9890	9894	9899	9903	9908
98	9912	9917	9921	9926	9930	9934	9939	9943	9948	9952
99	9956	9961	9965	9969	9974	9978	9983	9987	9991	9996
N	0	1	2	3	4	5	6	7	8	9

definition 3.6 If $a^m = n$, then we say that the m is the *logarithm* of n (usually written as log n) to the *base* a, and we write $m = \log_a n$ to indicate this. The whole-number part of m is called the *characteristic* of the logarithm, and the fractional part of m is called the *mantissa* of the logarithm.

If, for example, $10^{1.1703} = 14.8$, we say that 1.1703 is the logarithm of 14.8 to the base 10. We shall be concerned only with base 10, so we will not bother hereafter to use the phrase "to the base 10."

In the example above, 1 is the characteristic of the logarithm, and 0.1703 is the mantissa of the logarithm. If we examine a list of negative powers of 10,

$10^{-1} = 0.1$ $\qquad$ $10^{-4} = 0.0001$

$10^{-2} = 0.01$ $\qquad$ $10^{-5} = 0.00001$

$10^{-3} = 0.001$ $\qquad$ $10^{-6} = 0.000001$

as well as the one with the positive powers of 10, we see that the following rule may be formulated for determining the characteristic of a logarithm:

> Call the position right after the first nonzero digit the *starting point*. Count the number of places from the starting point to the decimal point. This is the characteristic—positive if you go to the right and negative if you go to the left.

The characteristic of 857, for example, is 2, the characteristic of 25.794 is 1, the characteristic of 7.96 is 0, and the characteristic of 0.00647 is -3. Notice that 0.00647 is between 10^{-2} and 10^{-3}, and when we add a positive mantissa between 0 and 1 to -3, we will get a logarithm between -2 and -3. *The mantissa as given by Table 3.1 is always positive.*

Let us see now how to use Table 3.1 to look up the mantissa of a logarithm assuming that we have already determined its characteristic. First, we use only the first three digits of the number, ignoring any zero at the beginning of the number and rounding off to three digits if there are more than three digits. If the number is 0.87492, for example, we use the digits 875. If there are less than 3 digits, we add enough 0's to get 3 digits. For example, we consider 6 as 6.00 and use the digits 600. We then find the first two digits in the column at the left, headed N, and then come over in that row until we reach the column headed by the third digit. In that spot, we will find the mantissa of the logarithm of the number.

For example, to find the logarithm of 418, we determine first that the characteristic is 2. Then we locate 41 in the left-hand column and come over to the column headed by 8. There we find 6212. This means that the mantissa is 0.6212 and that the logarithm of 418 is therefore 2.6212. This is another way of saying that $10^{2.6212} = 418$. By using the same procedure, we find that the logarithm of 650 is 2.8129. The logarithm of 418×650 is therefore $2.6212 + 2.8129$ or 5.4341. To find the product 418×650 we must now determine which number has 5.4341 as its logarithm. This number has a special name.

definition 3.7 If m is the logarithm of the number n, then we say that n is the *antilogarithm* of m (usually written as antilog m).

To find the antilogarithm of 5.4341 we reverse the above procedure. Look through the table of mantissas and find the one closest to 0.4341. This will be 0.4346, which is in the row headed by 27 and the column headed by 2. We therefore write down 272. The characteristic is 5, so we start right after the first 2, count off five places (adding 0's if necessary), and place the decimal there to obtain 272,000 as our answer. This is much closer than the answer previously obtained and shows that logarithms give a close approximation to the answer. Remember that the use of logarithms gives us only a close approximation to the answer. If we wish to get a better approximation, we would have to get a more detailed table of logarithms.

examples

1 Find the logarithm of 2980.

► We begin at the starting point (between the 2 and the 9) and count three places to the right to reach the decimal point (which is considered to be at the end of the number), so the characteristic is 3. We then find 29 in the column headed N in the table of logarithms and come over in this row until we are under 8, where we find 4742. The logarithm of 2980 is therefore 3.4742.

2 Find the logarithm of 18.279.

► We begin at the starting point (between the 1 and the 8) and count off one place to the right to reach the decimal point, so the characteristic is 1. We round off the first three digits to 183 and look for 18 in the column headed N in the table of logarithms and come over in this row until we are under 3, where we find 2625. The logarithm of 18.279 (rounded off to 18.3) is therefore 1.2625.

3 Find the antilogarithm of 1.9716.

► We look up the closest number to 9716 in the logarithm table and find 9717 in the row headed by 93 and in the column headed by 7, so we write down 937. The characteristic is 1, so we move one place to the right from the starting point and place the decimal between the 3 and the 7. The antilogarithm of 1.9716 is therefore 93.7.

For division problems, if we are satisfied with answers rounded off to three digits, we will find that logarithms provide an answer in much better agreement with the actual answer, if not in exact agreement. Let us consider the problem $859.8 \div 3.84$. The characteristic of 859.8 is 2; if we round off 859.8 to 860 and look in the row headed by 86 and in the column headed by 0, we find its mantissa to be 0.9345. The logarithm of 860 is therefore 2.9345. Similarly, the logarithm of 3.84 is 0.5843. The logarithm of $859.8 \div 3.84$ is therefore $2.9345 - 0.5843$ or 2.3502. We find the antilogarithm of 2.3502 by locating the closest mantissa to 0.3502, which is 0.3502, and noting that this is in the row headed by 22 and in the column headed by 4. We write down 224, and since the characteristic is 2, we count off two decimal places to the right from the position directly after the first 2 to determine the final answer, which is 224. Actual computation shows the answer to be 223.9, which rounds off to 224.

Properties I, II, III, and IV for exponents imply immediately the following rules for computation with logarithms:

1. $\log (ab) = \log a + \log b$
2. $\log (a/b) = \log a - \log b$
3. $\log a^m = m \log a$
4. $\log \sqrt[m]{a} = 1/m \log a$

If we agree that m may be any rational number, then rule 4 above is really a repetition of rule 3.

Translated into words, these rules say:

1. To multiply two numbers, determine their logarithms, add these logarithms, and then look up the antilogarithm of the sum.
2. To divide one number by another, subtract the logarithm of the divisor from that of the dividend and then look up the antilogarithm of the difference.
3. To raise a number to the mth power, determine its logarithm, multiply this logarithm by m, and then look up the antilogarithm.
4. To take the mth root of a number, determine its logarithm, divide this logarithm by m, and then look up the antilogarithm.

examples

4 Multiply 518.2×1.74.

$$\log 518.2 = 2.7143$$
$$\log 1.74 = \underline{0.2405}$$
$$2.9548$$
$$\text{antilog } 2.9548 = 901$$

Check how close this is to the actual answer.

5 Divide 7988.3 by 26.4.

$$\begin{aligned} \log 7988.3 &= 3.9025 \\ \log 26.4 &= \underline{1.4216} \\ & 2.4809 \\ \text{antilog } 2.4809 &= 303 \end{aligned}$$

Check how close this is to the actual answer.

6 Evaluate $(1.08)^{10}$.

$$\begin{aligned} \log 1.08 &= 0.0334 \\ 10 \log 1.08 &= 0.3340 \\ \text{antilog } 0.3340 &= 2.16 \end{aligned}$$

Therefore,

$$(1.08)^{10} = 2.16$$

7 Find the square root of 8712.

$$\begin{aligned} \log 8712 &= 3.9400 \\ \tfrac{1}{2} \log 8712 &= 1.9700 \\ \text{antilog } 1.9700 &= 93.3 \end{aligned}$$

Find $\sqrt{8712}$ by the other method and compare answers.

Logarithms may also be used to simplify computations involving more than two numbers.

examples

8 Multiply $187 \times 21.6 \times 6.59$.

$$\begin{aligned} \log 187 &= 2.2718 \\ \log 21.6 &= 1.3345 \\ \log 6.59 &= \underline{0.8189} \\ & 4.4252 \\ \text{antilog } 4.4252 &= 26{,}600 \end{aligned}$$

9 Evaluate $(2114 \times 43.7)/29.48$.

$$\begin{aligned} \log 2114 &= 3.3243 \\ \log 43.7 &= \underline{1.6405} \\ \log 2114 + \log 43.7 &= 4.9648 \\ \log 29.48 &= 1.4698 \\ \log 2114 + \log 43.7 - \log 29.48 &= 3.4950 \\ \text{antilog } 3.4950 &= 3130 \end{aligned}$$

10 Evaluate $(46.9)^3 \times (\sqrt{162})$.

$$\begin{aligned} \log 46.9 &= 1.6712 \\ 3 \log 46.9 &= \qquad\qquad 5.0136 \\ \log 162 &= 2.2095 \\ \tfrac{1}{2} \log 162 &= \qquad\qquad \underline{1.1048} \\ 3 \log 46.9 + \tfrac{1}{2} \log 162 &= \qquad\qquad 6.1184 \\ \text{antilog } 6.1184 &= 1{,}310{,}000 \end{aligned}$$

11 Evaluate $\sqrt{74.9}/2.6$.

$$\begin{aligned} \log 74.9 &= 1.8745 \\ \tfrac{1}{2} \log 74.9 &= 0.9373 \\ \log 2.6 &= \underline{0.4150} \quad \text{(we look up the mantissa for 260)} \\ \tfrac{1}{2} \log 74.9 - \log 26 &= 0.5223 \\ \text{antilog } 0.5223 &= 3.33 \end{aligned}$$

When dealing with numbers whose logarithms have negative characteristics, it would be awkward and confusing if we were to write down a negative characteristic and a positive mantissa. For example, the logarithm of 0.0813 would be $(-2) + 0.9101$. It has been found that it is much simpler to write $8. \quad - 10$ for the characteristic instead of -2 and then to fill in the mantissa so that we write $8.9101 - 10$ for the logarithm of 0.0813. Similarly, we would write $9. \quad - 10$ instead of -1, $7. \quad - 10$ instead of -3, etc. The rest of the computation proceeds as before, but we must remember our rules for working with positive and negative integers.

/ examples /

12 Multiply 0.0038 × 49700.

$$\log 0.0038 = 7.5798 - 10$$
$$\log 49700 = 4.6964$$
$$\log 0.0038 + \log 49700 = 12.2762 - 10 = 2.2762$$
$$\text{antilog } 2.2762 = 189$$

13 Divide 0.012/473.

$$\log 0.012 = 8.0792 - 10$$
$$\log 473 = 2.6749$$
$$\log 0.082 - \log 473 = 5.4043 - 10$$
$$\text{antilog of } 5.4043 - 10 = 0.0000254$$

14 Divide 473/0.012.

$$\log 473 = 2.6749$$
$$\log 0.012 = 8.0792 - 10$$

If we try to subtract 8.0792 − 10 from 2.6749 we will find that we end up subtracting 8 from 2, which again leaves us with a negative whole number in front and a positive mantissa in the middle. To avoid this difficulty we write 12.6749 − 10 for the logarithm of 473 instead of 2.6749 and then subtract:

$$\log 473 = 12.6749 - 10$$
$$\log 0.012 = 8.0792 - 10$$
$$\log 473 - \log 0.012 = 4.5957$$
$$\text{antilog } 4.5957 = 39{,}400$$

15 Evaluate $\sqrt{0.642}$.

$$\log 0.642 = 9.8075 - 10$$
$$\tfrac{1}{2}(\log 0.642) = 4.9038 - 5$$

The characteristic here is 4 − 5 or −1. Therefore,

$$\text{antilog } 4.9038 - 5 = 0.801$$

16 Evaluate $\sqrt[3]{0.0256}$.

$$\log 0.0256 = 8.4082 - 10$$

If we try to divide $8.4082 - 10$ by 3 we will end up with $-10 \div 3$ and have a negative fraction left over. To avoid this, we write $28.4082 - 30$ for the logarithm of 0.0256 instead of $8.4082 - 10$ and then divide by 3.

$$\tfrac{1}{3}(\log 0.0256) = \tfrac{1}{3}(28.4082 - 30)$$
$$= 9.4694 - 10$$
$$\text{antilog } 9.4694 - 10 = 0.295$$

17 Evaluate $(39.2)^2/\sqrt[4]{0.0053}$.

►
$$\begin{aligned} \log 39.2 &= 1.5933 \\ 2(\log 39.2) &= \qquad\qquad 3.1866 \\ \log 0.0053 &= 37.7243 - 40 \\ \tfrac{1}{4}(\log 0.0053) &= \qquad\qquad \underline{9.4311 - 10} \end{aligned}$$

Notice that when we try to subtract, we run into the same difficulty as in Example 11. We therefore rewrite 2 log 39.2 as $13.1866 - 10$ and then subtract:

$$\begin{aligned} 2(\log 39.2) &= 13.1866 - 10 \\ \tfrac{1}{4}(\log 0.0053) &= \underline{\ 9.4311 - 10} \\ 2(\log 39.2) - \tfrac{1}{4}(\log 0.0053) &= 3.7555 \end{aligned}$$

The antilogarithm of 3.7555 is exactly halfway between 5690 and 5700, so we write

$$\text{antilog } 3.7555 = 5695$$

18 Evaluate $(\sqrt[3]{18.7} \times 26)/\sqrt{0.4851}$.

►
$$\begin{aligned} \log 18.7 &= 1.2718 \\ \tfrac{1}{3}(\log 18.7) &= 0.4239 \\ \log 26 &= \underline{1.4150} \\ \tfrac{1}{3}(\log 18.7) + \log 26 &= 1.8389 \qquad = 11.8389 - 10 \\ \log 0.4851 &= 9.6857 - 10 \\ \tfrac{1}{2}(\log 0.4851) &= \qquad\qquad \underline{4.8429 - 5} \\ \tfrac{1}{3}(\log 18.7) + \log 26 - \tfrac{1}{2}(\log 0.4851) &= \quad 6.9960 - 5 \end{aligned}$$

$$\text{antilog } 6.9960 - 5 = \text{antilog } 1.9960 = 99.1$$

exercises 3.4

1 Determine the following:

(a) What is the logarithm of 8 to the base 2?
(b) What is the logarithm of 81 to the base 3?
(c) What is the logarithm of 3 to the base 27?
(d) What is the logarithm of 4 to the base 16?

2 Determine the logarithms of the following numbers:

(a) 18.4 (b) 2.15 (c) 17000 (d) 7
(e) 4.29 (f) 0.573 (g) 0.00015 (h) 0.0193

3 Determine the antilogarithms of the following numbers:

(a) 1.7639 (b) 3.8508 (c) 0.5084
(d) 6.7498 (e) 8.4083 − 10 (f) 5.3968 − 10

4 Perform the following computations by means of logarithms:

(a) 186×21.7 (b) $15.2 \div 3.12$
(c) $(19.1)^3$ (d) $\sqrt[5]{1.87}$
(e) $26.2 \times 197 \times 0.114$ (f) $(18.4 \div 0.125) \times 16.3$
(g) 0.0065×98.7 (h) $9.74 \div 0.0924$
(i) $(0.659)^3$ (j) $\sqrt{0.04762}$
(k) $0.583 \div 75.9$ (l) $0.00758 \div 0.000528$

5 Perform the following computations by means of logarithms:

(a) $(12.7)^3 \times \sqrt{1.82}$ (b) $\sqrt{0.197} \times \sqrt[3]{14.2}$

(c) $\dfrac{(18.7)^3 \times \sqrt[4]{97}}{\sqrt{1.73}}$ (d) $\dfrac{\sqrt[3]{217}}{(1.37)^4 \times \sqrt{18.2}}$

(e) $(1.04)^{20}$ (f) $\sqrt[20]{1.74}$

(g) $\dfrac{\sqrt[3]{0.748} \times \sqrt{85.9}}{(0.00486)^2}$ (h) $\dfrac{(0.0486)^3}{26.9}$

6 Determine the square roots of the following numbers by means of logarithms; check your answers by the method described in Section 3.3.

(a) 1817.2 (b) 96147.36 (c) 18 (d) 0.0572

/review test 3/

NAME ______________

DATE ______________

1 Evaluate the following quantities:

(a) 5^3 __________ (b) 3^{-2} __________ (c) $(\frac{1}{3})^4$ __________

(d) $(\frac{1}{2})^{-3}$ __________ (e) $(-6)^2$ __________ (f) -6^2 __________

2 Simplify:

(a) $(3^{-2})(4^2)$ __________

(b) $(4^{1.5})(4^{2.5})$ __________

(c) $(4^2)(4^{-1.5})$ __________

(d) $(\frac{1}{2})^3(2^4)$ __________

(e) $8^{-2/3}$ __________

(f) $16^{-3/2}$ __________

3 Simplify:

(a) $(2x^3y^2z^3)^4(3xy^2z)^2$

(b) $(3x^{-2}yz^{-1})^{-2}(4x^{-1}y^{-2}z^4)^2$

(c) $\dfrac{(2x^{-1}y^2z)^2(x^{-1}y^3z^{-2})^{-3}}{(3xy^{-2}z^{-1})^2}$

(d) $\dfrac{(4x^{1/2}y^{2/3}z)^{-1/2}}{(8x^{-1}y^{1/4}z^{1/2})^{1/3}}$

4 Compute the square root of 61493.8 to two decimal places.

5 Simplify the following expressions:

(a) $\sqrt{32} + \sqrt{72} - \sqrt{50}$

(b) $\sqrt{28} + \sqrt{63} + \sqrt{175}$

(c) $\sqrt{\frac{3}{16}} + \sqrt{12}$

(d) $\sqrt{18} \cdot \sqrt{2}$

(e) $\frac{\sqrt{75}}{\sqrt{3}}$

6 Simplify each expression by rationalizing the denominator:

(a) $\dfrac{5}{\sqrt{3}}$

(b) $\dfrac{1}{\sqrt{3} + \sqrt{2}}$

(c) $\dfrac{4}{\sqrt{8} - \sqrt{2}}$

(d) $\dfrac{5\sqrt{2} - 1}{\sqrt{18}}$

(e) $\dfrac{2}{\sqrt{3} - \sqrt{2}}$

(f) $\dfrac{3\sqrt{2}}{\sqrt{2} + \sqrt{8}}$

7 Determine the logarithms of the following numbers:

(a) 18.7 ______________________ (b) 196.42 ______________________

(c) 0.0063 ______________________ (d) 8.2 ______________________

8 Determine the antilogarithms of the following numbers:

(a) 3.2480 ______________________ (b) 9.7959 − 10 ______________________

(c) 7.6085 − 10 ______________________ (d) 0.8915 ______________________

9 Perform the following computations by means of logarithms:

(a) 76.4×1.86

(b) $142 \div 0.089$

(c) $\dfrac{491.8 \times 17.3}{7.26}$

(d) $\sqrt{187.4}$

(e) $\sqrt[3]{217}$

(f) $\dfrac{(181.1)^3 \times \sqrt{296}}{(0.073)^2}$

(g) $\dfrac{\sqrt[4]{87} \times (12.1)^3}{\sqrt{1700}}$

10 Which is larger, $\sqrt[5]{4500}$ or $\sqrt[4]{300}$?

chapter four

ALGEBRAIC EXPRESSIONS

4.1/ operations on polynomials

Most of our interest up to this point has been focused on the real-number system and the problems that we have dealt with involved specific numbers. Occasionally, when we wanted to express a rule or formula that was valid for all numbers, such as $x^2x^3 = x^5$, we used the symbol x to stand for an arbitrary number. When we wished to express a relationship that was true for any pair of numbers, such as $(xy)^4 = x^4y^4$, we used the symbol x to stand for one number and the symbol y to stand for the other number. This process of using symbols or letters to stand for numbers leads us to a consideration of algebraic expressions. We might say that an algebraic expression is any expression involving numbers and letters or other symbols that stand for numbers. For example, we might consider the algebraic expression $2x^2y - 3xy^2z$. This expression represents a certain numerical value depending on what values we assign to the symbols x, y, and z. If $x = 1$, $y = 3$, and $z = 2$, for example, this expression assumes the value $2 \cdot 1^2 \cdot 3 - 3 \cdot 1 \cdot 3^2 \cdot 2 = 6 - 54 = -48$. Notice that the symbol x has the same value, 1, everywhere that it appears but that different symbols may have different values. In a different problem, the symbols x, y, and z may assume different values.

Algebraic expressions may be broken down into various types, just like the real numbers. Polynomials, which are certain types of algebraic expressions that have no denominators, correspond to the integers, and rational expressions correspond to the rational numbers, in that they may be considered as quotients of polynomials with nonzero denominators. Our purpose in this section is to develop the rules for adding, subtracting, and multiplying polynomials. As in the case of the integers the operation of division is not closed for the polynomials.

There is actually no limit to the number of different symbols that may appear in an algebraic expression, but for most practical purposes, one does not usually consider more than two or three different quantities at a time, so we shall, for the most part, limit ourselves to expressions containing not more than three different letters, say x, y, and z. Of course, we may just as well use a, b, and c; p, q, and r; or any other set of symbols that we please.

definition 4.1 A *polynomial in x over the rationals* is an expression of the form

$$a_nx^n + a_{n-1}x^{n-1} + a_{n-2}x^{n-2} + \cdots + a_1x + a_0$$

where n is some nonnegative integer, the a_i are rational numbers for $i = 0, 1, 2, \ldots, n$, and x is merely a symbol that one may consider as representing some number, although this last interpretation is not really necessary.

If we permit the a_i to be real numbers, then the above expression is called a *polynomial in x over the reals*. If $a_n \neq 0$, then n is called the *degree* of the polynomial. The number a_i is called the *coefficient* of x_i. Note that a polynomial of degree n has $n + 1$ terms, some of which may have coefficient 0.

We may define similarly a polynomial in y over the rationals. A polynomial of degree 0 over the rationals would be merely a rational number. We have special names for polynomials of the first few degrees as indicated in Table 4.1. Above degree 5 we would merely refer to polynomials of degree 6, and so on.

To add two polynomials, we *add like terms*. To add like terms we must remember again our rules for adding rational numbers. For example, $5x^3 + (-3)x^3$ gives $2x^3$ as an answer. If we have any doubts about this, we refer back to the distributive or factoring rule, which tells us that $5x^3 + (-3)x^3 =$

$[5 + (-3)]x^3 = 2x^3$. To add two polynomials, we merely pair off the like terms and then add them.

table 4.1

degree	*name*	*example*
1	linear	$3x - 5$
2	quadratic	$2x^2 - 7x + 4$
3	cubic	$5x^3 - 8x + 2x - 7$
4	quartic	$2x^4 + 3x^3 - 5x + 8$
5	quintic	$9x^5 + 2x^3 - 8x + 2$

/ examples /

1 Add the polynomials $7x^4 - 3x^3 + 5x^2 + 6x - 3$ and $2x^5 - 3x^4 + x^3 - 2x^2 + 7x + 1$.

► We pair off like terms and obtain

$$2x^5 + [7 + (-3)]x^4 + [(-3) + 1]x^3 + [5 + (-2)]x^2 + (6 + 7)x + (-3 + 1)$$
$$= 2x^5 + 4x^4 - 2x^3 + 3x^2 + 13x - 2$$

This means that whatever number we substitute for x

$$(7x^4 - 3x^3 + 5x^2 + 6x - 3) + (2x^5 - 3x^4 + x^3 - 2x^2 + 7x + 1)$$
$$= 2x^5 + 4x^4 - 2x^3 + 3x^2 + 13x - 2$$

For example, if $x = 2$, then

$$7x^4 - 3x^3 + 5x^2 + 6x - 3 = 7 \cdot 16 - 3 \cdot 8 + 5 \cdot 4 + 6 \cdot 2 - 3$$
$$= 117$$

$$2x^5 - 3x^4 + x^3 - 2x^2 + 7x + 1$$
$$= 2 \cdot 32 - 3 \cdot 16 + 8 - 2 \cdot 4 + 7 \cdot 2 + 1$$
$$= 31$$

$$2x^5 + 4x^4 - 2x^3 + 3x^2 + 13x - 2$$
$$= 2 \cdot 32 + 4 \cdot 16 - 2 \cdot 8 + 3 \cdot 4 + 13 \cdot 2 - 2$$
$$= 148$$

and

$$117 + 31 = 148$$

2 Determine the sum $(4x^3 - 7x^2 + 6x - 3) + (9x^3 - 5x^2 - 3x + 1)$.

► $(4x^3 - 7x^2 + 6x - 3) + (9x^3 - 5x^2 - 3x + 1) = 13x^3 - 12x^2 + 3x - 2$

3 Add the polynomials $2x^3 - \frac{5}{2}x^2 + \frac{7}{3}x - \frac{1}{8}$ and $\frac{1}{4}x^3 + \frac{1}{2}x^2 + \frac{3}{8}x + 1$.

► $(2x^3 - \frac{5}{2}x^2 + \frac{7}{3}x - \frac{1}{8}) + (\frac{1}{4}x^3 + \frac{1}{2}x^2 + \frac{3}{8}x + 1)$

$$= (2 + \tfrac{1}{4})x^3 + (-\tfrac{5}{2} + \tfrac{1}{2})x^2 + (\tfrac{7}{3} + \tfrac{3}{8})x + (-\tfrac{1}{8} + 1)$$
$$= \tfrac{9}{4}x^3 - 2x^2 + \tfrac{65}{24}x + \tfrac{7}{8}$$

4 Compute $(4x^3 - 7x + 3) + (2x^3 - 3x^2 + 1)$.

► This may be considered as the sum

$$(4x^3 + 0x^2 - 7x + 3) + (2x^3 - 3x^2 + 0x + 1)$$

which gives us

$$6x^3 - 3x^2 - 7x + 4$$

The rule for subtracting polynomials is practically the same as that for subtracting integers: Change the sign of the polynomial being subtracted, and then add. As we shall see, to change the sign of a polynomial means to change the sign of each of its terms. Suppose, for example, that we wish to take the difference

$$(4x^3 + x^2 - 8) - (2x^3 + 2x^2 - 7x + 1)$$

This is the same as

$$(4x^3 + x^2 - 8) + (-1)(2x^3 + 2x^2 - 7x + 1)$$

By the distributive property

$$(-1)(2x^3 + 2x^2 - 7x + 1) = -2x^3 - 2x^2 + 7x - 1$$

so

$$(4x^3 + x^2 - 8) - (2x^3 + 2x^2 - 7x + 1)$$
$$= (4x^3 + x^2 - 8) + (-2x^3 - 2x^2 + 7x - 1)$$
$$= 2x^3 - x^2 + 7x - 9$$

Another way of looking at this is to remember that subtraction means addition of the additive inverse, and the additive inverse of a polynomial (being that polynomial that must be added to it to obtain 0) is obtained by changing the signs of all the terms. As a working rule of thumb, we may say that when a $-$ sign appears in front of parentheses, we may change it to a $+$ sign, provided that we change the signs of all the terms inside the parentheses.

/ examples /

5 Evaluate $(2x^4 - 7x^2 + 3) - (5x^4 - 8x^3 + x^2 + x)$.

► $(2x^4 - 7x^2 + 3) - (5x^4 - 8x^3 + x^2 + x)$

$$= (2x^4 - 7x^2 + 3) + (-5x^4 + 8x^3 - x^2 - x)$$
$$= -3x^4 + 8x^3 - 8x^2 - x + 3$$

Notice that once we have a + sign connecting the two pairs of parentheses, we may drop the parentheses and combine like terms.

6 Evaluate $(4\frac{1}{2}x^2 - \frac{7}{3}x - 5) - (3x^2 + 5x - \frac{2}{3})$.

► $$(4\tfrac{1}{2}x^2 - \tfrac{7}{3}x - 5) - (3x^2 + 5x - \tfrac{2}{3}) = \tfrac{9}{2}x^2 - \tfrac{7}{3}x - 5 - 3x^2 - 5x + \tfrac{2}{3}$$
$$= \tfrac{3}{2}x^2 - \tfrac{22}{3}x - \tfrac{13}{3}$$

7 Evaluate $(x^2 + 5) - (3x^2 - \frac{4}{3}x + 1)$.

► $$(x^2 + 5) - (3x^2 - \tfrac{4}{3}x + 1) = x^2 + 5 - 3x^2 + \tfrac{4}{3}x - 1$$
$$= -2x^2 + \tfrac{4}{3}x + 4$$

To discover the rule for multiplying two polynomials, we refer back to the distributive rule, which tells us that $m(n + p) = mn + mp$. This implies that

$$(m + q)(n + p) = (m + q)n + (m + q)p = mn + qn + mp + qp$$

In a similar manner, we show that in any product each term of the first factor[1] must be multiplied by each term of the second factor. This rule is referred to as the *generalized distributive rule* and tells us how to multiply two polynomials. For example, if the first polynomial consists of four terms and the second consists of two terms, then the product will consist of eight terms.

/examples/

8 Determine the product $(2x - 3)(4x + 5)$.

► Each term in the first factor must be multiplied by each term in the second factor, so we have

$$(2x - 3)(4x + 5) = (2x)(4x) + (2x)(5) + (-3)(4x) + (-3)(5)$$
$$= 8x^2 + 10x - 12x - 15$$

The two middle terms may then be combined to give $8x^2 - 2x - 15$ as the final answer. In general, like terms should be combined wherever possible, as this usually simplifies the answer.

9 Multiply $(4x^3 - 7x^2 + 3)(2x^2 - 5x + 1)$.

► $$(4x^3 - 7x^2 + 3)(2x^2 - 5x + 1)$$
$$= 8x^5 - 20x^4 + 4x^3 - 14x^4 + 35x^3 - 7x^2 + 6x^2 - 15x + 3$$
$$= 8x^5 - 34x^4 + 39x^3 - x^2 - 15x + 3$$

[1]A term is an expression that is added or subtracted, whereas a factor is an expression that is multiplied. Thus the expression $(4x^2y + 2xy)(2x + xy - y^2)$ consists of two factors. The first factor consists of two terms and the second factor consists of three terms. The first term of the first factor consists of three factors, 4, x^2, and y.

To keep track of things, it is advisable to multiply the first term of the first factor by each term of the second factor, then the second term of the first factor by each term of the second factor, and so on.

10 Multiply $(\frac{1}{2}x^2 - \frac{3}{4}x + 1)(x + \frac{1}{2})$.

$$\begin{aligned} ► \quad (\tfrac{1}{2}x^2 - \tfrac{3}{4}x + 1)(x + \tfrac{1}{2}) &= \tfrac{1}{2}x^3 + \tfrac{1}{4}x^2 - \tfrac{3}{4}x^2 - \tfrac{3}{8}x + x + \tfrac{1}{2} \\ &= \tfrac{1}{2}x^3 - \tfrac{1}{2}x^2 + \tfrac{5}{8}x + \tfrac{1}{2} \end{aligned}$$

As has been noted previously, division of polynomials is not closed; however, just as in the case of integers, there is a division algorithm for polynomials that states that when one polynomial is divided by another we get a quotient and a remainder whose degree is less than that of the divisor. If the remainder is zero, we say that we have factored the original polynomial into two factors.

examples

11 Divide $3x^4 - 7x^3 + 4x^2 + 2x - 3$ by $x^2 - 5x + 1$.

► We set this problem up as we do ordinary long division of natural numbers, being careful to write the terms of each polynomial in descending order of exponents:

$$x^2 - 5x + 1 \overline{\big) 3x^4 - 7x^3 + 4x^2 + 2x - 3}$$

We then divide the first term, $3x^4$, of the dividend by the first term, x^2, of the divisor and write the answer $3x^2$ above the first term of the dividend:

$$\begin{array}{r|l} & 3x^2 \\ \hline x^2 - 5x + 1 & 3x^4 - 7x^3 + 4x^2 + 2x - 3 \end{array}$$

Next, we multiply the $3x^2$ by the divisor and write this under the dividend, being careful to line up like powers beneath one another. Then we subtract:

$$\begin{array}{r|l} & 3x^2 \\ \hline x^2 - 5x + 1 & 3x^4 - 7x^3 + 4x^2 + 2x - 3 \\ & 3x^4 - 15x^3 + 3x^2 \\ \cline{2-2} & 8x^3 + x^2 \end{array}$$

We then bring down the next term and repeat the process by dividing the $8x^3$ of the difference by the x^2 of the quotient and writing the answer $8x$ above the next term of the dividend:

$$\begin{array}{r|l} & 3x^2 + 8x \\ \hline x^2 - 5x + 1 & 3x^4 - 7x^3 + 4x^2 + 2x - 3 \\ & 3x^4 - 15x^3 + 3x^2 \\ \cline{2-2} & 8x^3 + x^2 + 2x \end{array}$$

We then multiply $8x$ by the divisor and continue the process as above until we get a remainder whose degree is less than that of the divisor; in this case we continue until we get a remainder that is linear:

$$
\begin{array}{r|rrrrr}
 & 3x^2 & +\ 8x & +\ 41 & & \\
\hline
x^2 - 5x + 1 & 3x^4 & -\ 7x^3 & +\ 4x^2 & +\ 2x & -\ 3 \\
 & 3x^4 & -\ 15x^3 & +\ 3x^2 & & \\
\cline{2-4}
 & & 8x^3 & +\ x^2 & +\ 2x & \\
 & & 8x^3 & -\ 40x^2 & +\ 8x & \\
\cline{3-5}
 & & & 41x^2 & -\ 6x & -\ 3 \\
 & & & 41x^2 & -\ 205x & +\ 41 \\
\cline{4-6}
 & & & & 199x & -\ 44
\end{array}
$$

We thus end up with a quotient of $3x^2 + 8x + 41$ and a remainder of $199x - 44$. This means that if we multiply $3x^2 + 8x + 41$ by $x^2 - 5x + 1$ and add $199x - 44$ to the product, we should come out with $3x^4 - 7x^3 + 4x^2 + 2x - 3$.

12 Divide $4x^3 - 7x^2 + 3$ by $2x^2 + 4x - 1$.

►

$$
\begin{array}{r|rrrr}
 & 2x & -\ \frac{15}{2} & & \\
\hline
2x^2 + 4x - 1 & 4x^3 & -\ 7x^2 & & +\ 3 \\
 & 4x^3 & +\ 8x^2 & -\ 2x & \\
\cline{2-4}
 & & -\ 15x^2 & +\ 2x & +\ 3 \\
 & & -\ 15x^2 & -\ 30x & +\ \frac{15}{2} \\
\cline{3-5}
 & & & 32x & -\ \frac{9}{2}
\end{array}
$$

Notice that when we wrote the dividend we left space for the missing x term so that we could line things up properly for subtraction. The quotient here is $2x - \frac{15}{2}$, and the remainder is $32x - \frac{9}{2}$.

13 Evaluate $(6x^5 + 3x^4 - 2x^3 + 2x^2 - 1) \div (2x^3 + x^2 + 1)$.

►

$$
\begin{array}{r|rrrrr}
 & 3x^2 & -\ 1 & & & \\
\hline
2x^3 + x^2 + 1 & 6x^5 & +\ 3x^4 & -\ 2x^3 & +\ 2x^2 & -\ 1 \\
 & 6x^5 & +\ 3x^4 & & +\ 3x^2 & \\
\cline{2-6}
 & & & -\ 2x^3 & -\ x^2 & -\ 1 \\
 & & & -\ 2x^3 & -\ x^2 & -\ 1 \\
\cline{4-6}
 & & & & & 0
\end{array}
$$

Here the remainder is 0, so $6x^5 + 3x^4 - 2x^3 + 2x^2 - 1$ factors into $(2x^3 + x^2 + 1)(3x^2 - 1)$.

14 Evaluate $(2x^2 - 7x + 1) \div (x^3 - 1)$.

► Since the degree of the dividend is less than that of the divisor, the quotient is 0 and the remainder is $2x^2 - 7x + 1$. In other words, $2x^2 - 7x + 1 = 0(x^3 - 1) + (2x^2 - 7x + 1)$.

We have been concerned thus far with polynomials in one symbol, x.[1] A polynomial in several indeterminates or variables, say x, y, and z over the rationals, is an expression that could be obtained by taking the rational numbers along with the symbols x, y, and z and adding, subtracting, and multiplying them in any way that we wish; in other words, it is an expression with no denominators (except for the denominators of the rational number coefficients). The expression $6x^2yz^3 - \frac{4}{3}x^3y + \frac{2}{5}xyz^2 + 2xyz$, for example, is a polynomial in x, y, and z over the rationals. The rules for addition, subtraction, and multiplication of polynomials in several variables are exactly the same as those for polynomials in one variable:

1. To add two polynomials, combine like terms.
2. To subtract one polynomial from another, change the sign of each term of the polynomial being subtracted and then add.
3. To multiply two polynomials, multiply each term of the first factor by each term of the second factor.

It is seldom necessary to divide one polynomial in several variables by another, so we shall not be concerned with this except for the method of simplifying a quotient of two polynomials, which will be discussed in Section 4.4.

/examples/

15 Add $(5x^2y^3z - 3xyz^2 + 2xy^2z^3) + (3x^2yz - 4xy^2z^3 + 2x^2y^3z)$.

► We can only combine like terms, so we obtain

$$(5x^2y^3z - 3xyz^2 + 2xy^2z^3) + (3x^2yz - 4xy^2z^3 + 2x^2y^3z) = 7x^2y^3z - 3xyz^2 - 2xy^2z^3 + 3x^2yz$$

16 Subtract $(2xy^2z + 3xy^3z^2 - 4xyz^2) - (3xy^2z - 5xyz^2)$.

► $$\begin{aligned}(2xy^2z + 3xy^3z^2 - 4xyz^2) - (3xy^2z - 5xyz^2) &= 2xy^2z + 3xy^3z^2 - 4xyz^2 - 3xy^2z + 5xyz^2\\ &= -xy^2z + 3xy^3z^2 + xyz^2\end{aligned}$$

17 Multiply $(3xyz^2 - 2xy^2z + 5xyz)(2xy - 3yz)$.

► $$\begin{aligned}&(3xyz^2 - 2xy^2z + 5xyz)(2xy - 3yz)\\ &= 6x^2y^2z^2 - 9xy^2z^3 - 4x^2y^3z + 6xy^3z^2 + 10x^2y^2z - 15xy^2z^2\end{aligned}$$

No two terms of the answer are alike, so we cannot combine any terms here.

[1]The symbol x is sometimes also referred to as an *indeterminate* or a *variable*.

Each term of a polynomial in several variables has a degree that is the sum of the exponents of the variables, and the degree of a polynomial in several variables is the highest of the degrees of its terms. The degrees of the terms of the polynomial $2xy^2z - 5x^3y^2z^3 + 2xyz$, for example, are 4, 8, and 3, respectively, and the degree of the polynomial is therefore 8.

exercises 4.1

1 Find the following sums:

(a) $(3x^4 - 7x^3 + 2x - 5) + (x^2 - 7x + 3)$
(b) $(x^2 + 9x + 3) + (7x^3 - 2x^2 + 4x - 7)$
(c) $(3x^5 + x^4 + x^3 - 5x^2 - 2x + 1) + (x^3 - 7x^2 + 3x + 4)$
(d) $(2x^2 + 7x - 1) + (x^3 - 5x^2 + 3x - 2)$
(e) $(2\frac{1}{2}x^3 + 7\frac{2}{3}x + \frac{1}{2}) + (3x^2 - 1\frac{5}{6}x + \frac{3}{4})$

2 Perform the following subtractions:

(a) $(2x^4 - 7x^3 + 9x^2 - 3) - (2x^3 + 7x^2 - 8x + 1)$
(b) $(2x^2 - 9x + 1) - (5x^3 - 7x^2 + 9x - 3)$
(c) $(8x^7 - 3x^5 + 2x^4 - 5) - (x^6 + x^3 - 3x^2 + 5x)$
(d) $(9x^2 - 9x + 5) - (7x^4 - 3x^2 - 8x + 1)$
(e) $(4\frac{2}{3}x^3 - 7x^2 + \frac{1}{2}x - 3) - (3\frac{1}{5}x^3 + 2\frac{1}{2}x^2 + 3x - \frac{1}{3})$

3 Multiply the following polynomials:

(a) $(2x^2 - 7x + 1)(3x - 5)$
(b) $(3x^2 - 5x + 3)(2x^2 - 5x - 3)$
(c) $(x^3 - x + 1)(5x^2 - 2x - 1)$
(d) $(x^4 - x^2 + 3)(x^3 - 7x)$

4 Perform the following divisions:

(a) $(x^2 + 5x - 2) \div (x - 1)$
(b) $(5x^3 + 3x^2 + 4x - 5) \div (x + 3)$
(c) $(8x^2 - 7x + 3) \div (x - 5)$
(d) $(4x^4 - 7x^3 + 2x^2 - 4x - 3) \div (x^2 - 4)$

5 Add $(3x^2yz^2 - 2x^3yz^4 + x^3yz) + (5x^3yz^4 - 2xyz + x^3yz)$.

6 Subtract $(5xyz^2 - 2xy^2z + 4x^2yz) - (3xy^2z + x^2yz + 4xyz^2)$.

7 Multiply $(2xy^2z + x^2y^3z)(3xyz - 5x^2y^2z^2)$.

4.2/operations on rational expressions

In Chapter Two we found that the integers were closed with respect to addition, subtraction, and multiplication, but not with respect to division; to remedy the situation we extended the integers to the system of rational numbers, which were then closed with respect to all four operations (except, of course, for division by zero). We have the same situation with the polynomials, which are closed with respect to addition, subtraction, and multiplication, but not with respect to division; we will remedy the situation in the same manner, therefore, by forming a new system that will consist of the set of all quotients of polynomials with nonzero denominator. We refer to each member of this set as a *rational expression*. As before, if we agree that any polynomial is the same as the fraction that has that polynomial on top and a 1 on the bottom, we see that the set of rational expressions includes the set of polynomials properly, regardless of the number of variables that we are allowing for our polynomials.

Because rational expressions consist of numbers and symbols, which in turn may be thought of as standing for numbers, we may think of a rational expression as representing a fraction, the value of which depends upon the values that we substitute for the symbols. For example, if in the expression $(2x^2y - yz)/(xyz + z^2)$ we substitute $x = 2$, $y = 1$, and $z = -1$, we see that the expression is equal to $9/-1$ or -9. Since every rational expression may be thought of as representing a fraction, we would want the same rules to apply to rational expressions as applied to fractions.

If we use capital letters P, Q, R, S, etc., to stand for polynomials, then two rational expressions will be equal if and only if their cross products are equal; that is,

$$\frac{P}{Q} = \frac{R}{S} \quad \text{if and only if} \quad PS = QR$$

The cancelation law states that for any nonzero polynomial R,

$$\frac{PR}{QR} = \frac{P}{Q}$$

Let us consider the rational expressions $(x - y)/2y$ and $(x^2 - y^2)/(2xy + 2y^2)$. If we form cross products $(x - y)(2xy + 2y^2)$ and $(2y)(x^2 - y^2)$ and multiply them out, we come up with $2x^2y - 2y^3$ in both cases, so the two original rational expressions are equal.

The rules for addition, subtraction, multiplication, and division of rational expressions are exactly the same as the rules for operations on fractions and are summarized below with capital letters standing for polynomials. It is of course necessary to know how to add, subtract, and multiply polynomials to be able to apply these rules:

1. Addition:

$$\frac{P}{Q} + \frac{R}{S} = \frac{PS + QR}{QS}$$

2. Subtraction:

$$\frac{P}{Q} - \frac{R}{S} = \frac{PS - QR}{QS}$$

3. Multiplication:

$$\frac{P}{Q} \cdot \frac{R}{S} = \frac{PR}{QS}$$

4. Division:

$$\frac{P}{Q} \div \frac{R}{S} = \frac{P}{Q} \cdot \frac{S}{R} = \frac{PS}{QR}$$

Each of the letters P, Q, R, and S represents a polynomial that usually consists of several terms; consequently, computations with rational expressions usually take longer than computations with ordinary fractions.

/ examples /

1 $\dfrac{2xy - y^2}{3x} + \dfrac{x - y}{x + y}$

► $$\frac{2xy - y^2}{3x} + \frac{x - y}{x + y} = \frac{(2xy - y^2)(x + y) + (3x)(x - y)}{(3x)(x + y)}$$

$$= \frac{2x^2y + 2xy^2 - xy^2 - y^3 + 3x^2 - 3xy}{3x^2 + 3xy}$$

$$= \frac{2x^2y + xy^2 - y^3 + 3x^2 - 3xy}{3x^2 + 3xy}$$

2 $\dfrac{3x}{x - y} - \dfrac{2x + y}{y}$

► $$\frac{3x}{x - y} - \frac{2x + y}{y} = \frac{(3x)(y) - (x - y)(2x + y)}{(x - y)(y)}$$

$$= \frac{3xy - (2x^2 + xy - 2xy - y^2)}{xy - y^2}$$

$$= \frac{3xy - 2x^2 - xy + 2xy + y^2}{xy - y^2}$$

$$= \frac{4xy - 2x^2 + y^2}{xy - y^2}$$

3 $\dfrac{4x - y}{x + y} \cdot \dfrac{x^2 - y^2}{2y}$

$$\blacktriangleright \quad \frac{4x - y}{x + y} \cdot \frac{x^2 - y^2}{2y} = \frac{(4x - y)(x^2 - y^2)}{(x + y)(2y)}$$

$$= \frac{4x^3 - 4xy^2 - x^2y + y^3}{2xy + 2y^2}$$

4 $\dfrac{2x^3y^2 - x}{xy - y^2} \div \dfrac{x + y}{x - y}$

$$\blacktriangleright \quad \frac{2x^3y^2 - x}{xy - y^2} \div \frac{x + y}{x - y} = \frac{2x^3y^2 - x}{xy - y^2} \cdot \frac{x - y}{x + y}$$

$$= \frac{(2x^3y^2 - x)(x - y)}{(xy - y^2)(x + y)}$$

$$= \frac{2x^4y^2 - 2x^3y^3 - x^2 + xy}{x^2y + xy^2 - xy^2 - y^3}$$

$$= \frac{2x^4y^2 - 2x^3y^3 - x^2 + xy}{x^2y - y^3}$$

5 $\dfrac{1}{x} + \dfrac{1}{y}$

$$\blacktriangleright \quad \frac{1}{x} + \frac{1}{y} = \frac{y + x}{xy}$$

The last example tells us that to find the sum of two fractions whose numerators are both 1, we look at their denominators and form a fraction whose numerator is their sum and whose denominator is their product. Thus we can determine immediately that

$$\frac{1}{2} + \frac{1}{3} = \frac{5}{6}, \quad \frac{1}{4} + \frac{1}{7} = \frac{11}{28}, \quad \text{etc.}$$

/ example /

6 $\dfrac{1}{x} - \dfrac{1}{y}$

$$\blacktriangleright \quad \frac{1}{x} - \frac{1}{y} = \frac{y - x}{xy}$$

This example allows us to compute the difference of two fractions whose numerators are both 1. For example, we can determine immediately that

$$\frac{1}{2} - \frac{1}{3} = \frac{1}{6}, \quad \frac{1}{5} - \frac{1}{9} = \frac{4}{45}, \quad \frac{1}{3} - \frac{1}{8} = \frac{5}{24}, \quad \text{etc.}$$

Just as in the case of fractions, we may save ourselves some work in adding or subtracting rational expressions if we can find a common denominator that is less than their product. This is especially true when both fractions have the same denominator. If we look at the sum $(P/Q) + (R/Q)$, for example, and follow the rule for addition, we get

$$\frac{P}{Q} + \frac{R}{Q} = \frac{PQ + RQ}{Q^2} = \frac{(P + R)Q}{Q^2}$$

By the cancelation rule, this is the same as $(P + R)/Q$. Thus when adding two fractions with the same denominator, we add their numerators to get the numerator of the sum and take their common denominator for the denominator of the sum. Similarly, when subtracting fractions with a common denominator, we take the difference of the numerators as the numerator of the answer and take their common denominator as the denominator of the answer.

/ examples /

7 $\dfrac{2x^2y - y^2}{x + y} + \dfrac{x^2 + y^2}{x + y}$

► $$\frac{2x^2y - y^2}{x + y} + \frac{x^2 + y^2}{x + y} = \frac{(2x^2y - y^2) + (x^2 + y^2)}{x + y}$$

$$= \frac{2x^2y - y^2 + x^2 + y^2}{x + y}$$

$$= \frac{2x^2y + x^2}{x + y}$$

8 $\dfrac{3xy - y}{x^2 + y} - \dfrac{2x - xy}{x^2 + y}$

► $$\frac{3xy - y}{x^2 + y} - \frac{2x - xy}{x^2 + y} = \frac{(3xy - y) - (2x - xy)}{x^2 + y}$$

$$= \frac{3xy - y - 2x + xy}{x^2 + y}$$

$$= \frac{4xy - y - 2x}{x^2 + y}$$

9 $\dfrac{1}{x}\left(\dfrac{1}{y}+\dfrac{1}{z}\right)$

$$\blacktriangleright\ \frac{1}{x}\left(\frac{1}{y}+\frac{1}{z}\right)=\frac{1}{x}\left(\frac{z+y}{yz}\right)=\frac{z+y}{xyz}$$

10 $\dfrac{1}{x}+\dfrac{1}{y}+\dfrac{1}{z}$

$$\blacktriangleright\ \frac{1}{x}+\frac{1}{y}+\frac{1}{z}=\left(\frac{1}{x}+\frac{1}{y}\right)+\frac{1}{z}=\frac{y+x}{xy}+\frac{1}{z}$$

$$=\frac{(y+x)(z)+(xy)(1)}{xyz}=\frac{yz+xz+xy}{xyz}$$

11 $\dfrac{2}{2x+\dfrac{1}{x-\dfrac{2}{y}}}$

$$\blacktriangleright\ \frac{2}{2x+\dfrac{1}{x-\dfrac{2}{y}}}=\frac{2}{2x+\dfrac{1}{\dfrac{x}{1}-\dfrac{2}{y}}}=\frac{2}{\dfrac{2x}{1}+\dfrac{1}{\dfrac{xy-2}{y}}}$$

$$=\frac{2}{\dfrac{2x}{1}+\dfrac{y}{xy-2}}=\frac{2}{\dfrac{2x^2y-4x+y}{xy-2}}$$

$$=\frac{2}{1}\cdot\frac{xy-2}{2x^2y-4x+y}=\frac{2xy-4}{2x^2y-4x+y}$$

exercises 4.2

1 Simplify the following:

(a) $\dfrac{3x-2y}{x}+\dfrac{x+3y}{y}$

(b) $\dfrac{xy^2-y}{2x}+x^2+y$

(c) $\dfrac{3x+2y}{x}-\dfrac{x}{3x+y}$

(d) $\dfrac{x}{x+y}-\dfrac{x+y}{y}$

2 Simplify the following:

(a) $\dfrac{3x+y^2}{x}\cdot\dfrac{2x}{x-y}$

(b) $\dfrac{x^2+xy}{x-y}\cdot\dfrac{x+xy}{x+y}$

(c) $\dfrac{2x-y}{y}\div\dfrac{3x}{4x-y}$

(d) $\dfrac{3x^2-7x}{5}\div\dfrac{2xy^2-7x}{3y}$

3 Perform the following computations:

(a) $\frac{3x}{y}\left(\frac{2xy}{5} - \frac{6}{x+y}\right)$

(b) $\frac{1}{x}\left(\frac{1}{y} - \frac{1}{z}\right)$

(c) $\left(\frac{1}{x} - \frac{1}{y}\right)\frac{1}{z}$

(d) $\left(\frac{2x+y}{3x} + \frac{x+2y}{x-y}\right) \div \frac{x+y}{2x-y}$

(e) $\dfrac{1}{x + \frac{1}{y}}$

(f) $\dfrac{1}{x + \dfrac{1}{y + \frac{1}{z}}}$

4.3 / factorization of polynomials

If one were asked to reduce the fraction $\frac{35}{42}$ to lowest terms, he would see that 7 could be canceled from the numerator and the denominator and that $\frac{35}{42}$ was therefore reducible to $\frac{5}{6}$. To be able to reduce this fraction he would have to know that 35 was equal to 5×7 and that 42 was equal to 6×7; that is, he would have to know how to factor 35 and 42. If we are asked to multiply 5 times 7, we would answer 35. Conversely, if we were asked to factor 35, we would say 5×7.

The problem of factoring polynomials is exactly the same. When we are asked to factor a polynomial, we are really being asked to find two polynomials whose product is the given polynomial.

There is always one trivial factorization. If we were asked to factor $x^2 - y^2$, for example, we could write $(1)(x^2 - y^2)$. This is really a repetition of the original polynomial, so we will agree that when we factor a polynomial, both factors should be of degree at least 1.

It is simple to check a factoring problem. We merely multiply the two factors together, and if we get the original polynomial, then we have factored it correctly. Sometimes a polynomial cannot be factored; that is, there simply are not two polynomials each of degree at least 1 whose product is the given polynomial. In this case we say that the polynomial is *irreducible*. Irreducible polynomials are the counterparts of the prime numbers.

There are several approaches that may be employed in attempting to factor a polynomial. Two of these methods, however, will take care of most of the cases where a polynomial is factorable, and so we shall discuss only these two methods in this section and give brief indications of other approaches. These two methods are (1) factoring out a common factor and (2) factoring a trinomial.

To see how to factor a common factor out of a polynomial, let us go back to the distributive law in its more general form,

$$P(Q + R + S + T) = PQ + PR + PS + PT$$

Notice that when we multiply P by a sum of terms, each term of the answer has the factor P in it. In reverse, if each term in a polynomial has a factor P in it, it may be factored into P times the same sum of terms but with each term having P divided out of it.

As an example, consider the polynomial $6x^3y^2 - 9x^2y^3 + 15x^2y^2$. Note that each of the three terms of this polynomial has $3x^2y^2$ as a factor. We may therefore factor out $3x^2y^2$ and write this polynomial as $3x^2y^2$ times the sum of these terms, each with $3x^2y^2$ divided out; that is,

$$6x^3y^2 - 9x^2y^3 + 15x^2y^2 = 3x^2y^2(2x - 3y + 5)$$

We can check this very easily by multiplying the factors on the right side of this equation and verifying that we come up with the left side. Remember that if we wish to factor a common factor out of a polynomial, it must be a factor of each term of that polynomial. If it misses being a factor of even one term, then we cannot factor it out as a common factor.

/ examples /

1 Factor $16x^3y^2 + 24x^3y^4 - 8x^4y^3 + 12xy^3$.

► We note that $4xy^2$ is a common factor of all the terms of this polynomial. We therefore have

$$16x^3y^2 + 24x^3y^4 - 8x^4y^3 + 12xy^3 = 4xy^2(4x^2 + 6x^2y^2 - 2x^3y + 3y)$$

2 Factor $18p^3q + 12pq^2 + 3p - 6q$.

► The only common factor here is 3, so

$$18p^3q + 12pq^2 + 3p - 6q = 3(6p^3q + 4pq^2 + p - 2q)$$

3 Factor $15ab^2 + 4a^3 - 8b^2$.

► There is no factor common to the three terms of this polynomial, so we cannot factor out a common factor.

4 Factor $12ab^2 - 18a^2b + 9a^3b^3$.

► $12ab^2 - 18a^2b^2 + 9a^3b^3 = 3ab^2(4 - 6a + 3a^2b)$

We now turn to the factorization of a trinomial. A trinomial is a polynomial that consists of three terms. Let us go back and multiply together two polynomials each of which consists of two similar terms:

$$(3x - 2y)(5x + 6y) = 15x^2 + 18xy - 10xy - 12y^2$$

Notice that the product of the two middle terms gives us a term similar to the product of the two terms on the ends, and that these may then be combined in the final answer:

$$(3x - \overset{-10xy}{\overbrace{2y)(5x}} + \underset{+18xy}{\underbrace{6y)}} = 15x^2 + 8xy - 12y^2$$

This will always be the case when we multiply two polynomials each of which consists of two similar terms; that is, we will be able to combine the two middle terms of the answer so that the final answer will be a trinomial.

We will now try to reverse this procedure and write the trinomial, $15x^2 + 8xy - 12y^2$, as a product of two similar binomials (polynomials consisting of two terms) assuming that we do not know what the original factors were.

We write down the trinomial $15x^2 + 8xy - 12y^2$, and put two pairs of parentheses next to it in which we will try to fill in the correct factors:

$$15x^2 + 8xy - 12y^2 = (\quad)(\quad)$$

By the process of multiplication, we know that the product of the two first terms of the factors must be $15x^2$, so we ask what two expressions multiplied together gives us $15x^2$. This could be x and $15x$ or $3x$ and $5x$. Suppose that we try $3x$ and $5x$ first and fill these in as the first terms of the factors:

$$15x^2 + 8xy - 12y^2 \overset{?}{=} (3x \quad)(5x \quad)$$

We know that the product of the two second terms of the factors must be $-12y^2$, so we ask what gives us $-12y^2$. Here there are several possibilities: $12y$ and $-y$, $-12y$ and y, $4y$ and $-3y$, $-4y$ and $3y$, $6y$ and $-2y$, and $-6y$ and $2y$. Let us try $12y$ and $-y$ first and fill these in as the second terms of the factors:

$$15x^2 + 8xy - 12y^2 \overset{?}{=} (3x + \overset{+60xy}{\overbrace{12y)(5x}} - y) \qquad \text{(outer product: } -3xy\text{)}$$

This gives us $+60xy - 3xy = 57xy$ as the middle term of the product after combining terms, which is not what we want. If we reverse the positions of $12y$ and $-y$ we get

$$15x^2 + 8xy - 12y^2 \overset{?}{=} (3x - \overset{-5xy}{\overbrace{y)(5x}} + 12y) \qquad \text{(outer product: } +36xy\text{)}$$

which gives $-5xy + 36xy = 31xy$ for the middle term, which is also incorrect. We must now try all the other possibilities for the second terms of the factors until we get $+8xy$ for the middle term. If we carry out this trial-and-error process, we will eventually see that

$$15x^2 + 8xy - 12y^2 = (3x - \overset{-10xy}{\overbrace{2y)(5x}} + 6y) \qquad \text{(outer product: } +18xy\text{)}$$

is what we want because this gives us $-10xy + 18xy = +8xy$ for the middle term.

If none of these possibilities had worked, we would have gone back and tried $15x$ and x for the first terms of our answer. We see that factoring a trinomial is a trial-and-error process in which we try every possibility for the two first terms and the last two terms of the factors until we get the correct middle term. If nothing works, the trinomial is irreducible.

Let us try this approach on a problem where we do not know the answer beforehand by trying to factor $12x^2 - 7xy - 45y^2$. We write

$$12x^2 - 7xy - 45y^2 = (\quad)(\quad)$$

The various possibilities for the two first terms (which are supposed to give us $12x^2$ as a product) are $6x$ and $2x$, $4x$ and $3x$, and $12x$ and x. The possibilities for the two second terms are $45y$ and $-y$, $-45y$ and y, $3y$ and $-15y$, $-3y$ and $15y$, $5y$ and $-9y$, and $-5y$ and $9y$. We must try all these combinations and see which one, if any, gives us $-7xy$ as a middle term. The best procedure is to start with one pair for the two first terms and try all the possibilities for the two second terms. If none of these works, take the next possible pair for the first two terms and try all the possibilities for the two second terms again, and continue in this manner until you either find the correct factors or have exhausted all the possibilities.

If we apply this procedure to the problem above, we will find that the correct factorization is

$$12x^2 - 7xy - 45y^2 = (4x - 9y)(3x + 5y)$$

since this gives us $20xy - 27xy = -7xy$ for the middle term, which is what we want. Of course, if we had come up with

$$12x^2 - 7xy - 45y^2 = (3x + 5y)(4x - 9y)$$

this would have been just as good since the commutative law holds for polynomials as well as for numbers.

/ examples /

5 Factor $8x^2 - 26xy - 7y^2$.

$$\blacktriangleright\ 8x^2 - 26xy - 7y^2 = (2x - 7y)(4x + y)$$

6 Factor $6x^2 + 7xy - 10y^2$.

$$\blacktriangleright\ 6x^2 + 7xy - 10y^2 = (6x - 5y)(x + 2y)$$

7 Factor $12x^2 + x - 6$.

$$\blacktriangleright\ 12x^2 + x - 6 = (4x + 3)(3x - 2)$$

8 Factor $6a^2 - 19ab + 10b^2$.

$$\blacktriangleright\ 6a^2 - 19ab + 10b^2 = (3a - 2b)(2a - 5b)$$

In some factorization problems we can combine the two methods; that is, we can first factor out a common factor and then if the remaining factor is a trinomial, we may be able to factor this. For example, in the expression $24a^3b^2 + 22a^2b^3 - 30ab^4$, we can factor out $2ab^2$ to obtain

$$24a^3b^2 + 22a^2b^3 - 30ab^4 = 2ab^2(12a^2 + 11ab - 15b^2)$$

The second factor is a trinomial that may be factored into $(3a + 5b)(4a - 3b)$; therefore,

$$24a^3b^2 + 22a^2b^3 - 30ab^4 = 2ab^2(3a + 5b)(4a - 3b)$$

examples

9 Factor $12x^3 + 14x^2y - 20xy^2$.

$$\begin{aligned}\blacktriangleright\ 12x^3 + 14x^2y - 20xy^2 &= 2x(6x^2 + 7xy - 10y^2)\\ &= 2x(6x - 5y)(x + 2y)\end{aligned}$$

10 Factor $18x^3y + 15x^2y^2 - 18xy^3$.

$$\begin{aligned}\blacktriangleright\ 18x^3y + 15x^2y^2 - 18xy^3 &= 3xy(6x^2 + 5xy - 6y^2)\\ &= 3xy(2x + 3y)(3x - 2y)\end{aligned}$$

There is a special type of trinomial in which the middle term is 0 (so that it is really a binomial) and that is known as a difference of two squares. Notice that when we multiply two binomials that are the same except for the sign of the second term, the middle term drops out and we get a difference of two squares. For example, $(x + y)(x - y) = x^2 - y^2$. This means that a difference of two squares may always be factored into two binomials that are the same except for the sign of the second term.[1]

examples

11 Factor $16x^4 - 4y^2$.

$$\blacktriangleright\ 16x^4 - 4y^2 = (4x^2 + 2y)(4x^2 - 2y)$$

[1]A polynomial of the type $x^2 - 6$ is irreducible, for although it is technically correct to factor it into $(x + \sqrt{6})(x - \sqrt{6})$, we are restricting ourselves here to polynomials with rational coefficients and $\sqrt{6}$ is not a rational number. Similarly, $x - y$ is irreducible even though one might factor it into $(\sqrt{x} + \sqrt{y})(\sqrt{x} - \sqrt{y})$, because a polynomial includes only nonnegative integral powers of x and y.

12 Factor $25a^4b^2 - 16a^2b^6$.

$$\blacktriangleright\ 25a^4b^2 - 16a^2b^6 = (5a^2b + 4ab^3)(5a^2b - 4ab^3)$$

This rule for factoring a difference of two squares may be extended to a difference of two nth powers by the formula,

$$x^n - y^n = (x - y)(x^{n-1} + x^{n-2}y + x^{n-3}y^2 + \cdots + xy^{n-2} + y^{n-1})$$

When $n = 3$, for example, we have the factoring formula

$$x^3 - y^3 = (x - y)(x^2 + xy + y^2)$$

When $n = 4$, we have

$$(x^4 - y^4) = (x - y)(x^3 + x^2y + xy^2 + y^3), \text{ etc.}$$

We will close this section by mentioning one other possibility for factoring a polynomial. It is sometimes possible to break a polynomial up into two parts, factor a common factor out of each part, and if we find that the other factors are both the same, this may then be factored out as a common factor of the result. Consider, for example, the polynomial $xw + xz + yw + yz$. If we break this up into $(xw + xz) + (yw + yz)$ and factor x out of the first part and y out of the second part, we have $x(w + z) + y(w + z)$. We see that $(w + z)$ is a common factor of the result, which may be factored out to give us

$$xw + xz + yw + yz = (x + y)(w + z)$$

/examples/

13 Factor $6ac - 2ad + 9bc - 3bd$.

$$\begin{aligned}\blacktriangleright\ 6ac - 2ad + 9bc - 3bd &= (6ac - 2ad) + (9bc - 3bd)\\ &= 2a(3c - d) + 3b(3c - d)\\ &= (2a + 3b)(3c - d)\end{aligned}$$

14 Factor $8xp - 12xq - 6yp + 9yq$.

$$\begin{aligned}\blacktriangleright\ 8xp - 12xq - 6yp + 9yq &= (8xp - 12xq) - (6yp - 9yq)\\ &= 4x(2p - 3q) - 3y(2p - 3q)\\ &= (4x - 3y)(2p - 3q)\end{aligned}$$

/exercises 4.3/

1 Factor the following expressions:

(a) $6x^3y + 4x^2y^2 - 2x^2y$ (b) $12a^3b^3 - 16a^2b^3 + 20a^3b^2$

(c) $5pq^2 - 15p^2q + 20p$ (d) $9abc^2 - 15a^2bc - 12ab^2$

2 Factor the following expressions:

(a) $12x^2 + xy - 6y^2$ (b) $6p^2 + 7pq - 10q^2$
(c) $6x^2 - x - 12$ (d) $10s^2 + 9st - 9t^2$
(e) $s^2 + s - 30$ (f) $4x^2 - 4xy - 35y^2$

3 Factor the following expressions:

(a) $x^2y^2 - z^2$ (b) $16x^4 - 81y^4$
(c) $xw + 2yw - xz - 2yz$ (d) $6px + 4xq - 3py - 2qy$

4 Since $(x + y)(x - y) = x^2 - y^2$, we can multiply 72×68 very quickly by writing $(70 + 2)(70 - 2) = 4900 - 4 = 4896$. Use this principle to determine the following products:

(a) 84×76 (b) 59×61 (c) 67×73
(d) 48×52 (e) 35×25

4.4 / simplification of algebraic expressions

The operations of addition, subtraction, multiplication, and division are binary operations. This means that when we add, subtract, multiply, or divide numbers or algebraic expressions, we combine only two at a time. Thus in a seemingly complicated expression involving more than two numbers, we can, by combining the numbers two at a time, reduce it to a single number. Parentheses, brackets, and other symbols of grouping are used to tell us which two numbers to combine first, which two to combine next, and so on. Reduction of fractions permits us to reduce cumbersome fractions such as $\frac{4760}{6664}$ to more simple fractions like $\frac{5}{7}$, which are much easier to work with.

By similar procedures, we may reduce complicated algebraic expressions to simpler form. Because we are permitted to combine only like terms when adding or subtracting and we cannot always divide one expression by another evenly, it is not always possible to reduce one of these to a single term.

Given an algebraic expression, we usually try to simplify it by performing the indicated operations first. If we end up with a quotient, we then try to reduce this to the lowest terms. To do this we factor the numerator and denominator, if possible, and then see whether they have a common factor that may be canceled out.

/ examples /

1 Simplify $\frac{2x^2}{y}\left(\frac{1}{x^2} - \frac{1}{y}\right)$.

► $$\frac{2x^2}{y}\left(\frac{1}{x^2} - \frac{1}{y}\right) = \frac{2x^2}{y} \cdot \frac{y - x^2}{x^2y} = \frac{2x^2(y - x^2)}{x^2y^2}$$

We may cancel x^2 from the numerator and denominator to give

$$\frac{2(y - x^2)}{y^2} \quad \text{or} \quad \frac{2y - 2x^2}{y^2}$$

2 Simplify $2x - \dfrac{x^2 + y}{x}$.

► $$2x - \frac{x^2 + y}{x} = \frac{2x}{1} - \frac{x^2 + y^2}{x} = \frac{2x^2 - (x^2 + y^2)}{x}$$

$$= \frac{2x^2 - x^2 - y^2}{x} = \frac{x^2 - y^2}{x}$$

We could factor the numerator into $(x + y)(x - y)$, but that will not help us do any further simplification here.

3 Simplify $\dfrac{1 + x}{2y} \Big/ \dfrac{2 + y}{x}$.

► $$\frac{\dfrac{1 + x}{2y}}{\dfrac{2 + y}{x}} = \frac{1 + x}{2y} \div \frac{2 + y}{x} = \frac{1 + x}{2y} \cdot \frac{x}{2 + y}$$

$$= \frac{(1 + x)x}{2y(2 + y)} = \frac{x + x^2}{4y + 2y^2}$$

4 Simplify $\dfrac{1}{x}\left(\dfrac{\sqrt{x + 1}}{3} + \dfrac{1}{2\sqrt{x + 1}}\right)$.

► $$\frac{1}{x}\left(\frac{\sqrt{x + 1}}{3} + \frac{1}{2\sqrt{x + 1}}\right) = \frac{1}{x} \cdot \frac{(\sqrt{x + 1})(2\sqrt{x + 1}) + 3}{6\sqrt{x + 1}}$$

$$= \frac{1}{x} \cdot \frac{2(x + 1) + 3}{6\sqrt{x + 1}} = \frac{1}{x} \cdot \frac{2x + 2 + 3}{6\sqrt{x + 1}}$$

$$= \frac{1}{x} \cdot \frac{2x + 5}{6\sqrt{x + 1}} = \frac{2x + 5}{6x\sqrt{x + 1}}$$

If we wish, we may rationalize the denominator and write

$$\frac{2x + 5}{6x\sqrt{x + 1}} = \frac{(2x + 5)\sqrt{x + 1}}{6x\sqrt{x + 1}\,\sqrt{x + 1}} = \frac{(2x + 5)\sqrt{x + 1}}{6x(x + 1)}$$

$$= \frac{(2x + 5)\sqrt{x + 1}}{6x^2 + 6x}$$

5 Simplify $\dfrac{\sqrt{x} - \sqrt{y}}{\sqrt{x} + \sqrt{y}} \cdot \dfrac{x - y}{4}$.

► If we first rationalize the denominator of the first factor, we get

$$\frac{\sqrt{x} - \sqrt{y}}{\sqrt{x} + \sqrt{y}} \cdot \frac{x - y}{4} = \left(\frac{\sqrt{x} - \sqrt{y}}{\sqrt{x} + \sqrt{y}} \cdot \frac{\sqrt{x} - \sqrt{y}}{\sqrt{x} - \sqrt{y}}\right) \frac{x - y}{4}$$

$$= \frac{x - 2\sqrt{x}\sqrt{y} + y}{x - y} \cdot \frac{x - y}{4} = \frac{x - 2\sqrt{xy} + y}{4}$$

since $x - y$ may be canceled out before multiplying.

6 Add $\dfrac{3x + 4}{3x} + \dfrac{x - 1}{x^2}$.

► Here we see that $3x^2$ would be a common denominator, so we may write

$$\frac{3x + 4}{3x} + \frac{x - 1}{x^2} = \frac{(3x + 4)x}{3x^2} + \frac{(x - 1)(3)}{3x^2}$$

$$= \frac{3x^2 + 4x + 3x - 3}{3x^2} = \frac{3x^2 + 7x - 3}{3x^2}$$

7 Subtract $\dfrac{2x}{3x^2 - 3x} - \dfrac{3}{x^2 - 1}$.

► If we factor the denominators, we may write the problem as

$$\frac{2x}{3x(x - 1)} - \frac{3}{(x + 1)(x - 1)}$$

so a common denominator is $3x(x + 1)(x - 1)$. We then have

$$\frac{2x}{3x^2 - 3x} - \frac{3}{x^2 - 1} = \frac{2x(x + 1) - 3(3x)}{3x(x + 1)(x - 1)}$$

$$= \frac{2x^2 + 2x - 9x}{3x(x + 1)(x - 1)} = \frac{2x^2 - 7x}{3x^3 - 3x}$$

We may then cancel x from the numerator and denominator to obtain

$$\frac{2x - 7}{3x^2 - 3}$$

In the last two problems we could have followed the rule for adding or subtracting fractions in a straightforward manner and gotten the same answer, but we would have had more work and we would have had to reduce the final answer to lowest terms.

8 Simplify $\dfrac{2x^2 + 5x - 3}{5x^2 + 17x + 6}$.

$$\blacktriangleright \quad \frac{2x^2 + 5x - 3}{5x^2 + 17x + 6} = \frac{(2x - 1)(x + 3)}{(5x + 2)(x + 3)} = \frac{2x - 1}{5x + 2}$$

9 Simplify $\dfrac{x^2 - 2x}{x^2 - 4}$.

$$\blacktriangleright \quad \frac{x^2 - 2x}{x^2 - 4} = \frac{x(x - 2)}{(x + 2)(x - 2)} = \frac{x}{x + 2}$$

10 Simplify $\dfrac{x^3 - y^3}{x^2 - y^2}$.

$$\blacktriangleright \quad \frac{x^3 - y^3}{x^2 - y^2} = \frac{(x - y)(x^2 + xy + y^2)}{(x - y)(x + y)} = \frac{x^2 + xy + y^2}{x + y}$$

To apply what you have learned, you might try the following mind-reading trick on some of your friends.[1] Tell someone to pick a number without revealing his choice. Since you have no idea what he chose, write down the symbol x on a piece of paper. The feature that sets this mind-reading trick apart from others is that you do not tell your friend what to do to the number; rather, he may perform any operations he wishes with it as long as he lets you in on what he is doing. You might suggest omitting division in order to eliminate fractions and make the computation easier, although division is certainly not prohibited.

Suppose that he tells you first, "Add 5." You add 5 to x, so you have $x + 5$ on your paper. Suppose that he says next, "Multiply by 3." You then have $3(x + 5)$. If he then says, "Subtract 4," you should have $3(x + 5) - 4$, and if he says finally, "Multiply by 2," you should have $2[3(x + 5) - 4]$. You simplify this to obtain

$$2[3(x + 5) - 4] = 2(3x + 15 - 4) = 2(3x + 11) = 6x + 22$$

You then say that you would like to add a couple of steps of your own. Tell him to subtract 22 so that you will have $6x$ left; then tell him to divide by the number that he started with (the value of which you need never discover) and tell him that his answer is 6 (since $6x/x = 6$). Thus by simplifying an algebraic expression, you discover the number your friend ends up with, although you have no idea of what number he started with.

As another example of this mind-reading trick, suppose that you ask your friend to take a number and he then gives you the following sequence of steps:

1. Add 6.
2. Multiply by 2.

[1]Adapted from "Take a Number," by L. M. Weiner, *The Mathematics Teacher*, Vol. 48, No. 4 (April 1955), p. 203.

3. Subtract 4.
4. Add 9.
5. Multiply by 5.
6. Subtract 100.

In accordance with these instructions, you should write the following sequence of algebraic expressions on your paper:

1. $x + 6$
2. $2(x + 6)$
3. $2(x + 6) - 4$
4. $2(x + 6) - 4 + 9$
5. $5[2(x + 6) - 4 + 9]$
6. $5[2(x + 6) - 4 + 9] - 100$

If you simplify this last expression, you would obtain

$$\begin{aligned} 5[2(x + 6) - 4 + 9] - 100 &= 5(2x + 12 + 5) - 100 \\ &= 5(2x + 17) - 100 \\ &= 10x + 85 - 100 \\ &= 10x - 15 \end{aligned}$$

To simplify this tell your friend to add 15 so that you then have $10x$, and then tell him to divide by the number that he started with, and you can announce confidently that his answer must be 10, provided that he made no errors in computation.

exercises 4.4

1 Perform the following computations:

(a) $\dfrac{2x + y}{x^2} - \dfrac{8y}{7x}$ (b) $\dfrac{3x}{x^2 - 1} + \dfrac{2}{x + 1} - \dfrac{y}{x - 1}$

(c) $\dfrac{5xy}{x^2 - xy} + \dfrac{2x + y}{x^2 - y^2}$ (d) $\dfrac{p + q}{p^2 - q^2} + \dfrac{2p + 3q}{2p^2 + 3pq + q^2}$

2 Simplify the following expressions:

(a) $\dfrac{2x^2 + 5xy - 3y^2}{4x^2 + 4xy - 3y^2}$ (b) $\dfrac{x^2 - y^2}{2x^2 + xy - y^2}$

(c) $\dfrac{x^4 - y^4}{x^2 - y^2}$ (d) $\dfrac{x^4 - y^4}{x - y}$

(e) $\dfrac{x^2 - y^2}{2x + y} \cdot \dfrac{2x^2 + 3xy + y^2}{x - y}$ (f) $\dfrac{3x - y}{x + y} \div \dfrac{3x^2 - 4xy + y^2}{x^2 + 3xy + 2y^2}$

(g) $\dfrac{x^2 + x}{x - 1} \cdot \dfrac{x^3 - y^3}{4x^2 - y^2} \cdot \dfrac{2x - y}{x^2 - y^2}$

review test 4

NAME ____________

DATE ____________

1 Add $(9x^4 - 3x^3 - 7x^2 + 2x - 1) + (x^3 - 4x^2 + 5x - 1)$.

2 Subtract $(7x^3 - 4x^2 - 9x + 3) - (9x^3 + 4x^2 + 2x - 1)$.

3 Multiply $(x^3 - 7x^2 + 3x - 2)(x^2 - 5x + 1)$.

4 Divide $(4x^3 - 7x^2 + 3x + 4) \div (x - 3)$.

5 Simplify $(3x^2y - 7xy^2 + x^2y^2) + (2xy^2 - 7xy + y) - (3x^2y + 4xy - 3x^2y^2)$.

6 Multiply $(2x^2y^2 - 7xy^2)(3x^2y - 4xy)$.

7 Simplify the following:

(a) $\dfrac{2x - 3y}{2x} + \dfrac{1}{x + y}$

(b) $\dfrac{3x + y}{x - y} - \dfrac{1 + x}{y}$

(c) $\dfrac{1}{x + y} + \dfrac{1}{x - y}$

(d) $\dfrac{x + y}{1} - \dfrac{1}{x + y}$

(e) $\dfrac{1}{x} + \dfrac{1}{y}$

8 Perform the following computations:

(a) $\left(\dfrac{1}{x} + \dfrac{1}{y}\right)(x + y)$

(b) $\dfrac{1}{x}\left(\dfrac{1}{x} + \dfrac{1}{y}\right)$

(c) $\frac{1}{x} + \frac{1}{y} + \frac{1}{z}$

(d) $\dfrac{x + y}{\frac{1}{x} + \frac{1}{y}}$

(e) $\dfrac{2x - y}{x + 2y} \div \dfrac{3x + y}{x - 3y}$

(f) $\dfrac{1}{\left(\frac{1}{x} + y\right)\left(x + \frac{1}{y}\right)}$

9 Factor the following polynomials:

(a) $3x^2y^3 - 15x^3y^2 + 12x^2y^2$

(b) $12s^2t - 8s^2t^2 + 20st^2$

(c) $6p^2q - 9p^3q^2 + 4pq^2$

10 Factor the following polynomials:

(a) $6x^2 - 11xy - 10y^2$

(b) $6s^2 - 7st - 10t^2$

(c) $12p^2 + pq - 6q^2$

(d) $4x^2y^2 - 9$

(e) $256x^4 - 81y^4$

11 Using the fact that $(x + y)(x - y) = x^2 - y^2$, determine the following products:
(a) 62×58

(b) 86×94

(c) 83×77

12 Perform the following computations:

(a) $\dfrac{x - 3y}{3x^2} + \dfrac{2x - y}{5x}$

(b) $\dfrac{2x}{x^2 - 4} + \dfrac{3x}{x + 2}$

(c) $\dfrac{2x}{x^2 - y^2} + \dfrac{3y}{x^2 + xy}$

(d) $\dfrac{s + t}{2s^2 + st - t^2} + \dfrac{s - t}{s^2 - st - 2t^2}$

13 Simplify the expression

$$\frac{2x^2 + 3xy - 9y^2}{6x^2 - 13xy + 6y^2}$$

14 Simplify the expression

$$\frac{2x^2 + xy - y^2}{x^2 + 2x} \cdot \frac{xy + 2y}{x^2 - y^2}$$

15 Write the expression

$$\frac{x^6 - y^6}{x^5 - y^5}$$

in a different form by factoring the numerator and the denominator and then canceling.

chapter five

SOLVING EQUATIONS

5.1/equations and identities

Any statement that tells us that one thing is equal to another is an *equation.* The statement $3 + 4 = 7$, for example, is an equation. We may interpret this to mean that $3 + 4$ is the same thing as 7 or that $3 + 4$ is another name for 7. If we write $3 + 4 = 8$, we again have an equation, but one that is not a true statement.

When we see an equation such as $3 + 4 = 7$ or $3 + 4 = 8$, we can look at it and decide whether or not it is true. Some equations, however, contain symbols or indeterminates (usually x). Equations such as $3 + x = 7$ and $x(x + 1) = x^2 + x$ are examples of this type. What are we supposed to do with equations like these?

Equations that contain indeterminates may be considered as questions rather than statements. We may view them as asking: What value of the indeterminate, if any, will make this a true equation? or Find the set of all numbers that will satisfy this equation.

This answer set may consist of the null set or it may consist of the set of all real numbers. An equation that is true no matter what number we substitute for x is called an *identity*. Thus $x(x + 1) = x^2 + x$ is an identity. Equations whose answer set consists of a finite set of numbers are called *conditional equations* or, more simply, just *equations*. It is this latter type in which we shall be interested in this chapter and we shall discuss methods for finding the set of answers or *roots*, as they are usually referred to, of these equations.

Equations containing just one indeterminate or *unknown* are referred to as *equations in one unknown*, those with two indeterminates are called *equations in two unknowns*, and so on. We shall concentrate first on equations in one unknown and then, later in the chapter, on equations in two unknowns.

An equation that states that a certain polynomial in x is equal to zero is called a *polynomial equation in one unknown*. As before, if the degree of the polynomial is 1, the equation is called *linear*, if it is 2, the equation is called *quadratic*, and so on. The equation $4x + 3 = 0$ would thus be an example of a linear equation, whereas $5x^2 - 7x + 2 = 0$ would be an example of a quadratic equation.

One way to find the roots of an equation is to employ the trial-and-error approach; that is, keep trying different numbers until you find the one or ones that work. Unless the equation is very simple and you can practically spot the answer, this method is not very practical because there is an infinite set of numbers that could be answers. We shall start with the simplest case, linear equations in one unknown, then proceed to quadratic equations in one unknown, and finally to linear equations in two unknowns and linear equations in three unknowns. To demonstrate their practical value, we shall show how solutions of these equations may be used to solve certain word problems.

exercises 5.1

1 Determine whether each of the following is an identity or a conditional equation:

(a) $5x + 2 = 3x - 1$

(b) $2(x + 4) = 2x + 8$

(c) $2x + 1 = 5x - 3 - 3x + 4$

(d) $\dfrac{x + 1}{2} = \dfrac{3x + 3}{6}$

(e) $\dfrac{2x}{7} = \dfrac{x + 1}{3}$

2 Explain why the equation $2x + 5 = 2x + 1$ does not have any roots.

3 Find the two roots of the equation $x^2 - 10x + 21$ by trial and error. (*Hint:* The roots are integers between 1 and 10.)

5.2/linear equations in one unknown

Our basic objective in solving an equation in one unknown is to find those values of x that will satisfy the equation. In other words, starting with the original equation, we want to end up with an equation of the form $x =$ the answer. The original equation tells us that two quantities are equal, and that if we do the same thing to these quantities, the resulting quantities should be equal. Thus our basic working rule for the solution of equations will be that, with certain exceptions, if we do the same thing to both sides of an equation, the resulting equation will have the same answer set as the original. We must be careful, however, for there are certain exceptions to this rule. For example, if we were to multiply or divide both sides by something involving x, we could enlarge or diminish the original answer set.

A lincar equation in one unknown can involve only terms of the first degree in x and numbers. Thus it must be similar to the equation

$$5x - 7 + 2x = 3 + 4x - 4$$

Using this equation as a guide, notice that if we add 7 to both sides of the equation we would get

$$5x - 7 + 2x + 7 = 3 + 4x - 4 + 7$$

The -7 would cancel the $+7$ on the left and we would have

$$5x + 2x = 3 + 4x - 4 + 7$$

Comparing this with the original equation shows that the -7 on the left went over to the right and became $+7$. Now if we subtract $4x$ from both sides we have

$$5x + 2x - 4x = 3 + 4x - 4 + 7 - 4x$$

or

$$5x + 2x - 4x = 3 - 4 + 7$$

so the net effect is that the $+4x$ went over to the left and became $-4x$.

This example illustrates the fact that we may pick up a term from either side of an equation and move it over to the other side provided that we change its sign. This procedure is known as *transposing terms*. *Transposing terms of an equation does not change its answer set.*

Our purpose in transposing terms is to get all the terms containing x on one side and all the other terms on the other side so that we may combine terms on each side. If we do this in the equation above, we have

$$3x = 6$$

Our objective is to end up with $x =$ the answer, and we obtain x on the left side if we divide by 3. What we do to one side, however, we must do to the other side, so we divide both sides by 3 to get the answer,

$$x = 2$$

We may check this answer by substituting 2 for x in the original equation and seeing that we come out with a true equation—that is, that $x = 2$ satisfies the equation.

By way of summary, we may list the following steps for solving linear equations in one unknown:

1. Transpose all the terms containing x to the left side and all the other terms to the right side. (Those terms already on the proper side are left untouched.)
2. Combine terms on each side.
3. Divide both sides by the coefficient of x. (This should yield $x =$ the answer.)

Some equations are somewhat more complex than the example above. Consider, for example,

$$\frac{3x + 1}{2x} - 5 = \frac{2}{3}$$

Our objective is to simplify this equation, and we can best do this by eliminating the denominators. We can eliminate the denominator $2x$ if we multiply both sides by $2x$ (we must multiply each term on both sides by $2x$) and then cancel the $2x$. We would then have

$$\cancel{2x} \cdot \frac{3x + 1}{\cancel{2x}} - 2x \cdot 5 = \frac{2}{3} \cdot 2x$$

Next we eliminate the denominator 3 by multiplying each term on both sides by 3 and canceling. This yields

$$3 \cdot \cancel{2x} = \frac{3x + 1}{\cancel{2x}} - 3 \cdot 2x \cdot 5 = \frac{2}{\cancel{3}} \cdot 2x \cdot \cancel{3}$$

We then write out what we have left:

$$3(3x + 1) - 3 \cdot 2x \cdot 5 = 2 \cdot 2x$$

Multiplying out yields

$$9x + 3 - 30x = 4x$$

and then the three-step procedure just described yields

$$9x - 30x - 4x = -3$$

$$-25x = -3$$

$$x = \frac{3}{25}$$

Thus $x = \frac{3}{25}$ is the root of the original equation. It would be a lot more work to check this answer than in the previous example, but substituting $x = \frac{3}{25}$ in the original equation will yield a true equation.

For equations with denominators, we must therefore apply the following two steps first before determining x by the three steps listed before:

1. Eliminate each denominator by multiplying each term on both sides by it and then canceling.

2. Multiply out what you have left.

This reduces the equation to the form indicated above, which we can then solve as before.

examples

1 $3x + 2 - 7 = 4x + 9 - 2 - 5x + 8$

► Transposing all the terms involving x to the left and the others to the right yields

$$3x - 4x + 5x = 9 - 2 + 8 - 2 + 7$$

Then

$$4x = 20$$

$$x = 5$$

2 $8 - 4x - 7 = 2x + 5 + 6x$

► Following the three-step procedure above yields

$$-4x - 2x - 6x = 5 - 8 + 7$$

$$-12x = 4$$

$$x = -\frac{4}{12} = -\frac{1}{3}$$

3 $\dfrac{2x - 5}{7} - 5 = \dfrac{2x}{5}$

► Multiplying through by 7 and by 5 yields

$$5 \cdot \cancel{7}\,\frac{2x - 5}{\cancel{7}} - 5 \cdot 7 \cdot 5 = \frac{2x}{\cancel{5}} \cdot 7 \cdot \cancel{5}$$

$$5(2x - 5) - 5 \cdot 7 \cdot 5 = 2x \cdot 7$$

$$10x - 25 - 175 = 14x$$

$$10x - 14x = 25 + 175$$

$$-4x = 200$$

$$x = -\frac{200}{4} = -50$$

4 $\dfrac{3x+1}{6} + x = \dfrac{4}{3}$

► If we multiply through by 6, we will eliminate both denominators in one step:

$$\cancel{6} \cdot \frac{3x+1}{\cancel{6}} + x \cdot 6 = \frac{4}{\cancel{3}} \cdot \cancel{6}^{2}$$

$$3x + 1 + 6x = 4 \cdot 2$$

$$3x + 1 + 6x = 8$$

$$3x + 6x = 8 - 1$$

$$9x = 7$$

$$x = \frac{7}{9}$$

5 $\dfrac{2x-3}{10x} - 3 = \dfrac{7}{5}$

► If we multiply through by $10x$, we will eliminate both denominators in one step:

$$\cancel{10x} \cdot \frac{2x-3}{\cancel{10x}} - 3 \cdot 10x = \frac{7}{\cancel{5}} \cdot \cancel{10x}^{2x}$$

$$2x - 3 - 3 \cdot 10x = 7 \cdot 2x$$

$$2x - 3 - 30x = 14x$$

$$2x - 30x - 14x = 3$$

$$-42x = 3$$

$$x = -\frac{3}{42} = -\frac{1}{14}$$

6 $\dfrac{5x-1}{3} = \dfrac{2x+7}{5}$

► Multiplying through first by 3 and then by 5 yields

$$5 \cdot \cancel{3} \cdot \frac{5x-1}{\cancel{3}} = \frac{2x+7}{\cancel{5}} \cdot 3 \cdot \cancel{5}$$

$$5(5x - 1) = (2x + 7)(3)$$

$$25x - 5 = 6x + 21$$

$$19x = 26$$

$$x = \frac{26}{19}$$

Notice that after eliminating the denominators, we come up with

$$(5)(5x + 1) = (2x + 7)(3)$$

That is, the numerator of the left side times the denominator of the right side equals the numerator of the right side times the denominator of the left side. Whenever each side of an equation consists of a single fraction we may *cross-multiply* as above and proceed from there. In other words, if the equation is of the form

$$\frac{A}{B} = \frac{C}{D}$$

we may write

$$AD = BC$$

and proceed as before.

examples

7 $\dfrac{6}{2x - 1} = \dfrac{5}{8x + 3}$

► Applying the cross-multiplication procedure yields

$$\begin{aligned} 6(8x + 3) &= 5(2x - 1) \\ 48x + 18 &= 10x - 5 \\ 48x - 10x &= -5 - 18 \\ 38x &= -23 \\ x &= -\frac{23}{38} \end{aligned}$$

8 $\dfrac{3}{2x + 3} = \dfrac{6}{4x + 6}$

► Cross-multiplying yields

$$\begin{aligned} 3(4x + 6) &= 6(2x + 3) \\ 12x + 18 &= 12x + 18 \end{aligned}$$

Applying the usual procedure from here on will yield $0 = 0$. We see that this equation is actually an identity.

9 $\dfrac{3}{2x + 3} = \dfrac{6}{4x + 1}$

► Proceeding as before,

$$3(4x + 1) = 6(2x + 3)$$
$$12x + 3 = 12x + 18$$
$$12x - 12x = 18 - 3$$
$$0 = 15$$

This contradiction tells us that the original equation has no roots, because $12x - 12x = 18 - 3$ has the same answer set as the original equation, and there is no number that will satisfy the last equation.

10 $\dfrac{\frac{4}{5} - x}{3x + 2} = \dfrac{1}{2}$

► Multiplying both sides by $3x + 2$ yields

$$\cancel{(3x + 2)} \cdot \frac{\frac{4}{5} - x}{\cancel{3x + 2}} = \frac{1}{2} \cdot (3x + 2)$$

which may be written

$$\frac{4}{5} - x = \frac{3x + 2}{2}$$

Proceeding as before yields

$$2 \cdot \cancel{5} \cdot \frac{4}{\cancel{5}} - x \cdot 5 \cdot 2 = \frac{3x + 2}{\cancel{2}} \cdot 5 \cdot \cancel{2}$$
$$2 \cdot 4 - x \cdot 5 \cdot 2 = (3x + 2)(5)$$
$$8 - 10x = 15x + 10$$
$$-10x - 15x = 10 - 8$$
$$-25x = 2$$
$$x = -\frac{2}{25}$$

exercises 5.2

1 Solve the following equations:

(a) $3x + 7 - 2x = 4 - 8x + 3$
(b) $2 + 5x + 3 - 1 = 7x + 5$
(c) $3x - 5x + 2 + 7 = 4 - 7x + 8x - 1$
(d) $2 + 7 - 8x = 9x + 2x + 1$

2 Solve the following equations:

(a) $5(2x + 4) - 2 = 7(x + 3) - 1$
(b) $3 - 4(8x + 1) = 2(4x - 1) + 3x$
(c) $2x - 5(4x + 1) = 2 - 8(x - 3)$
(d) $5x - 3(2x + 5) = 9 + 6(3x - 1)$

3 Solve the following equations:

(a) $\frac{5}{3x} - 4 = \frac{2}{5}$ (b) $\frac{5x}{3} - 7 = 2x$

(c) $\frac{2x + 1}{5x} - 4 = \frac{3}{2}$ (d) $\frac{1}{x - 1} + 5 = \frac{4}{7}$

4 Solve the following equations:

(a) $\frac{4x - 2}{3} = \frac{3x + 1}{7}$ (b) $\frac{3x + 5}{2} = \frac{x - 3}{4}$

(c) $\frac{5}{6x + 1} = \frac{3}{2x - 3}$ (d) $\frac{x + \frac{1}{2}}{3x - \frac{2}{3}} = \frac{5}{7}$

5.3 / word problems

The material covered in Section 5.2 may be used to solve word problems. As a typical word problem, suppose that John has \$12 more than Bill. If John gives Bill \$14, Bill will have three times as much money as John. We are asked to find how much each has to begin with.

To find the answer, we must write an equation that says the same thing as the problem says in words, and then we must solve this equation. The task of writing an equation that says the same thing as the words is generally difficult, and it is just as difficult to give a detailed set of instructions that will permit one to do this. Instead, we will give a general set of guidelines. First determine what it is you are looking for, and let x stand for this quantity. If you are looking for more than one quantity, determine what expression in x must stand for the other quantity. In the problem above we are asked to find Bill's original amount and John's original amount, so we write

Let x = Bill's original amount (in dollars)

Since John started with \$12 more than Bill, we write

Let $x + 12$ = John's original amount

(If we had decided to let x = John's original amount, we would have let $x - 12$ = Bill's original amount.) After John gives Bill \$14, John will have left $x + 12 - 14$ or $x - 2$, whereas Bill will have $x + 14$. We then wish to state that Bill's final amount is 3 times John's final amount, so we write

$$x + 14 = 3(x - 2)$$

which says the same thing as the problem says in words. Solving this yields

$$x + 14 = 3x - 6$$

$$x - 3x = -6 - 14$$

$$-2x = -20$$

$$x = \$10 \qquad \text{Bill's original amount}$$

$$x + 12 = \$22 \qquad \text{John's original amount}$$

Let us examine another word problem of the same type. John has four times as much money as Bill. If John gives Bill \$20, Bill will have \$13 more than John. How much did each have originally? To solve the problem, we write

$$\text{Let } x = \text{Bill's original amount}$$

$$\text{Let } 4x = \text{John's original amount}$$

After John gives Bill \$20, John has $4x - 20$ and Bill has $x + 20$. We wish to say that Bill's amount is then John's amount plus \$13, so we write

$$x + 20 = 4x - 20 + 13$$

Solving this yields

$$x - 4x = -20 + 13 - 20$$

$$-3x = -27$$

$$x = \$9 \qquad \text{Bill's original amount}$$

$$4x = \$36 \qquad \text{John's original amount}$$

As a general set of guidelines,

1. Read the problem and determine what you are looking for.
2. Let x equal what you are looking for.
3. If more than one quantity is sought, determine what the other must be in terms of x.
4. Write an equation that says the same thing as the problem.
5. Solve the equation.

As a further aid in solving word problems, they may generally be classified into types, and if you study the method of setting up the equation for any one type, you will probably be able to write the equations for other word problems of the same type. We will present several examples of different types of word problems, starting out with two that are similar to the type discussed above, and then giving two examples of each of various other types of word problems. If you encounter a problem that does not fall into one of these types, your best approach will probably be to follow the guidelines set forth above.

examples

1 John has \$2 more than Bill. If Bill gives John \$9, John will have six times as much money as Bill. How much did each have to start with?

► Let x = Bill's original amount

Let $x + 2$ = John's original amount

After the exchange, John will have $x + 2 + 9$, Bill will have $x - 9$, and John's amount will be six times Bill's amount, so we write

$$\begin{aligned} x + 2 + 9 &= 6(x - 9) \\ x + 11 &= 6x - 54 \\ x - 6x &= -54 - 11 \\ -5x &= -65 \\ x &= \$13 \quad \text{Bill's original amount} \\ x + 2 &= \$15 \quad \text{John's original amount} \end{aligned}$$

2 John has three times as much money as Bill. If John gives Bill half of what he has, Bill will have \$12 more than John. How much did each have originally?

► Let x = Bill's original amount

Let $3x$ = John's original amount

Half of John's amount is $\frac{3}{2}x$, so after the exchange John has $3x - \frac{3}{2}x$ or $\frac{3}{2}x$, whereas Bill has $x + \frac{3}{2}x$ or $\frac{5}{2}x$, and we wish to say that Bill's amount is \$12 more than John's amount, so we write

$$\frac{5}{2}x = \frac{3}{2}x + 12$$

Multiplying through by 2 yields

$$\cancel{2} \cdot \frac{5}{\cancel{2}}x = \cancel{2} \cdot \frac{3}{\cancel{2}}x + 12 \cdot 2$$

$$\begin{aligned} 5x &= 3x + 24 \\ 2x &= 24 \\ x &= \$12 \quad \text{Bill's original amount} \\ 3x &= \$36 \quad \text{John's original amount} \end{aligned}$$

3 If 12 is added to three times a certain number, the result is five times the original number. What is the original number?

► Let x = the number

$$12 + 3x = 5x$$
$$3x - 5x = -12$$
$$-2x = -12$$
$$x = 6 \qquad \text{the original number}$$

4 The sum of three consecutive positive integers is 36. What are the integers?

► Let x = the first integer

Let $x + 1$ = the second integer

Let $x + 2$ = the third integer

$$x + (x + 1) + (x + 2) = 36$$
$$x + x + x = 36 - 1 - 2$$
$$3x = 33$$
$$x = 11 \qquad \text{the first integer}$$
$$x + 1 = 12 \qquad \text{the second integer}$$
$$x + 2 = 13 \qquad \text{the third integer}$$

5 Mary is 15 years older than Jane. Two years from now Mary will be four times as old as Jane. How old is each now?

► Let x = Jane's age now

Let $x + 15$ = Mary's age now

Two years from now Jane's age will be $x + 2$, whereas Mary's age will be $x + 15 + 2$; we wish to say that Mary's age will be four times Jane's age, so we write

$$x + 15 + 2 = 4(x + 2)$$
$$x + 17 = 4x + 8$$
$$x - 4x = 8 - 17$$
$$-3x = -9$$
$$x = 3 \qquad \text{Jane's age now}$$
$$x + 15 = 18 \qquad \text{Mary's age now}$$

6 A man is twice as old as his son. Fourteen years ago he was four times as old. How old is each now?

$$\blacktriangleright \quad \text{Let } x = \text{son's age now}$$

$$\text{Let } 2x = \text{father's age now}$$

Fourteen years ago the son's age was $x - 14$, the father's age was $2x - 14$, and the father's age was four times the son's age, so we write

$$2x - 14 = 4(x - 14)$$

$$2x - 14 = 4x - 56$$

$$2x - 4x = -56 + 14$$

$$-2x = -42$$

$$x = 21 \qquad \text{son's age now}$$

$$2x = 42 \qquad \text{father's age now}$$

7 I have 20 coins in my pocket consisting of nickels and dimes, and their total value is \$1.45. How many of each do I have?

$$\blacktriangleright \quad \text{Let } x = \text{the number of nickels}$$

Since I have 20 coins altogether, the number of dimes must be 20 minus the number of nickels, so we write

$$\text{Let } 20 - x = \text{the number of dimes}$$

Each nickel is worth 5 cents, so x nickels are worth $5x$ cents, and each dime is worth 10 cents, so $(20 - x)$ dimes are worth $10(20 - x)$ cents. The total value is \$1.45 or 145 cents (the units of measure must be the same on both sides of the equation), so we write

$$5x + 10(20 - x) = 145$$

$$5x + 200 - 10x = 145$$

$$5x - 10x = 145 - 200$$

$$-5x = -55$$

$$x = 11 \qquad \text{the number of nickels}$$

$$20 - x = 9 \qquad \text{the number of dimes}$$

8 I have 30 coins in my pocket consisting of dimes and quarters, and their total value is \$4.80. How many of each do I have?

► Let x = the number of dimes

Let $30 - x$ = the number of quarters

$$10x + 25(30 - x) = 480$$
$$10x + 750 - 25x = 480$$
$$-15x = -270$$
$$x = 18 \quad \text{the number of dimes}$$
$$30 - x = 12 \quad \text{the number of quarters}$$

9 A store sells one brand of coffee at \$0.85 per pound and another brand at \$0.60 per pound. A man wants a 10-lb mixture of these two brands and wishes to pay a total of \$6.75. How many pounds of each brand should he get?

► Let x = the number of pounds of the \$0.85 brand

Since he wants 10 lb altogether, he will get $10 - x$ lb of the \$0.60 brand, so we write

Let $10 - x$ = the number of pounds of the \$0.60 brand

x lb at \$0.85 per pound will cost $85x$, and $(10 - x)$ lb at \$0.60 per pound will cost $60(10 - x)$, and the total must be \$6.75 or 675 cents:

$$85x + 60(10 - x) = 675$$
$$85x + 600 - 60x = 675$$
$$85x - 60x = 675 - 600$$
$$25x = 75$$
$$x = 3 \quad \text{the number of pounds of the \$0.85 brand}$$
$$10 - x = 7 \quad \text{the number of pounds of the \$0.60 brand}$$

10 One brand of nuts sells for \$0.50 per pound whereas another brand sells for \$0.65 per pound. I wish to buy a 12-lb mixture of these two brands and pay a total of \$6.60. How many pounds of each should I get?

► Let x = number of pounds of the \$0.50 brand

Let $12 - x$ = number of pounds of the \$0.65 brand

$$50x + 65(12 - x) = 660$$
$$50x + 780 - 65x = 660$$
$$50x - 65x = 660 - 780$$
$$-15x = -120$$
$$x = 8 \quad \text{the number of pounds of the \$0.50 brand}$$
$$12 - x = 4 \quad \text{the number of pounds of the \$0.65 brand}$$

exercises 5.3

1 John has \$11 more than Bill. If John gives Bill \$18, Bill will have six times as much money as John. How much money does each one have?

2 Tom has three times as much money as John. If John gives Tom \$1, Tom will have four times as much money as John. How much money does each one have?

3 Two times a certain integer plus three times the succeeding integer is three times the first integer plus 13. What are the integers?

4 The sum of three consecutive positive integers is 18. What are the integers?

5 Mary is six years older than Jane. Five years ago, Mary was twice as old as Jane. How old is each one now?

6 A man is three times as old as his son. Six years from now he will be twice as old. How old is each one now?

7 I have 30 coins in my pocket consisting of dimes and quarters, and their total value is \$5.85. How many of each do I have?

8 I have 15 coins in my pocket consisting of nickels and dimes and their total value is \$1.05. How many of each do I have?

9 A store sells one brand of coffee at \$0.70 per pound and another brand at \$0.55 per pound. How many pounds of each brand should be included in a 12-lb mixture of these that is to sell for \$7.95?

10 One brand of candy sells for \$0.60 per pound, whereas another brand sells for \$0.40 per pound. An 8-lb mixture of these two is to sell for \$4.00. How many pounds of each brand should be included in the 8-lb mixture?

11 If 18 is added to twice a certain number, the result is three times that number. What is the number?

12 Mary is three times as old as Jane was two years ago. The sum of their ages is 30 years. How old is each now?

5.4/quadratic equations in one unknown

A linear equation in one unknown cannot involve any x^2 terms (after all denominators have been eliminated and terms combined). If it contains an x^2 term, but no term of higher degree, it is classified as a quadratic equation. When one first looks at an equation that contains denominators, it is usually difficult to tell whether or not it will lead to a linear equation. Consider, for example, the equations

$$\frac{4x - 3}{2x} + 1 = \frac{5}{3}$$

and

$$\frac{4x - 3}{3} + 1 = \frac{5}{2x}$$

If we follow the simplification procedures described previously for both equations, we get

$$3(4x - 3) + 1 \cdot 2x \cdot 3 = 5 \cdot 2x$$
$$12x - 9 + 6x = 10x$$
$$8x = 9$$

for the first equation, which is obviously linear, and

$$2x(4x - 3) + 1 \cdot 2x \cdot 3 = 5 \cdot 3$$
$$8x^2 - 6x + 6x = 15$$
$$8x^2 = 15$$

for the second, which is obviously quadratic.

Unless one had some previous experience with such equations, the two original equations would probably look pretty much alike to him. Fortunately, it is not necessary to classify an equation right at the start. This can be done after the denominators have been eliminated and the factors multiplied out. Thus the first two steps will remain the same as before: First we eliminate denominators, then we multiply out what we have left. At this point we look at the equation and determine whether it is linear or quadratic. If there are no x^2 or higher-degree terms, then it is linear and we apply the three-step procedure previously described to solve it. If there are x^2 terms but none of higher degree, then it is quadratic and we must apply a different procedure to solve it. In case the equation is a cubic, quartic, or one of higher degree, there are other methods for solving it or obtaining approximate solutions for it, but these are beyond the scope of this book.

For the case of a quadratic equation, we transpose all the terms to the left, leaving 0 on the right, and then combine terms on the left. This should leave the equation in the form

$$ax^2 + bx + c = 0$$

where a, b, and c are any numbers. This is called the *general form* of the quadratic equation, and we shall now present two methods for completing the solution of a quadratic equation, once it has been reduced to general form.

The first method is *by factoring*. To illustrate this method, let us choose a specific problem, say

$$2x^2 + 5x - 12 = 0$$

We factor the left side to obtain

$$(x + 4)(2x - 3) = 0$$

Remember that we are looking for a value that, when substituted for x in the left side of the equation, will give 0. Since 0 times anything is equal to 0, we can make the left side 0 either by making the first factor 0 or by making the second factor 0. This means that we must solve the two simple equations $x + 4 = 0$ and $2x - 3 = 0$, which yield $x = -4$ and $x = \frac{3}{2}$, respectively. These are the two answers to the original quadratic equation $2x^2 + 5x - 12 = 0$, as may be verified by direct substitution. You will note that a quadratic equation has two roots, whereas a linear equation only has one.

We may summarize the steps for solving a quadratic equation by factoring, once it has been reduced to general form:

1. Factor the left-hand side.

2. Set each factor equal to 0 and solve the resulting equation for x. (This will give the two roots of the original equation.)

examples

1 $x^2 - 5x + 6 = 0$

$$x^2 - 5x + 6 = 0$$

$$(x - 2)(x - 3) = 0$$

$$x - 2 = 0 \qquad x = 2$$

$$x - 3 = 0 \qquad x = 3$$

2 $6x^2 - 13x - 5 = 0$

$$6x^2 - 13x - 5 = 0$$

$$(3x + 1)(2x - 5) = 0$$

$$3x + 1 = 0 \qquad 3x = -1 \qquad x = -\frac{1}{3}$$

$$2x - 5 = 0 \qquad 2x = 5 \qquad x = \frac{5}{2}$$

3 $4x^2 - 28x + 49 = 0$

$$4x^2 - 28x + 49 = 0$$

$$(2x - 7)(2x - 7) = 0$$

$$2x - 7 = 0 \qquad 2x = 7 \qquad x = \frac{7}{2}$$

$$2x - 7 = 0 \qquad 2x = 7 \qquad x = \frac{7}{2}$$

Here the two roots are the same. This shows that it is possible for a quadratic equation to have two equal roots.

4 $x^2 - 4x - 9 = 0$

► The left-hand side of this equation cannot be factored, so this quadratic equation cannot be solved by factoring.

The last example indicated that not every quadratic equation can be solved by factoring. For these equations there is the so-called *quadratic formula*, which provides the values of the roots in terms of the coefficients a, b, and c. To derive this formula, we start with the general quadratic

$$ax^2 + bx + c = 0$$

Remember that we are permitted to do whatever we please to both sides of this equation as long as we do the same thing to both sides.

Transpose c:

$$ax^2 + bx = -c$$

Divide both sides by a:

$$x^2 + \frac{b}{a}x = -\frac{c}{a}$$

Add $b^2/4a^2$ to both sides:

$$x^2 + \frac{b}{a}x + \frac{b^2}{4a^2} = \frac{b^2}{4a^2} - \frac{c}{a}$$

Factor the left side:

$$\left(x + \frac{b}{2a}\right)^2 = \frac{b^2}{4a^2} - \frac{c}{a}$$

Put the right side over the common denominator:

$$\left(x + \frac{b}{2a}\right)^2 = \frac{b^2 - 4ac}{4a^2}$$

Take the square root of each side:

$$x + \frac{b}{2a} = \frac{\pm\sqrt{b^2 - 4ac}}{2a}$$

We have written $\pm\sqrt{b^2 - 4ac}$ because both the positive value of the square root of $b^2 - 4ac$ and the negative value will give $b^2 - 4ac$ when squared.

Transpose $b/2a$:

$$x = -\frac{b}{2a} \pm \frac{\sqrt{b^2 - 4ac}}{2a}$$

Simplify the right side:

$$x = \frac{-b \pm \sqrt{b^2 - 4ac}}{2a}$$

The formula in the box is the desired quadratic formula. To use it, we merely substitute the number values for the coefficients a, b, and c and simplify. We use the + sign to get one root and the − sign to get the other root.

You may wonder at this point whether there is any way to tell when a quadratic equation of the form $ax^2 + bx + c = 0$ can be factored. It is possible to prove that if a, b, and c are integers and $b^2 - 4ac$ is a perfect square, then $ax^2 + bx + c$ can be factored over the integers; otherwise, it cannot be factored over the integers, and you must use the quadratic formula to solve the equation.

examples

5 $x^2 - 5x + 6 = 0$

► First determine the values of a, b, and c. a is the coefficient of x^2, b is the coefficient of x, and c is the constant term, so in the present example $a = 1$, $b = -5$, and $c = 6$. Substituting these values in the quadratic formula yields

$$x = \frac{-(-5) \pm \sqrt{(-5)^2 - 4 \cdot 1 \cdot 6}}{2 \cdot 1} = \frac{5 \pm \sqrt{25 - 24}}{2}$$

$$= \frac{5 \pm \sqrt{1}}{2} = \frac{5 \pm 1}{2}$$

If we use the + sign,

$$x = \frac{5 + 1}{2} = \frac{6}{2} = 3$$

If we use the − sign,

$$x = \frac{5 - 1}{2} = \frac{4}{2} = 2$$

These are the same answers that we obtained for this problem before by factoring.

6 $15x^2 + x - 2 = 0$

► $a = 15 \quad b = 1 \quad c = -2$

$$x = \frac{-1 \pm \sqrt{1^2 - 4 \cdot 15 \cdot (-2)}}{2 \cdot 15}$$

$$= \frac{-1 \pm \sqrt{1 + 120}}{30} = \frac{-1 \pm \sqrt{121}}{30} = \frac{-1 \pm 11}{30}$$

$$x = \frac{-1 + 11}{30} = \frac{10}{30} = \frac{1}{3} \qquad x = \frac{-1 - 11}{30} = \frac{-12}{30} = -\frac{2}{5}$$

7 $2x^2 - 8x + 5 = 0$

► $a = 2 \quad b = -8 \quad c = 5$

$$x = \frac{-(-8) \pm \sqrt{(-8)^2 - 4 \cdot 2 \cdot 5}}{2 \cdot 2} = \frac{8 \pm \sqrt{64 - 40}}{4}$$

$$= \frac{8 \pm \sqrt{24}}{4}$$

Since 24 is not a perfect square, we approximate the square root of 24:

$$\sqrt{24} \approx 4.90$$

$$x \approx \frac{8 \pm 4.90}{4}$$

$$x \approx \frac{8 + 4.90}{4} = \frac{12.90}{4} = 3.23 \qquad x \approx \frac{8 - 4.90}{4} = \frac{3.10}{4} = 0.78$$

8 $3x^2 + 7x - 2 = 0$

► $a = 3 \quad b = 7 \quad c = -2$

$$x = \frac{-7 \pm \sqrt{7^2 - 4 \cdot 3 \cdot (-2)}}{2 \cdot 3} = \frac{-7 \pm \sqrt{49 + 24}}{6}$$

$$= \frac{-7 \pm \sqrt{73}}{6}$$

$$\sqrt{73} \approx 8.54$$

$$x \approx \frac{-7 \pm 8.54}{6}$$

$$x \approx \frac{-7 + 8.54}{6} = \frac{1.54}{6} = 0.26 \qquad x \approx \frac{-7 - 8.54}{6} = \frac{-15.54}{6} = -2.59$$

9 $2x^2 + 4x + 7 = 0$

► $a = 2 \qquad b = 4 \qquad c = 7$

$$x = \frac{-4 \pm \sqrt{4^2 - 4 \cdot 2 \cdot 7}}{2 \cdot 2} = \frac{-4 \pm \sqrt{16 - 56}}{4} = \frac{-4 \pm \sqrt{-40}}{4}$$

Since there is no real number whose square is -40, there is no real number that satisfies this equation; that is, this equation has no real roots.[1]

10 $x^2 + x + 1 = 0$

► $a = 1 \qquad b = 1 \qquad c = 1$

$$x = \frac{-1 \pm \sqrt{1^2 - 4 \cdot 1 \cdot 1}}{2 \cdot 1} = \frac{-1 \pm \sqrt{1 - 4}}{2} = \frac{-1 \pm \sqrt{-3}}{2}$$

Again, since $\sqrt{-3}$ is not a real number, this equation has no real roots.

In Section 5.3 it was shown how word problems may be solved by setting up equations that say the same thing as the words and then solving them. Occasionally, we may come across a word problem the equation of which is quadratic. As an example, consider the following problem.

A rectangular field is twice as long as it is wide. If 300 yd were added to the length and 100 yd to the width, the area would be multiplied by 5. What are the dimensions of the field? Proceeding as before, we write

Let x = the width[2]

Let $2x$ = the length

As we shall show in Chapter Six, the area of a rectangle is the product of its length and width. The present area is therefore $2x \cdot x$ or $2x^2$. If the length were increased by 300, it would become $2x + 300$, and if the width were increased by 100, it would become $x + 100$, so the new area would be $(2x + 300)(x + 100)$. We wish to say that the new area is 5 times the original area, so we write

$$(2x + 300)(x + 100) = 5 \cdot 2x^2$$

Simplifying this yields

$$2x^2 + 500x + 30{,}000 = 10x^2$$

$$-8x^2 + 500x + 30{,}000 = 0$$

[1]We could follow our previous practice and enlarge the system of real numbers by defining square roots of negative numbers and thus obtain roots for this equation. This enlarged system is called the *complex numbers*. A discussion of the complex numbers is beyond the scope of this text, so we shall merely say that this equation has no roots among the real numbers.

[2]If we were to let x = the length, then the width would be $x/2$. By letting x = the width, we avoid fractions.

We can simplify this quadratic equation by dividing both sides by -4 to obtain

$$2x^2 - 125x - 7500 = 0$$

This can be solved by factoring:

$$(2x + 75)(x - 100) = 0$$

Setting each factor equal to zero and solving for x yields

$$2x + 75 = 0 \qquad 2x = -75 \qquad x = -\tfrac{75}{2} = -37\tfrac{1}{2}$$

$$x - 100 = 0 \qquad x = 100$$

The width of a field cannot be negative, so we must discard the first value of x.[1] The answer to the problem is therefore

$$x = 100 \text{ yd} = \text{width of field}$$

$$2x = 200 \text{ yd} = \text{length of field}$$

/examples/

11 If a certain number is increased by 4 and then multiplied by 3 times the original number, the result is 33 times the original number. What is the original number?

► Let x = the original number

The equation may then be written

$$(x + 4)(3x) = 33x$$

$$3x^2 + 12x = 33x$$

$$3x^2 - 21x = 0$$

This can be factored as follows:

$$3x(x - 7) = 0$$

$$3x = 0 \qquad x = 0$$

$$x - 7 = 0 \qquad x = 7$$

Here the equation yields two answers, both of which satisfy the original word problem.

[1]This case of an equation that yields a root that does not satisfy the original word problem shows that it is a good idea to check each answer against the original word problem.

12 A rectangular field is 60 yd longer than it is wide. If the length is increased by 100 yd and the width is doubled, the area is increased by 79200 sq yd. What are the original dimensions?

► Let x = the original width of the field

$x + 60$ = the original length of the field

The original area is therefore $x(x + 60)$. After the alterations, we have

$x + 60 + 100$ = new length

$2x$ = new width

so the new area is $(x + 160)(2x)$. The statement of the problem tells us that the new area is the old area plus 79200 sq yd, so we write

$$(x + 160)(2x) = x(x + 60) + 79200$$

$$2x^2 + 320x = x^2 + 60x + 79200$$

$$x^2 + 260x - 79200 = 0$$

This can be factored as follows:

$$(x - 180)(x + 440) = 0$$

$$x - 180 = 0 \qquad x = 180$$

$$x + 440 = 0 \qquad x = -440$$

Since the width cannot be negative, we discard the second answer and are left with

$x = 180$ yd = original width of field

$x + 60 = 240$ yd = original length of field

exercises 5.4

1 Solve the following equations by factoring:

(a) $x^2 - 11x + 18 = 0$
(b) $5x^2 - 24x - 5 = 0$
(c) $6y^2 - 7y - 3 = 0$
(d) $15s^2 - s - 2 = 0$
(e) $9t^2 - 9t - 4 = 0$

2 Solve the following equations by the quadratic formula:

(a) $2x^2 - 7x + 1 = 0$
(b) $x^2 - 11x + 15 = 0$
(c) $6x^2 - 13x - 5 = 0$
(d) $2p^2 - p - 5 = 0$
(e) $3x^2 + x - 1 = 0$

3 Solve the following equations:

(a) $2x^2 - 5x + 2 = 0$ (b) $x^2 - 7x + 2 = 0$

(c) $3x^2 - x - 1 = 0$ (d) $x^2 + x - 1 = 0$

(e) $\frac{2x}{5} - \frac{1}{x} = 3$ (f) $\frac{2x - 1}{3} = \frac{1}{x + 1}$

4 A rectangle is twice as long as it is wide. If 10 ft were added to the length and 5 ft added to the width, the area would be 128 sq ft. What are its dimensions?

5 If a certain number is decreased by 4 and then multiplied by twice the original number, the result is 42. What is the original number? (*Hint:* Two answers are possible.)

6 If twice a certain number is multiplied by 3 more than the number, the result is 140. What is the number? (*Hint:* Two answers are possible.)

7 Mary is 6 years older than Ann. The product of their ages is 16 years more than 12 times Ann's age. What are their ages?

8 Tom is 3 times as old as Bill. The product of their ages is 21 years less than 8 times Tom's age. What are their ages?

5.5 / systems of linear equations

Thus far we have been concerned only with equations in one unknown, and we have shown how to solve linear and quadratic equations of this type. We turn now to equations in two unknowns. Consider the equation $3x - 2y = 1$. To give an answer to this equation, we must find a value for x and a value for y that will satisfy it. In other words, a solution consists of two numbers, one for x and one for y. We see immediately that $x = 1$ and $y = 1$ will be an answer to this equation. If we agree that we will always write the value for x first, we could write this answer in the compact form (1, 1). Using this compact form for expressing answers, we can verify that (3, 4), (5, 7), (7, 10), $(-1, -2)$, $(-3, -5)$, $(2, 2\frac{1}{2})$, $(\frac{1}{2}, \frac{1}{4})$, and $(-\frac{1}{2}, -\frac{5}{4})$ are also answers to this equation. In fact, we can choose any value whatsoever for x, substitute this in the equation, and solve the resulting equation for y to get a pair of numbers that will satisfy the original equation, $3x - 2y = 1$. We see that the equation $3x - 2y = 1$ has an infinite number of answers, and the same holds true for any linear equation in two unknowns.[1] For this reason we usually do not try to solve one equation in two unknowns; instead, we consider a set of two equations in two unknowns and try to find a pair of values, one for x and one for y, that will satisfy the two equations

[1]As before, a linear equation in two unknowns is one in which neither unknown appears to a power higher than 1 and that contains no xy term; that is, one that can be reduced to the form $ax + by = c$, where a, b, and c are real numbers.

simultaneously. Such a set of equations is usually referred to as a *system of two equations in two unknowns.*

If we consider, along with the above equation, $3x - 2y = 1$, the equation $2x - y = 3$, we will find that the only answer listed above for the first equation that also satisfies the second equation is (5, 7). We say, therefore, that the *ordered pair*[1] (5, 7) is the solution to the system of equations

$$3x - 2y = 1$$

$$2x - y = 3$$

It is true, as we shall prove in Section 5.6, that a system of two linear equations in two unknowns must have exactly one solution, an infinite number of solutions, or no solutions. The problems in which we shall be most interested are those with one solution.

As in the case of equations in one unknown, it is not practical to attempt to find the solution of a system of two equations in two unknowns by a trial-and-error method. We shall describe in this section two methods for finding the solutions of such systems, one known as the *method of elimination* and the other known as the *method of substitution*, and in Sections 5.6 and 5.7 we shall describe two additional methods.

The method of elimination will be illustrated by a specific example. Consider the system

$$5x - 3y = 1$$

$$3x + y = 9$$

The same principle applies as before; that is, if we do the same thing to both sides of either or both of these equations, we do not change the answer set of the system. If we multiply both sides of the first equation by 3 and both sides of the second equation by 5, we come up with the equivalent system

$$15x - 9y = 3$$

$$15x + 5y = 45$$

in which the coefficients of x are the same. Now, if we make use of the axiom that tells us that when equals are added to or subtracted from equals, the results are equal, and if we subtract the second equation from the first, the x drops out[2] and we are left with

$$-14y = -42$$

[1]*Ordered pair* means that the 5 must be written first and the 7 written second. This is not the same as the *ordered pair* (7, 5). The *pair* (5, 7), however, is the same as the *pair* (7, 5).

[2]This is why this is called the method of elimination; we have eliminated one of the unknowns.

which is a linear equation in one unknown, and which we can proceed to solve for y to obtain

$$y = 3$$

We now take either of the two original equations, substitute 3 for y, and solve the resulting equation for x. If we do this in the first equation, we get

$$5x - 3 \cdot 3 = 1$$

$$5x - 9 = 1$$

$$5x = 10$$

$$x = 2$$

The solution of our system is thus (2, 3).

We could have solved this system in another way. If we multiply the first equation by 1 (in other words, leave it unaltered) and multiply the second equation by 3, we get

$$5x - 3y = 1$$

$$9x + 3y = 27$$

in which the coefficients of y are the same except for the sign. If we now add these equations, we get

$$14x = 28$$

in which the y has been eliminated. (If we were to subtract the second equation from the first, we would not eliminate the y; we would get $-4x - 6y = -26$, which is not what we want.) We can now solve this to get

$$x = 2$$

Substituting this in the first equation gives us

$$5 \cdot 2 - 3y = 1$$

$$10 - 3y = 1$$

$$-3y = -9$$

$$y = 3$$

so the solution is again (2, 3).

As a general procedure for solving a system of two equations in two unknowns by elimination, we have the following steps:

1. Multiply both sides of the first equation by some number and both sides of the second equation by some number so that either the x's or the y's will have the same coefficient, except possibly for the sign. (These multipliers are easy to determine by inspection.)

2. If the signs of these coefficients are different, add the two equations; if they are the same, subtract one equation from the other. This should eliminate one unknown and leave one equation in one unknown.

3. Solve this equation for the unknown.

4. Substitute this value for the unknown in either of the two original equations and solve for the other unknown.

examples

1 Solve

$$2x - 4y = 7$$
$$3x + 3y = 2$$

► Multiplying the first equation by 3 and the second equation by 2 gives us

$$6x - 12y = 21$$
$$6x + 6y = 4$$

Subtracting the second equation from the first gives

$$-18y = 17$$
$$y = -\frac{17}{18}$$

Substituting this value for y in the first of the original set yields

$$2x - 4\left(-\frac{17}{18}\right) = 7$$

We multiply through by 18 to eliminate the denominator:

$$18 \cdot 2x - 4\left(-\frac{17}{\cancel{18}}\right) \cdot \cancel{18} = 7 \cdot 18$$
$$36x + 68 = 126$$
$$36x = 58$$
$$x = \frac{58}{36} = \frac{29}{18}$$

The solution is thus $(\frac{29}{18}, -\frac{17}{18})$.

For the alternative solution, multiply the first equation by 3 and the second one by 4:

$$6x - 12y = 21$$
$$12x + 12y = 8$$

Add the two equations and solve the resulting one for x:

$$18x = 29$$

$$x = \frac{29}{18}$$

Substitute this for x in the first of the original equations and solve for y.

$$2 \cdot \frac{29}{18} - 4y = 7$$

$$\cancel{18} \cdot 2 \cdot \frac{29}{\cancel{18}} - 4y \cdot 18 = 7 \cdot 18$$

$$58 - 72y = 126$$

$$-72y = 68$$

$$y = -\frac{68}{72} = -\frac{17}{18}$$

so the solution is $(\frac{29}{18}, -\frac{17}{18})$, as before.

2 Solve

$$x + 5y = 6$$

$$3x - 2y = 4$$

► Multiply the first equation by 3:

$$3x + 15y = 18$$

$$3x - 2y = 4$$

Subtract the second equation from the first:

$$17y = 14$$

$$y = \frac{14}{17}$$

Substitute in the first equation:

$$x + 5 \cdot \frac{14}{17} = 6$$

$$17 \cdot x + 5 \cdot \frac{14}{\cancel{17}} \cdot \cancel{17} = 6 \cdot 17$$

$$17x + 70 = 102$$

$$17x = 32$$

$$x = \frac{32}{17}$$

The solution is thus $(\frac{32}{17}, \frac{14}{17})$.

3 Solve

$$8x + 3y = 1$$
$$5x - 2y = 6$$

$$16x + 6y = 2$$
$$15x - 6y = 18$$
$$31x = 20$$
$$x = \frac{20}{31}$$

$$8 \cdot \frac{20}{31} + 3y = 1$$

$$\cancel{31} \cdot 8 \cdot \frac{20}{\cancel{31}} + 3y \cdot 31 = 1 \cdot 31$$

$$160 + 93y = 31$$
$$93y = -129$$
$$y = -\frac{129}{93} = -\frac{43}{31}$$

The solution is thus $(\frac{20}{31}, -\frac{43}{31})$.

The method of substitution for solving a set of two linear equations in two unknowns consists of solving one of the equations for one of the unknowns in terms of the other and then substituting this in the other equation. This leaves one equation in one unknown, which may then be solved. The other unknown is found as before by going back to either of the original equations.

This method will be illustrated by means of the example

$$3x - 2y = 1$$
$$2x - y = 3$$

Solve the first equation for x by transposing all terms not containing x to the right. This gives

$$3x = 1 + 2y$$
$$x = \frac{1 + 2y}{3}$$

Then substitute this value for x in the second equation,

$$2\left(\frac{1 + 2y}{3}\right) - y = 3$$

which gives us one equation in one unknown. Solving this gives

$$\not{3} \cdot 2 \left(\frac{1 + 2y}{\not{3}}\right) - y \cdot 3 = 3 \cdot 3$$

$$2 + 4y - 3y = 9$$

$$y = 7$$

We solve for x the same way as before by substituting 7 for y in either of the original equations. If we use the first one, we have

$$3x - 2 \cdot 7 = 1$$

$$3x = 15$$

$$x = 5$$

so the solution to the system is (5, 7).

If one of the unknowns in one of the equations has either 1 or -1 as its coefficient, it is best to solve for this unknown in terms of the other first, as this will avoid denominators if the original coefficients are whole numbers.

/examples/

4 Solve

$$2x - 5y = 2$$

$$3x + y = 1$$

► Solve the second equation for y in terms of x:

$$y = 1 - 3x$$

Substitute this back into the first equation:

$$2x - 5(1 - 3x) = 2$$

Solve for x:

$$2x - 5 + 15x = 2$$

$$17x = 7$$

$$x = \frac{7}{17}$$

Substitute this value in the first equation and solve for y.

$$2\left(\frac{7}{17}\right) - 5y = 2$$

$$\cancel{17} \cdot 2\left(\frac{7}{\cancel{17}}\right) - 17 \cdot 5y = 2 \cdot 17$$

$$14 - 85y = 34$$

$$-85y = 20$$

$$y = -\frac{20}{85} = -\frac{4}{17}$$

The answer is therefore $(\frac{7}{17}, -\frac{4}{17})$.

5 Solve

$$8x - 3y = 5$$

$$2x + 4y = 7$$

► $$8x = 5 + 3y$$

$$x = \frac{5 + 3y}{8}$$

$$2\left(\frac{5 + 3y}{8}\right) + 4y = 7$$

$$\cancel{8} \cdot 2\left(\frac{5 + 3y}{\cancel{8}}\right) + 4y \cdot 8 = 7 \cdot 8$$

$$10 + 6y + 32y = 56$$

$$38y = 46$$

$$y = \frac{46}{38} = \frac{23}{19}$$

Substitute this into the first equation:

$$8x - 3 \cdot \frac{23}{19} = 5$$

$$19 \cdot 8x - 3 \cdot \frac{23}{\cancel{19}} \cdot \cancel{19} = 5 \cdot 19$$

$$152x - 69 = 95$$

$$152x = 164$$

$$x = \frac{164}{152} = \frac{41}{38}$$

The answer is thus $(\frac{41}{38}, \frac{23}{19})$.

Word problems in which we are looking for two quantities and that were solved previously by using one linear equation in one unknown may also be solved by using two linear equations in two unknowns by letting x stand for one of the quantities and y for the other. Example 1 of Section 5.3 could have been solved as follows:

Let x = Bill's original amount

Let y = John's original amount

Since John has \$2 more than Bill, our first equation reads

$$y = x + 2$$

If Bill gives John \$9, Bill has $x - 9$ and John has $y + 9$, and we say that John has six times as much as Bill by writing

$$y + 9 = 6(x - 9)$$

If we simplify these two equations and write the x and y terms on the left and the constants on the right, we get

$$-x + y = 2$$
$$-6x + y = -63$$

We can solve these by subtracting the second from the first:

$$5x = 65$$

$x = 13$ Bill's original amount

$y = x + 2 = 15$ John's original amount

examples

6 Do Example 5 of Section 5.3 using two equations in two unknowns.

► Let x = Jane's age now

Let y = Mary's age now

$$y = x + 15$$
$$y + 2 = 4(x + 2)$$

Rewriting these equations yields

$$-x + y = 15$$
$$-4x + y = 6$$
$$3x = 9$$

$x = 3$ Jane's age now

$y = x + 15 = 18$ Mary's age now

7 Do Example 7 of Section 5.3 using two equations in two unknowns.

► Let x = the number of nickels

Let y = the number of dimes

$$x + y = 20$$
$$5x + 10y = 145$$
$$5x + 5y = 100$$
$$5x + 10y = 145$$
$$-5y = -45$$
$$y = 9 \qquad \text{the number of dimes}$$
$$x = 20 - y = 11 \qquad \text{the number of nickels}$$

The method of substitution may also be used to solve three equations in three unknowns.

example

8 Solve

$$2x + 3y - z = 1$$
$$3x + y - 2z = 4$$
$$x + 5y + 4z = 6$$

► Solve the first equation for z:

$$-z = 1 - 2x - 3y$$
$$z = -1 + 2x + 3y$$

Substitute this into the other two equations:

$$3x + y - 2(-1 + 2x + 3y) = 4$$
$$x + 5y + 4(-1 + 2x + 3y) = 6$$

This results in two equations in two unknowns, which may be simplified down to

$$3x + y + 2 - 4x - 6y = 4$$
$$x + 5y - 4 + 8x + 12y = 6$$
$$-x - 5y = 2$$
$$9x + 17y = 10$$

Solve this set of equations:

$$-9x - 45y = 18$$
$$9x + 17y = 10$$
$$-28y = 28$$
$$y = -1$$
$$-x - 5(-1) = 2$$
$$-x + 5 = 2$$
$$-x = -3$$
$$x = 3$$
$$\begin{aligned} z &= -1 + 2x + 3y \\ &= -1 + 2 \cdot 3 + 3(-1) \\ &= -1 + 6 - 3 = 2 \end{aligned}$$

The solution of the original system of three equations in three unknowns is thus $x = 3$, $y = -1$, and $z = 2$, which may be written in compact form as before $(3, -1, 2)$.

exercises 5.5

1 Solve the following systems of equations by elimination:

(a) $\begin{aligned} 2x - y &= 3 \\ x - 5y &= 1 \end{aligned}$ (b) $\begin{aligned} 5x - 3y &= 1 \\ 2x + 7y &= 5 \end{aligned}$

(c) $\begin{aligned} 2x - 5y &= 1 \\ 4x - 9y &= 2 \end{aligned}$ (d) $\begin{aligned} 3x - y &= 7 \\ 2x + 4y &= 3 \end{aligned}$

2 Solve the following systems of equations by substitution:

(a) $\begin{aligned} x - 5y &= 3 \\ 2x + 5y &= 1 \end{aligned}$ (b) $\begin{aligned} 3x + 2y &= 7 \\ 4x - y &= 1 \end{aligned}$

(c) $\begin{aligned} 4x - 7y &= 3 \\ 2x + 5y &= 2 \end{aligned}$ (d) $\begin{aligned} 5x - 9y &= 1 \\ 4x + 3y &= 7 \end{aligned}$

3 Solve the following systems of equations:

(a) $\begin{aligned} x - 3y &= 1 \\ 2x + y &= 7 \end{aligned}$ (b) $\begin{aligned} 2s - 3t &= 7 \\ 4s + 7t &= 5 \end{aligned}$

(c) $\begin{aligned} 3p - 2q &= 1 \\ 4p + 3q &= 5 \end{aligned}$ (d) $\begin{aligned} 3a + 5b &= 9 \\ 2a - b &= 4 \end{aligned}$

4 Do Exercises 1, 2, 5, 6, 7, and 8 of Section 5.3 using two equations in two unknowns.

5 Solve the system of equations

$$\begin{aligned} x + 3y - 2z &= 4 \\ 3x - y + 5z &= 1 \\ 2x + y + z &= 3 \end{aligned}$$

6 Solve the system of equations

$$\begin{aligned} 5x + y - z &= 4 \\ x + y + z &= 7 \\ 3x + 2y + z &= 1 \end{aligned}$$

5.6 / solution of systems of equations by graphing

In this section we shall show that any equation in two unknowns corresponds to a geometric plane curve, and we shall show how this fact may be utilized to provide another method for solving two equations in two unknowns. To do this, we must first describe the *Cartesian coordinate system*, or *rectangular coordinate system*, as it is often referred to. This coordinate system consists of two perpendicular lines, one horizontal and one vertical. The horizontal line is called the x axis, the vertical line is called the y axis, and the point where they intersect is called the origin, denoted by 0 in Figure 5.1. We construct a number line on the x axis, using the origin as the zero point, and we construct a number line on the y axis with the same zero point and with the same-sized intervals as on the x axis, with the positive integers going upward and the negative integers going downward.

figure 5.1

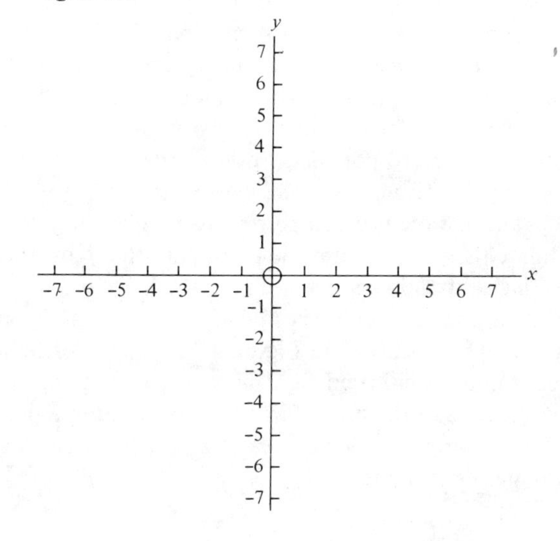

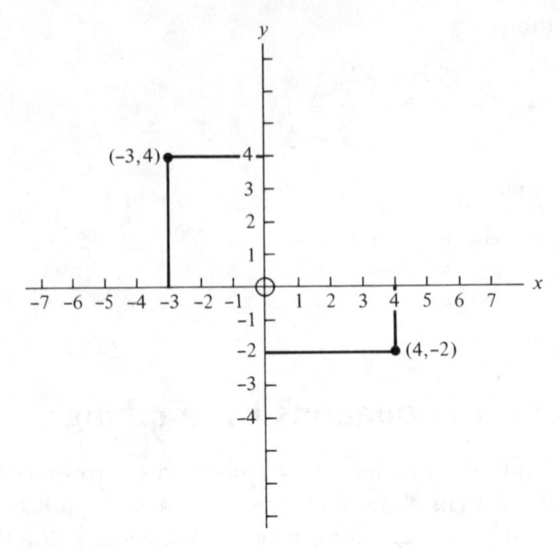

figure 5.2

This coordinate system permits us to set up a one-to-one correspondence between solutions of equations in two unknowns, expressed as ordered pairs and points in the plane, determined by the x and y axes. For example, to locate the point corresponding to the ordered pair $(4, -2)$, we start at the origin and move 4 units to the right and 2 units downward and mark the point $(4, -2)$ right there. The point $(-3, 4)$ is located by starting at the origin and moving 3 units to the left and 4 units upward (see Figure 5.2). The first number of the ordered pair is called the *x coordinate* and tells us how far to move to the right or left starting from the origin, whereas the second number of the ordered pair is called the *y coordinate* and tells us how far we then move upward or downward to locate the point that corresponds to the ordered pair. This procedure of locating and marking points corresponding to ordered pairs is referred to as *plotting* points.

By the definition of a one-to-one correspondence, every ordered pair corresponds to a unique point, and, conversely, every point corresponds to a unique ordered pair of numbers. To determine the ordered pair corresponding to a given point, we take for the x coordinate the distance that the point is to the right or left of the y axis (positive if it is to the right and negative if it is to the left), and for the y coordinate we take the distance that the point is above or below the x axis (positive if it is above and negative if it is below). In Figure 5.3 a number of points have been plotted and their coordinates indicated right next to them so that you may check to be sure that you understand the procedures both of plotting a point, given its coordinates, and of determining the coordinates of a point from its location with respect to the rectangular coordinate system.

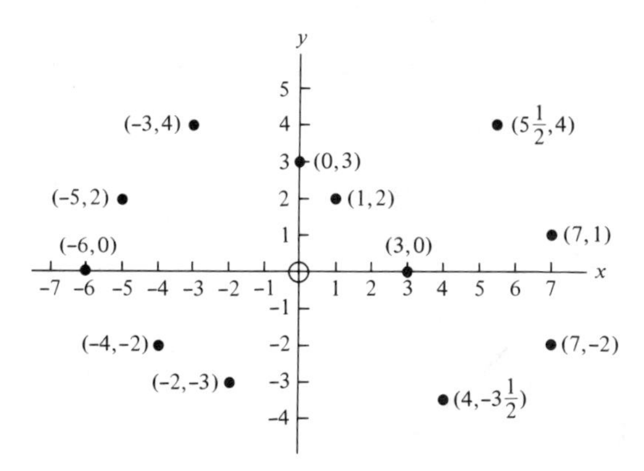

figure 5.3

We see that the rectangular coordinate system that we have described allows us to identify points in the plane by means of ordered pairs of real numbers and to plot a point in the plane for each ordered pair of real numbers. We will now show how this may be used to solve a set of two linear equations in two unknowns.

Consider the equations one at a time, and for each equation determine several ordered pairs that are solutions of the equation. This can be done easily enough by substituting any number for x and solving for y. Then plot the points corresponding to these ordered pairs. You will find that the set of points corresponding to each equation lies on a straight line. (This is why these equations are called linear equations.) Using a straightedge, draw the line determined by these points. (We refer to this as *plotting* the line.) This line is called *the line corresponding to the equation* because *every point on the line has coordinates that constitute a solution of the equation, and every solution of the equation corresponds to a point that lies on this line.*

By this procedure we get one line corresponding to each equation. The solutions of the first equation all lie on one line, whereas the solutions of the second equation all lie on the other line, so, since we are looking for the ordered pair that satisfies both equations simultaneously, we locate that point that lies on both lines simultaneously, that is, their point of intersection. The coordinates of this point of intersection constitute the solution of the system of equations.

As a practical matter, since we know that each equation will result in a straight line and a straight line is determined by two points, it is only necessary to plot two points for each equation in order to plot the line to which it corresponds.

By way of illustration, let us solve the system of equations

$$\begin{aligned} x + y &= 1 \\ 2x + 3y &= 5 \end{aligned}$$

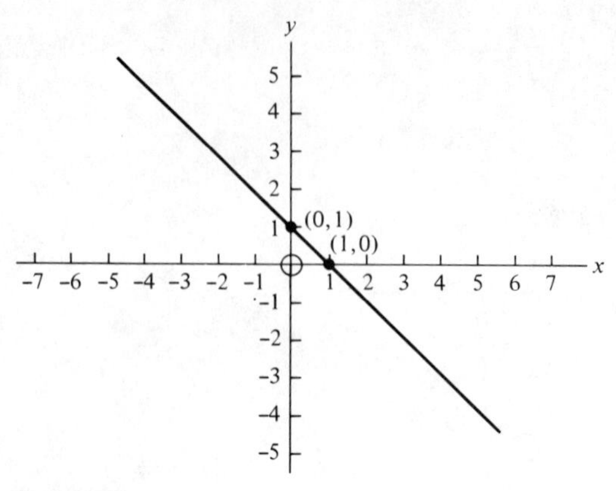

figure 5.4

Starting with the equation $x + y = 1$, we can make a small table in which we list two of its solutions:

x	y
1	0
0	1

In this table we have indicated that (1, 0) and (0, 1) are solutions. There are many more solutions, but we need only two to plot the line corresponding to this equation. We next plot the line (see Figure 5.4).

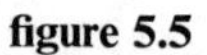

figure 5.5

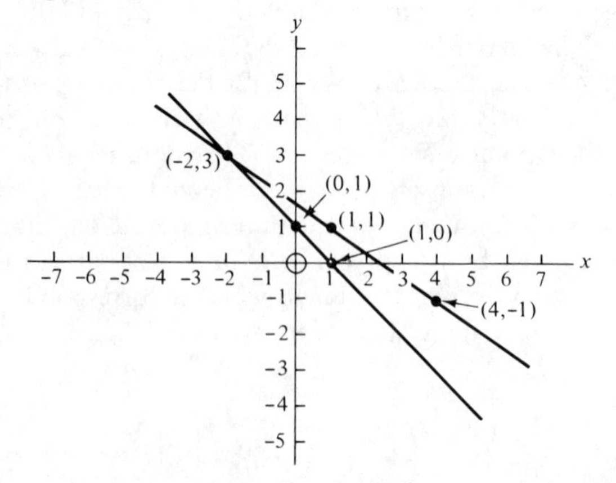

Similarly, for the equation $2x + 3y = 5$ we set up a table of two solutions,

x	y
1	1
4	-1

and plot the corresponding line (see Figure 5.5). The point where these two lines intersect has coordinates $(-2, 3)$, so the solution of the set of equations is $x = -2$, $y = 3$.

/examples/

1 Solve the following system of equations by graphing:

$$\begin{aligned} 3x - y &= 7 \\ 9x + 2y &= 1 \end{aligned}$$

► We plot the line of each equation:

$3x - y = 7$

x	y
2	-1
3	2

$9x + 2y = 1$

x	y
-1	5
0	$\frac{1}{2}$

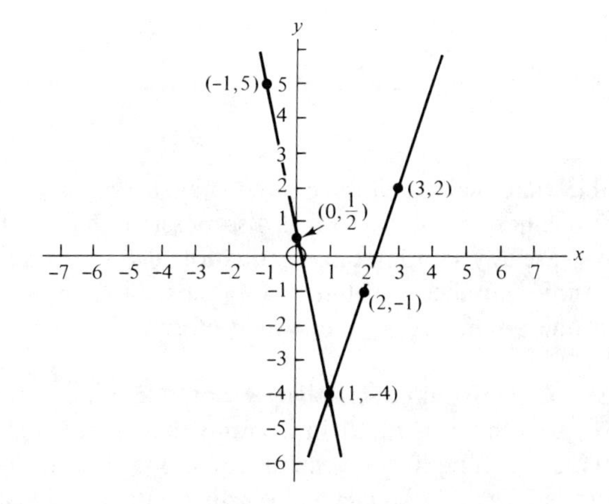

The lines intersect at $(1, -4)$ on the graph, so the solution is $x = 1$, $y = -4$.

2 Solve by graphing:

$$2x + 3y = 1$$
$$4x + \quad y = -8$$

► We plot the line of each equation:

$2x + 3y = 1$

x	y
-1	1
-4	3

$4x + y = -8$

x	y
-2	0
-1	-4

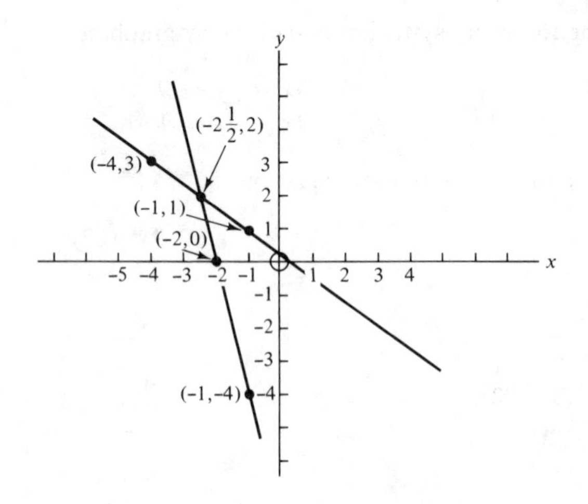

These lines intersect at $(-2\frac{1}{2}, 2)$ on the graph, so the solution is $x = -2\frac{1}{2}$, $y = 2$.

It is possible that the two lines plotted may be parallel. In this case the system of equations has no solution. It is also possible that the two lines may be identical. In this case any solution of one equation also satisfies the other, and the system has an infinite number of solutions. In most cases, however, the two lines will intersect in one point, and so the system of equations will have exactly one solution.

An equation in two unknowns that is not linear may also be plotted as described above, and this can result in a graph that is not a straight line. The analysis of curves resulting from general equations and their properties is a separate study called *analytic geometry*. We shall give one illustration here of a curve resulting from a quadratic equation in two unknowns by plotting the curve corresponding to the equation $x^2 + y^2 = 25$.

We first set up a table of values:

x	y
0	± 5
± 5	0
± 3	± 4
± 4	± 3
± 1	± 4.9
± 2	± 4.5

The entry $(\pm 3, \pm 4)$ in the table stands for the four points $(3, 4)$, $(-3, 4)$, $(3, -4)$, and $(-3, -4)$; the same is true of the other $\pm$ signs. We have plotted more than two points because this curve is not a straight line, and we must plot enough points to be able to see what the curve looks like. If we plot all the points in the table above we have the array shown in Figure 5.6. If we draw a smooth curve through these points, we come up with a circle whose radius is 5 and whose center is at the origin, as shown in Figure 5.7. This circle is the curve of the equation $x^2 + y^2 = 25$ in the same sense as before; that is, every solution of the equation corresponds to a point that lies on this circle, and every point on the circle has coordinates that satisfy the equation.

figure 5.6

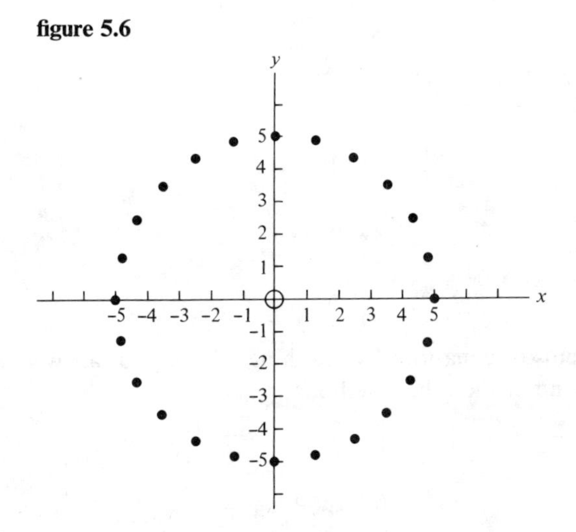

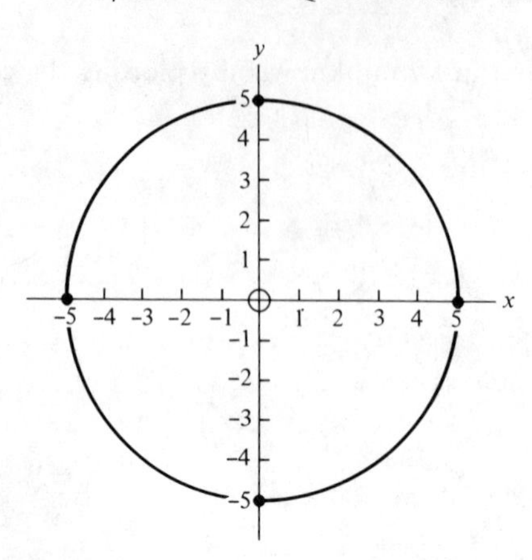

figure 5.7

exercises 5.6

1 Plot the following points on a rectangular coordinate system:

(a) $(-1, 2)$ (b) $(3, -4)$ (c) $(-5, -1)$

(d) $(0, 6)$ (e) $(-4, 0)$ (f) $(-1, -6)$

2 Give the coordinates of the points A, B, C, and D plotted on the graph.

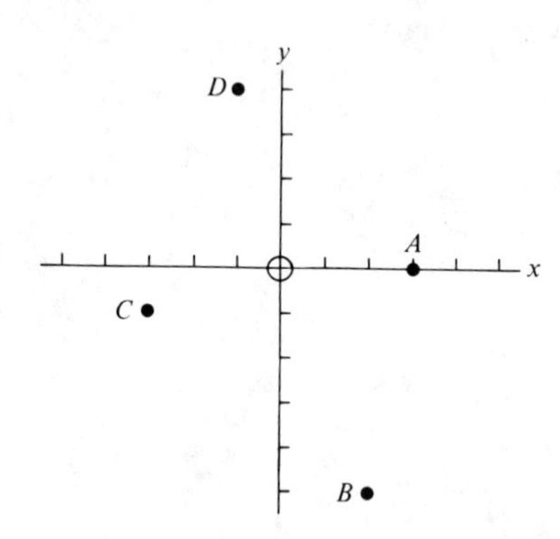

3 Solve the following systems of equations by graphing; check your answers by solving each system by one of the other methods:

(a) $x + 2y = 5$
$3x - 2y = -9$

(b) $6x - y = 3$
$2x + 3y = 11$

(c) $2x + 3y = -1$
$x - 2y = 10$

(d) $2x + y = 8$
$x - 6y = 4$

4 Plot the curve corresponding to the equation $9x^2 + 16y^2 = 144$.

5 Plot the curve corresponding to the equation $y^2 = 8x$.

6 Plot the curve corresponding to the equation $x^2 + 4y^2 - 2x - 4 = 0$.

5.7 / solution of systems of equations by determinants

There is yet another method for solving a system of two linear equations in two unknowns, one that makes use of *determinants*. The method of determinants provides us with a formula for the solution of the equations in terms of the coefficients much in the same way as the quadratic formula does for quadratic equations.

To derive the formula, we write a general set of two equations in two unknowns with letters to stand for the coefficients:

$$ax + by = c$$
$$dx + ey = f$$

Multiplying the first equation by e and the second by b yields

$$aex + bey = ce$$
$$bdx + bey = bf$$

Subtracting gives

$$(ae - bd)x = ce - bf$$
$$x = \frac{ce - bf}{ae - bd}$$

Instead of substituting this to solve for y, it is easier here to multiply the original first equation by d and the second by a to obtain

$$adx + bdy = cd$$
$$adx + aey = af$$

Then subtracting gives

$$(bd - ae)y = cd - af$$
$$y = \frac{cd - af}{bd - ae}$$

If we multiply the numerator and denominator by -1, we have

$$y = \frac{af - cd}{ae - bd}$$

and so the solution of the original equation is

$$x = \frac{ce - bf}{ae - bd} \quad \text{and} \quad y = \frac{af - cd}{ae - bd}$$

There is a simple scheme for memorizing this formula. If we write the four coefficients on the left sides of the original equations in an array in the order in which they appear, we have the *determinant of coefficients*, which appears as follows:

$$\begin{vmatrix} a & b \\ d & e \end{vmatrix}$$

This determinant is described as a 2×2 determinant because it has two rows and two columns.[1] We attach a certain value to this determinant, which is defined as the product of the two elements on the diagonal that runs from the upper left to the lower right minus the product of the two elements on the other diagonal. In other words, we have Definition 5.1.

definition 5.1

$$\begin{vmatrix} a & b \\ d & e \end{vmatrix} = ae - bd$$

Notice that

$$\begin{vmatrix} a & b \\ d & e \end{vmatrix} = ae - bd$$

is the denominator of each of the fractions that give the value of x and y in the boxed formula above. We therefore write

$$x = \frac{\begin{vmatrix} & \\ & \end{vmatrix}}{\begin{vmatrix} a & b \\ d & e \end{vmatrix}} \qquad y = \frac{\begin{vmatrix} & \\ & \end{vmatrix}}{\begin{vmatrix} a & b \\ d & e \end{vmatrix}}$$

To write the numerators as determinants, we use the numbers c and f on the right side of the equations and put them in the first column of the numerator for x and in the second column of the numerator for y. This gives

$$x = \frac{\begin{vmatrix} c & \\ f & \end{vmatrix}}{\begin{vmatrix} a & b \\ d & e \end{vmatrix}} \qquad y = \frac{\begin{vmatrix} & c \\ & f \end{vmatrix}}{\begin{vmatrix} a & b \\ d & e \end{vmatrix}}$$

The blank columns in the numerators are now filled in by copying the corresponding column from each denominator. This gives us the formulas above for the solution of two linear equations in two unknowns in terms of determinants:

$$x = \frac{\begin{vmatrix} c & b \\ f & e \end{vmatrix}}{\begin{vmatrix} a & b \\ d & e \end{vmatrix}} \qquad y = \frac{\begin{vmatrix} a & c \\ d & f \end{vmatrix}}{\begin{vmatrix} a & b \\ d & e \end{vmatrix}}$$

[1]The rows run horizontally and the columns run vertically.

If you will recall the definition of a determinant, you will see that this is just another way of writing

$$x = \frac{ce - bf}{ae - bd} \qquad y = \frac{af - cd}{ae - bd}$$

examples

1 Solve the following set of equations by determinants:

$$3x - 5y = 3$$

$$2x + 7y = 1$$

► We first set up the solution for determinants by writing

$$x = \frac{\begin{vmatrix} & \\ & \end{vmatrix}}{\begin{vmatrix} & \\ & \end{vmatrix}} \qquad y = \frac{\begin{vmatrix} & \\ & \end{vmatrix}}{\begin{vmatrix} & \\ & \end{vmatrix}}$$

For the denominators we fill in the four coefficients on the left sides in the same positions as they appear, so we have

$$x = \frac{\begin{vmatrix} & \\ & \end{vmatrix}}{\begin{vmatrix} 3 & -5 \\ 2 & 7 \end{vmatrix}} \qquad y = \frac{\begin{vmatrix} & \\ & \end{vmatrix}}{\begin{vmatrix} 3 & -5 \\ 2 & 7 \end{vmatrix}}$$

In the numerators we fill in the numbers on the right sides of the equation in the first column for x and in the second column for y:

$$x = \frac{\begin{vmatrix} 3 & \\ 1 & \end{vmatrix}}{\begin{vmatrix} 3 & -5 \\ 2 & 7 \end{vmatrix}} \qquad y = \frac{\begin{vmatrix} & 3 \\ & 1 \end{vmatrix}}{\begin{vmatrix} 3 & -5 \\ 2 & 7 \end{vmatrix}}$$

The blank columns are now filled in by copying what is directly below them, and we then have

$$x = \frac{\begin{vmatrix} 3 & -5 \\ 1 & 7 \end{vmatrix}}{\begin{vmatrix} 3 & -5 \\ 2 & 7 \end{vmatrix}} \qquad y = \frac{\begin{vmatrix} 3 & 3 \\ 2 & 1 \end{vmatrix}}{\begin{vmatrix} 3 & -5 \\ 2 & 7 \end{vmatrix}}$$

We complete the solution by evaluating these determinants:

$$\begin{vmatrix} 3 & -5 \\ 1 & 7 \end{vmatrix} = 3 \cdot 7 - (-5)(1) = 21 + 5 = 26$$

$$\begin{vmatrix} 3 & 3 \\ 2 & 1 \end{vmatrix} = 3 \cdot 1 - 3 \cdot 2 = 3 - 6 = -3$$

$$\begin{vmatrix} 3 & -5 \\ 2 & 7 \end{vmatrix} = 3 \cdot 7 - (-5)(2) = 21 + 10 = 31$$

The solution is therefore

$$x = \frac{\begin{vmatrix} 3 & -5 \\ 1 & 7 \end{vmatrix}}{\begin{vmatrix} 3 & -5 \\ 2 & 7 \end{vmatrix}} = \frac{26}{31} \qquad y = \frac{\begin{vmatrix} 3 & 3 \\ 2 & 1 \end{vmatrix}}{\begin{vmatrix} 3 & -5 \\ 2 & 7 \end{vmatrix}} = \frac{-3}{31}$$

which may be written in the compact form $(\frac{26}{31}, \frac{-3}{31})$.

2 Solve by determinants

$$3y = 4 - 7x$$
$$8 = x - 5y$$

► We must first write these equations in the standard form

$$7x + 3y = 4$$
$$-x + 5y = -8$$

Then solving these by determinants, we have

$$x = \frac{\begin{vmatrix} 4 & 3 \\ -8 & 5 \end{vmatrix}}{\begin{vmatrix} 7 & 3 \\ -1 & 5 \end{vmatrix}} = \frac{4 \cdot 5 - (3)(-8)}{7 \cdot 5 - (3)(-1)} = \frac{20 + 24}{35 + 3} = \frac{44}{38} = \frac{22}{19}$$

$$y = \frac{\begin{vmatrix} 7 & 4 \\ -1 & -8 \end{vmatrix}}{\begin{vmatrix} 7 & 3 \\ -1 & 5 \end{vmatrix}} = \frac{(7)(-8) - (4)(-1)}{7 \cdot 5 - (3)(-1)} = \frac{-56 + 4}{35 + 3} = \frac{-52}{38} = -\frac{26}{19}$$

The solution may then be written in the form $(\frac{22}{19}, -\frac{26}{19})$.

3 Solve by determinants

$$4x - 10y = 1$$
$$2x - 5y = 3$$

► $$x = \frac{\begin{vmatrix} 1 & -10 \\ 3 & -5 \end{vmatrix}}{\begin{vmatrix} 4 & -10 \\ 2 & -5 \end{vmatrix}} = \frac{-5 - (-30)}{(-20) - (-20)} = \frac{25}{0}$$

$$y = \frac{\begin{vmatrix} 4 & 1 \\ 2 & 3 \end{vmatrix}}{\begin{vmatrix} 4 & -10 \\ 2 & -5 \end{vmatrix}} = \frac{12 - 2}{(-20) - (-20)} = \frac{10}{0}$$

Since we cannot divide by zero, we cannot solve this set of equations by determinants. In general, when the denominator determinant is zero, it means that the lines corresponding to the equations are either parallel or identical; that is, there is either no solution or there are an infinite number of solutions.

Determinants may also be used to solve a system of three linear equations in three unknowns. The procedure is similar to the one described above. To illustrate, consider the system of equations

$$\begin{aligned} 2x - 3y + 5z &= 4 \\ 3x + y - 2z &= 1 \\ 7x + 3y - 4z &= 2 \end{aligned}$$

The determinant of coefficients here is the 3×3 determinant

$$\begin{vmatrix} 2 & -3 & 5 \\ 3 & 1 & -2 \\ 7 & 3 & -4 \end{vmatrix}$$

and so we write

$$x = \frac{\begin{vmatrix} & & \\ & & \\ & & \end{vmatrix}}{\begin{vmatrix} 2 & -3 & 5 \\ 3 & 1 & -2 \\ 7 & 3 & -4 \end{vmatrix}} \qquad y = \frac{\begin{vmatrix} & & \\ & & \\ & & \end{vmatrix}}{\begin{vmatrix} 2 & -3 & 5 \\ 3 & 1 & -2 \\ 7 & 3 & -4 \end{vmatrix}} \qquad z = \frac{\begin{vmatrix} & & \\ & & \\ & & \end{vmatrix}}{\begin{vmatrix} 2 & -3 & 5 \\ 3 & 1 & -2 \\ 7 & 3 & -4 \end{vmatrix}}$$

For the numerators, we fill the numbers 4, 1, and 2 on the right side of the equations in the first column for x, in the second column for y, and in the third column for z, so we then have

$$x = \frac{\begin{vmatrix} 4 & & \\ 1 & & \\ 2 & & \end{vmatrix}}{\begin{vmatrix} 2 & -3 & 5 \\ 3 & 1 & -2 \\ 7 & 3 & -4 \end{vmatrix}} \qquad y = \frac{\begin{vmatrix} & 4 & \\ & 1 & \\ & 2 & \end{vmatrix}}{\begin{vmatrix} 2 & -3 & 5 \\ 3 & 1 & -2 \\ 7 & 3 & -4 \end{vmatrix}} \qquad z = \frac{\begin{vmatrix} & & 4 \\ & & 1 \\ & & 2 \end{vmatrix}}{\begin{vmatrix} 2 & -3 & 5 \\ 3 & 1 & -2 \\ 7 & 3 & -4 \end{vmatrix}}$$

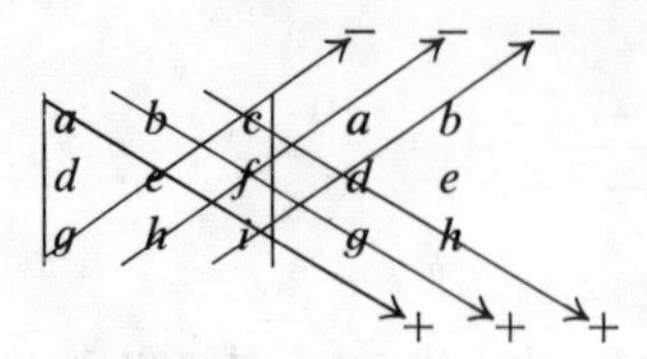

figure 5.8

The blank columns in the numerators are completed by copying the columns directly below them; this gives the setup for the solution of the set of equations by determinants:

$$x = \frac{\begin{vmatrix} 4 & -3 & 5 \\ 1 & 1 & -2 \\ 2 & 3 & -4 \end{vmatrix}}{\begin{vmatrix} 2 & -3 & 5 \\ 3 & 1 & -2 \\ 7 & 3 & -4 \end{vmatrix}} \qquad y = \frac{\begin{vmatrix} 2 & 4 & 5 \\ 3 & 1 & -2 \\ 7 & 2 & -4 \end{vmatrix}}{\begin{vmatrix} 2 & -3 & 5 \\ 3 & 1 & -2 \\ 7 & 3 & -4 \end{vmatrix}} \qquad z = \frac{\begin{vmatrix} 2 & -3 & 4 \\ 3 & 1 & 1 \\ 7 & 3 & 2 \end{vmatrix}}{\begin{vmatrix} 2 & -3 & 5 \\ 3 & 1 & -2 \\ 7 & 3 & -4 \end{vmatrix}}$$

It remains to describe how to evaluate the determinant of a 3×3 matrix in order to complete the solution. This is done by Definition 5.2.

definition 5.2

$$\begin{vmatrix} a & b & c \\ d & e & f \\ g & h & i \end{vmatrix} = aei + bfg + cdh - (ceg + afh + bdi)$$

A simple way to evaluate a 3×3 determinant is to write the first two columns over again as in Figure 5.8 and then to take the six products as indicated by the arrows and put the proper sign in front of each term.[1]

We may now complete the solution to our problem:

$$x = \frac{\begin{vmatrix} 4 & -3 & 5 \\ 1 & 1 & -2 \\ 2 & 3 & -4 \end{vmatrix} \begin{matrix} 4 & -3 \\ 1 & 1 \\ 2 & 3 \end{matrix}}{\begin{vmatrix} 2 & -3 & 5 \\ 3 & 1 & -2 \\ 7 & 3 & -4 \end{vmatrix} \begin{matrix} 2 & -3 \\ 3 & 1 \\ 7 & 3 \end{matrix}}$$

$$= \frac{-16 + 12 + 15 - (10 - 24 + 12)}{-8 + 42 + 45 - (35 - 12 + 36)} = \frac{11 - (-2)}{79 - (59)} = \frac{13}{20}$$

[1]The evaluation of a 4×4 determinant is much more complicated and involves 24 terms; in this book, however, we shall not be concerned with determinants larger than 3×3.

$$y = \frac{\begin{array}{|rrr|rr} 2 & 4 & 5 & 2 & 4 \\ 3 & 1 & -2 & 3 & 1 \\ 7 & 2 & -4 & 7 & 2 \end{array}}{\begin{array}{|rrr|rr} 2 & -3 & 5 & 2 & -3 \\ 3 & 1 & -2 & 3 & 1 \\ 7 & 3 & -4 & 7 & 3 \end{array}}$$

$$= \frac{-8 - 56 + 30 - (35 - 8 - 48)}{20} = \frac{-34 - (-21)}{20} = \frac{-13}{20}$$

$$z = \frac{\begin{array}{|rrr|rr} 2 & -3 & 4 & 2 & -3 \\ 3 & 1 & 1 & 3 & 1 \\ 7 & 3 & 2 & 7 & 3 \end{array}}{\begin{array}{|rrr|rr} 2 & -3 & 5 & 2 & -3 \\ 3 & 1 & -2 & 3 & 1 \\ 7 & 3 & -4 & 7 & 3 \end{array}}$$

$$= \frac{4 - 21 + 36 - (28 + 6 - 18)}{20} = \frac{19 - (16)}{20} = \frac{3}{20}$$

This solution may be written in compact form as $(\frac{13}{20}, -\frac{13}{20}, \frac{3}{20})$.

/example/

4 Solve by determinants

$$\begin{aligned} 2x - y - 7z &= 6 \\ 3x + 5y - z &= 2 \\ x + 4y - 6z &= 5 \end{aligned}$$

$$\blacktriangleright\ x = \frac{\begin{array}{|rrr|rr} 6 & -1 & -7 & 6 & -1 \\ 2 & 5 & -1 & 2 & 5 \\ 5 & 4 & -6 & 5 & 4 \end{array}}{\begin{array}{|rrr|rr} 2 & -1 & -7 & 2 & -1 \\ 3 & 5 & -1 & 3 & 5 \\ 1 & 4 & -6 & 1 & 4 \end{array}}$$

$$= \frac{-180 + 5 - 56 - (-175 - 24 + 12)}{-60 + 1 - 84 - (-35 - 8 + 18)}$$

$$= \frac{-231 - (-187)}{-143 - (-25)} = \frac{-44}{-118} = \frac{22}{59}$$

$$y = \frac{\begin{vmatrix} 2 & 6 & -7 \\ 3 & 2 & -1 \\ 1 & 5 & -6 \end{vmatrix} \begin{matrix} 2 & 6 \\ 3 & 2 \\ 1 & 5 \end{matrix}}{\begin{vmatrix} 2 & -1 & -7 \\ 3 & 5 & -1 \\ 1 & 4 & -6 \end{vmatrix} \begin{matrix} 2 & -1 \\ 3 & 5 \\ 1 & 4 \end{matrix}}$$

$$= \frac{-24 - 6 - 105 - (-14 - 10 - 108)}{-118}$$

$$= \frac{-135 - (-132)}{-118} = \frac{-3}{-118} = \frac{3}{118}$$

$$z = \frac{\begin{vmatrix} 2 & -1 & 6 \\ 3 & 5 & 2 \\ 1 & 4 & 5 \end{vmatrix} \begin{matrix} 2 & -1 \\ 3 & 5 \\ 1 & 4 \end{matrix}}{\begin{vmatrix} 2 & -1 & -7 \\ 3 & 5 & -1 \\ 1 & 4 & -6 \end{vmatrix} \begin{matrix} 2 & -1 \\ 3 & 5 \\ 1 & 4 \end{matrix}}$$

$$= \frac{50 - 2 + 72 - (30 + 16 - 15)}{-118}$$

$$= \frac{120 - (31)}{-118} = -\frac{89}{118}$$

We may write this solution in compact form $(\frac{44}{118}, \frac{3}{118}, -\frac{89}{118})$; we have kept all the denominators alike to simplify checking this solution in the original equation.

As in the case of two equations in two unknowns, if the determinant in the denominator should be zero, it means either that there are no solutions or that there are an infinite number of solutions.

exercises 5.7

1 Solve the following systems of equations by determinants; check your answers by one of the other methods:

(a) $x + 3y = 4$
$2x - y = 1$

(b) $5x - 7y = 3$
$2x + 8y = 5$

(c) $9x + y = 3$
$2x - 5y = 7$

(d) $4x - 6y = 1$
$7x + 2y = 5$

2 Solve the following systems of equations by determinants; check your answers by using the method of substitution:

(a) $3x + y + z = 1$
$2x + 5y - 3z = 5$
$4x - y - 7z = 2$

(b) $x + 5y + 4z = 2$
$3x - y + 7z = 5$
$2x + 5y + z = 1$

(c) $5x + 2y + 7z = 1$
$3x + y + z = 6$
$5x + y + 2z = 3$

3 Solve the following systems of equations by each of the four methods discussed in this chapter:

(a) $3x + 5y = 7$
$x - 5y = 9$

(b) $2x + 6y = 7$
$3x - 2y = 4$

review test 5

NAME ______________

DATE ______________

1 Determine whether each of the following is an identity or a conditional equation:

(a) $2x + 1 - 7x - 3 = 4x - 8 - 9x + 6$ ______________

(b) $3(x + 4) = 2x + 7 - 4x + 2$ ______________

(c) $\frac{2x + 1}{7} = \frac{3x - 1}{2}$ ______________

(d) $\frac{4x - 3}{5} = \frac{8x - 6}{10}$ ______________

2 Solve the following linear equations:

(a) $3(2x - 7) + 4 = 2(x + 1) - 8$

(b) $3x - 7(x + 1) = 3 - 5(x - 2)$

(c) $\frac{2x - 1}{7} + 5x = \frac{1}{3}$

(d) $\frac{3x - 2}{2x} - 6 = \frac{1}{2}$

(e) $\frac{2x - 3}{5} = \frac{3x + 2}{4}$

(f) $\frac{1}{x + 1} = \frac{3}{x + 2}$

(g) $\frac{x + 1/2}{x + 1/3} = \frac{1}{2}$

3 Solve the following quadratic equations:

(a) $x^2 - 8x + 15 = 0$

(b) $6x^2 - 7x - 3 = 0$

(c) $10x^2 + 3x + 5 = 2x + 8$

(d) $\frac{x}{4} + \frac{5}{x + 1} = 2$

(e) $\dfrac{3x + 1}{5} = \dfrac{7}{3x - 1}$

(f) $\dfrac{4}{x - 1} = \dfrac{5x - 1}{7}$

4 Solve the following equations:

(a) $2x - 7(x + 1) = 3 + 4(x + 2)$

(b) $x + x(1 + x) = 2 + 3(2x - 2)$

(c) $1 + 2x(x - 1) = x + 2(x + 1)$

(d) $2x^2 - 7x + 1 = 0$

(e) $\dfrac{2x + 1}{7} - \dfrac{x}{3} = 1$

(f) $\dfrac{3x - 1}{2} + \dfrac{1}{x} = 2$

(g) $\dfrac{x - 1}{x + 1} + \dfrac{1}{2} = 3$

5 John has \$8 more than Bill. If John gives Bill \$12, Bill will have three times as much money as John. How much money does each one have now?

6 Twice a certain integer plus 7 is equal to three times that integer minus 4. What is the integer?

7 A man is 25 years younger than his father. Three years ago he was one sixth as old as his father. How old is each one now?

8 I have 20 coins in my pocket consisting of nickels and dimes, and their total value is \$1.35. How many of each do I have?

9 One brand of coffee sells for \$0.80 per pound, whereas another brand sells for \$0.90 per pound. A 10-lb mixture of these two is worth \$8.30. How many pounds of each brand are in this mixture?

10 Mary is twice as old as Jane will be four years from now. The sum of their ages is 35 years. How old is each now?

11 A rectangle is three times as long as it is wide. If 5 ft were added to the width and 50 ft were added to the length, the area would be doubled. What are its dimensions?

12 If a certain number is increased by 4 and this then squared, the result is 16 times the original number. What is the number?

13 Solve by elimination:

$$2x - 3y = 4$$
$$3x + 5y = 1$$

14 Solve by substitution:

$$4x - y = 7$$
$$3x + 5y = 2$$

15 Solve by graphing:

$$\begin{aligned} 2x - 3y &= 5 \\ x + 2y &= 6 \end{aligned}$$

16 Solve by determinants:

$$\begin{aligned} 2x + 5y &= 4 \\ 3x - y &= 1 \end{aligned}$$

17 Solve by determinants:

$$\begin{aligned} 2x + 3y - z &= 4 \\ 5x + y + 2z &= 1 \\ 2x - 4y + 3z &= -3 \end{aligned}$$

18 Solve the system of equations in problem 17 by substitution.

19 Plot the curve corresponding to the equation $x^2 = 2y - 4$.

20 Plot the curve corresponding to the equation $xy = 8$.

chapter six

GEOMETRY AND TRIGONOMETRY

6.1/ areas of plane figures

In the first five chapters of this book we have been concerned with numbers and the solutions of problems involving numbers. In this chapter we shall discuss plane figures and the way in which numbers are used to express the size of a plane figure. A plane figure is one that may be drawn on a flat sheet of paper,

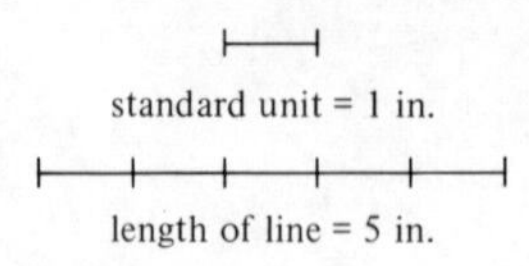

figure 6.1

or, as it is usually referred to, one that is two-dimensional. We shall be concerned with the space or area enclosed by simple, closed curves, which, in all but a few cases, will be composed of a set of straight-line segments. A simple closed curve is one that does not intersect itself and whose terminal point is the same as its initial point. To be more specific, we shall derive formulas for the areas of rectangles, parallelograms, triangles, circles, and combinations of these.

In order to have a system in which we use numbers to express the areas of figures, we must agree upon a unit of measurement. When we measure the length of a straight line, we use a standard unit of measurement such as the inch and then express the length of the line in terms of the number of inches that can be laid off on it (see Figure 6.1).

We could just as well use other units of linear measurement such as the foot, the yard, the meter, the centimeter, and so on. What we are doing is to see how many times the unit of measurement that we are using will fit on the line. In the case of two-dimensional figures, we adopt a square unit such as the square inch, square foot, or some similar unit, and see how many times it will fit into the area that we are measuring. A square inch is a square each of whose sides is one inch long, and any square unit of measurement is a square each of whose sides is one linear unit of measurement long.

We start with the rectangle, which is a four-sided figure each of whose angles is 90°; that is, each side is *perpendicular* to its adjacent side,[1] as shown in Figure 6.2. Suppose that the dimensions of the rectangle are whole numbers; that is, the length (usually the longer side) is m units long and the width (usually the shorter side) is n units long where m and n are positive integers. It is clear from Figure 6.3,

figure 6.2

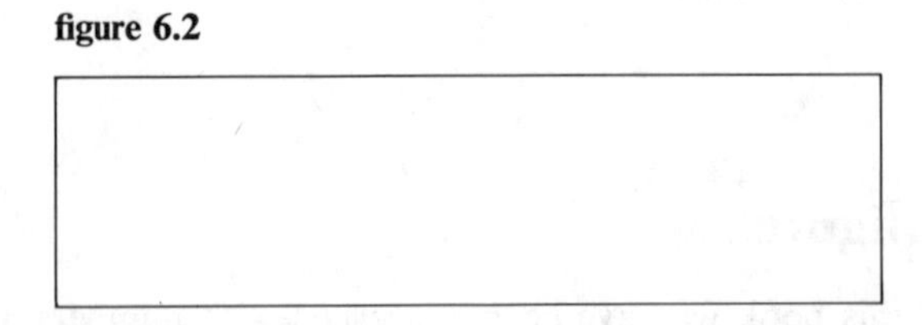

[1]A complete revolution is 360°, so 90° is one fourth of a complete revolution. This is what is meant by two lines or line segments being perpendicular. A 90° angle is usually referred to as a *right angle*.

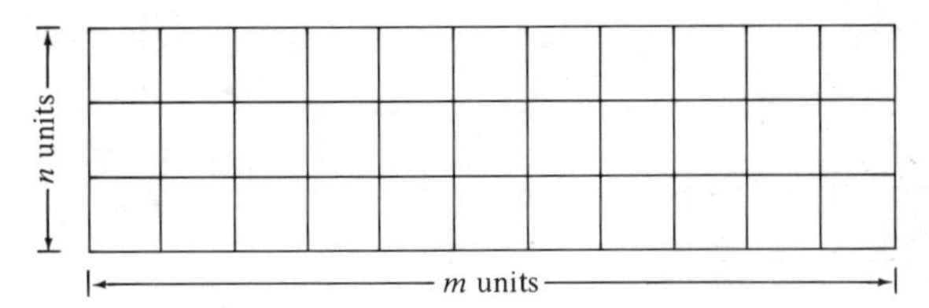

figure 6.3

where m is 11 and n is 3, that we can fit m square units along the bottom row of the rectangle and that we can fit in n such rows, each containing m square units, and thus completely fill up the rectangle. The area of the rectangle is therefore mn, or, if we let L stand for the length of the rectangle and W stand for its width, the area is the length times the width, which may be written

$$A = LW$$

where A stands for the area.

If the length is not an integer but a rational number of the form $m + (a/b)$, where m is a positive integer and a and b are positive integers with $b \neq 0$ and $a < b$, then the rectangle appears as in Figure 6.4. Each row contains m square units plus a fractional square unit that is the (a/b)th portion of one square unit, and there are n such rows; the area, therefore, is $[m + (a/b)]n$ and we still have the formula $A = LW$.

If the width is now made a rational number of the form $n + (c/d)$, where n, c, and d are positive integers, $d \neq 0$, and $c < d$, then the rectangle appears as in Figure 6.5. Each unit in the top row except the one on the right is c/d square unit, and there are m of these; and each unit in the right-hand column except the top one is a/b square unit, and there are n of these. The one in the upper right-hand corner is $(a/b) \cdot (c/d)$ square unit. The total area of the rectangle is therefore

$$mn + m \cdot \frac{c}{d} + n \cdot \frac{a}{b} + \frac{a}{b} \cdot \frac{c}{d}$$

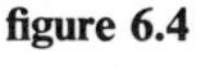

figure 6.4

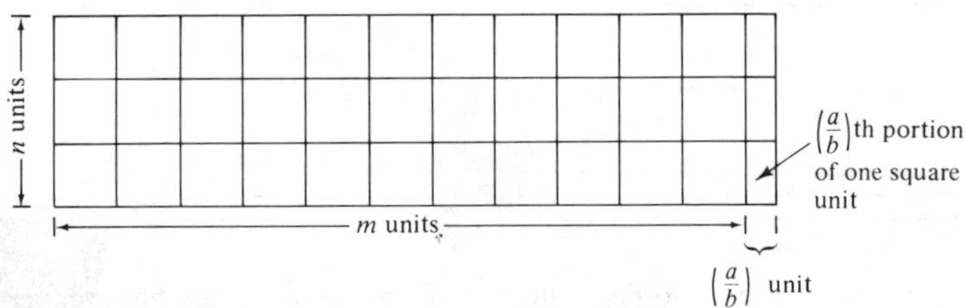

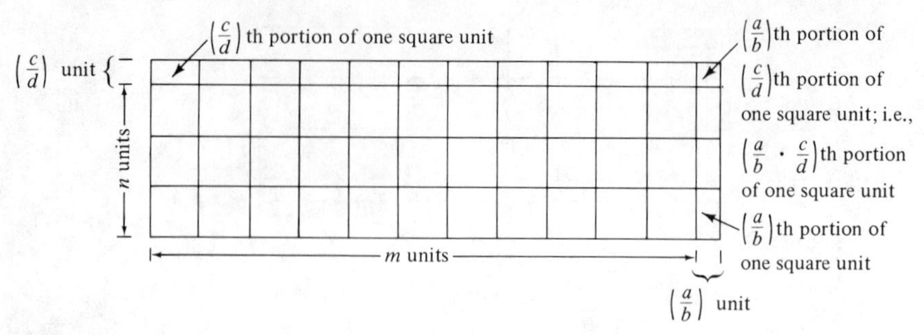

figure 6.5

and this expression can be factored into

$$\left(m + \frac{a}{b}\right)\left(n + \frac{c}{d}\right)$$

Thus when the dimensions are rational numbers, the formula $A = LW$ is still valid. It can be shown by a limiting process (since any real number may be considered as the limit of sequence of rational decimals) that this formula holds when the dimensions are any real numbers, so $A = LW$ is the formula for the area of a rectangle in any case.

We consider next the area of a parallelogram, which is a four-sided figure whose opposite sides are parallel, as shown in Figure 6.6. Note that every rectangle is a parallelogram but not every parallelogram a rectangle. If one of the sides is called the *base*, denoted by B, then the distance between it and its opposite side is called the *height*, denoted by H.

To derive the formula for the area of a parallelogram, we drop a perpendicular from one vertex[1] to one of the opposite sides, cut off the triangle thus formed and move it to the other side of the parallelogram; notice that we come up with a rectangle as in Figure 6.7. The length of this rectangle is the base of the original

figure 6.6

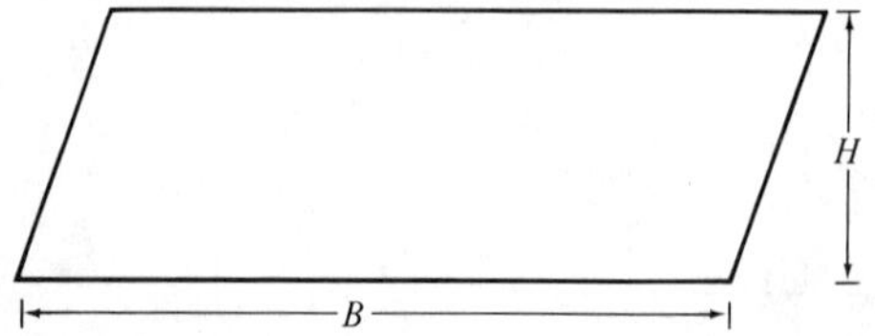

[1]A vertex is a point where two sides intersect. Each parallelogram has four vertices.

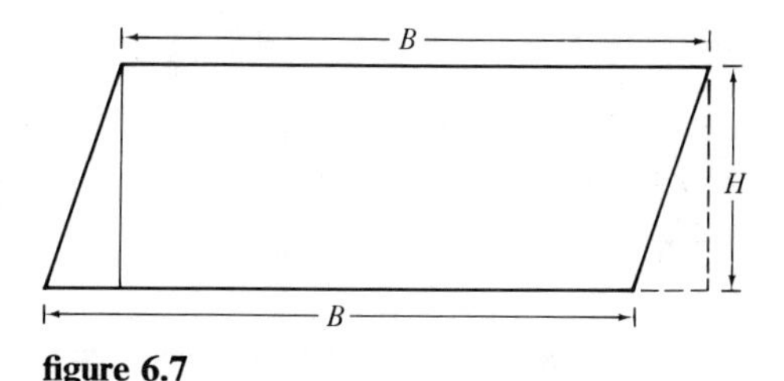

figure 6.7

parallelogram, and its width is the height of the original parallelogram. Since the area was unaltered in moving the triangle, the area of the original parallelogram is the same as the area of the rectangle; that is, we have the formula

$$A = BH$$

for the area of a parallelogram.

We could have chosen one of the other sides of the parallelogram of Figure 6.6 as the base, and the height would then be as shown in Figure 6.8, but the area would still be given by the formula $A = BH$ and would give the same number for the area as before.

We may now use the formula for the area of a parallelogram to determine the area of a triangle, which is the figure formed by three intersecting line segments (which do not all intersect in one point so that we will have some positive area to measure), as shown in Figure 6.9. If one side is called the *base*, then the distance

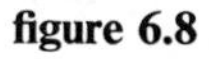

figure 6.8

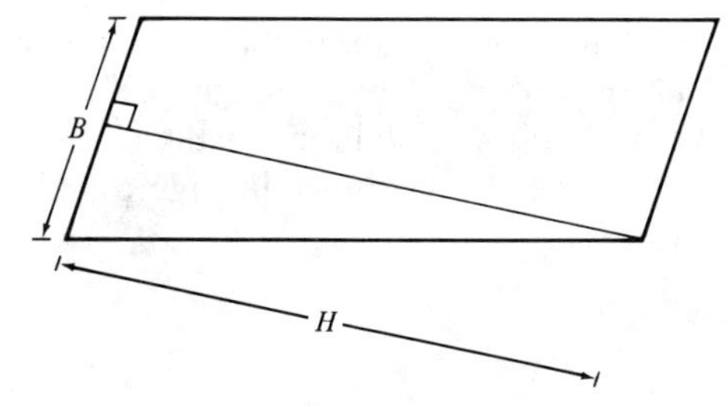

figure 6.9

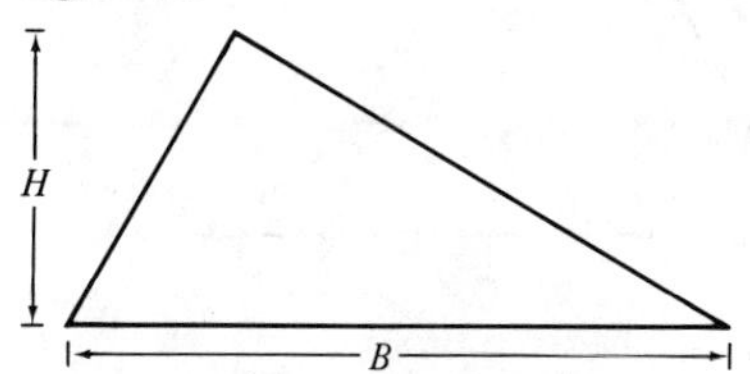

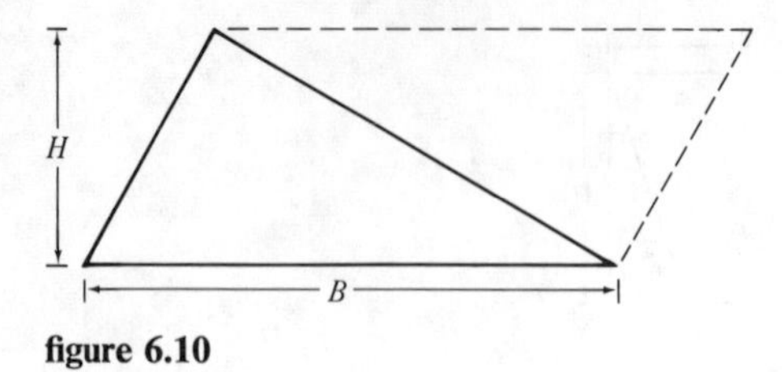

figure 6.10

from the opposite vertex to this side is called the height. To determine the area of the triangle, we draw a line from the vertex at the top parallel to the base and draw a line from the vertex at the right parallel to its opposite side and extend these lines to the point where they intersect; notice that we get a parallelogram, as shown in Figure 6.10. This parallelogram is made up of two triangles, the original one and a new one, which, if inverted, may be seen to be the same size and shape[1] as the original, and therefore has the same area. The area of our original triangle is therefore one half the area of the parallelogram, so we have the formula for the area of a triangle,

$$A = \tfrac{1}{2}BH$$

Any of three sides may be chosen as the base, so we could have the two possibilities shown in Figure 6.11 in addition to that of Figure 6.9. In both cases of Figure 6.11, it was necessary to extend the base somewhat in order to draw the height perpendicular to it. Regardless of which side we choose for the base, the area is always given by $A = \tfrac{1}{2}BH$ and always gives the same number for an answer.

We can now find the area of any closed figure whose boundary consists of straight line segments by breaking it up into triangles or rectangles or both and adding up the areas of its components to obtain the total area of the figure. The area in Figure 6.12, for example, may be broken up as shown in Figure 6.13.

figure 6.11

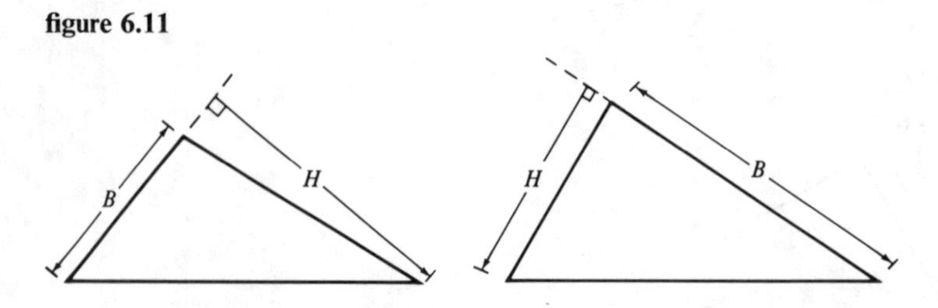

[1]In Section 6.2 such triangles that are identical will be defined to be *congruent*.

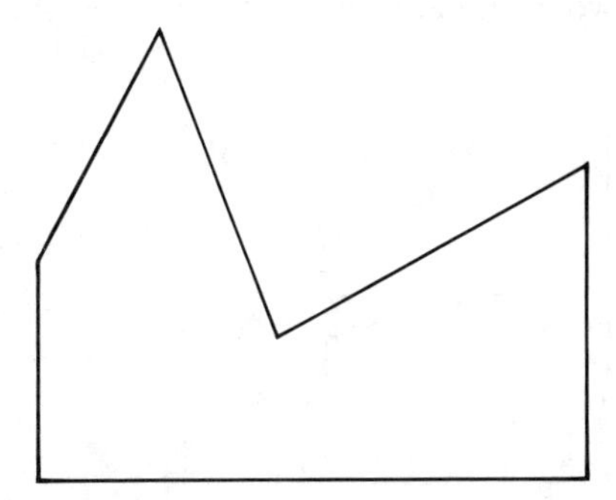

figure 6.12

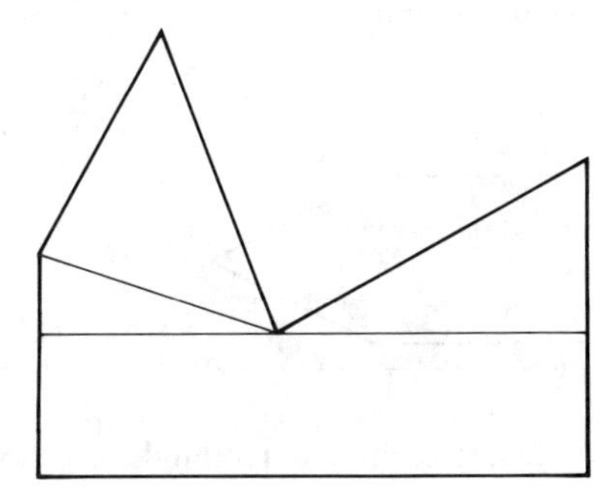

figure 6.13

examples

1 Find the area of the figure shown.

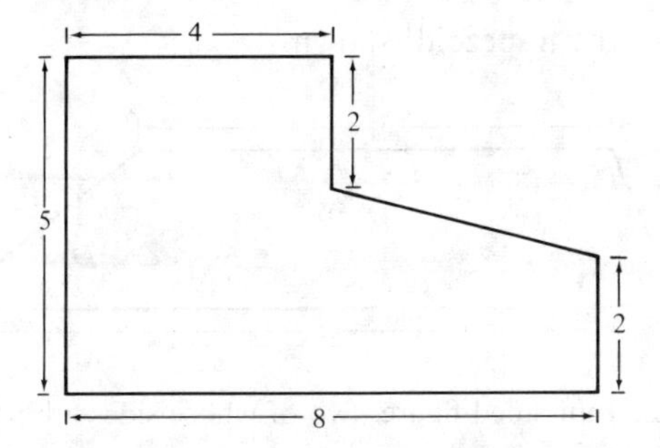

► We break the figure up into two rectangles and a triangle as shown.

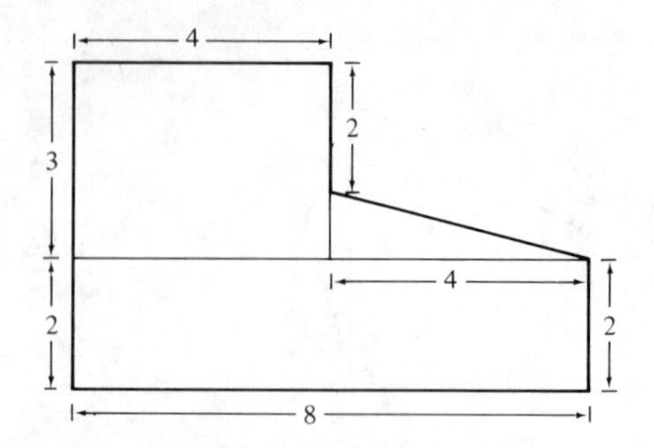

The areas of the rectangles are $4 \times 3 = 12$ square units and $8 \times 2 = 16$ square units, whereas the area of the triangle is $\frac{1}{2}(4 \times 1) = 2$ square units, so the total area is $12 + 16 + 2 = 30$ square units.

2 Find the area of the figure shown.

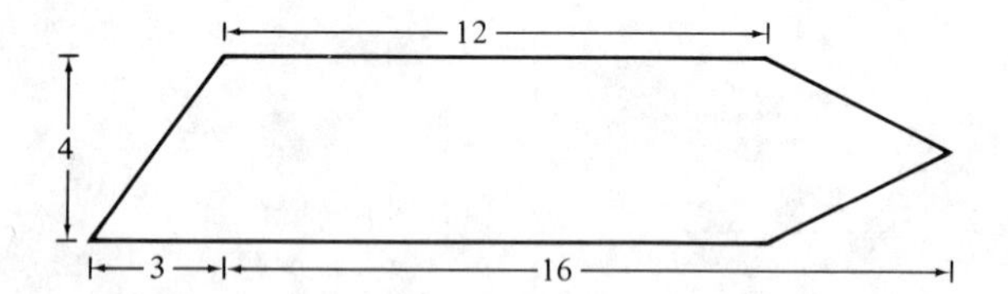

► We break the figure up into two triangles and a rectangle as shown.

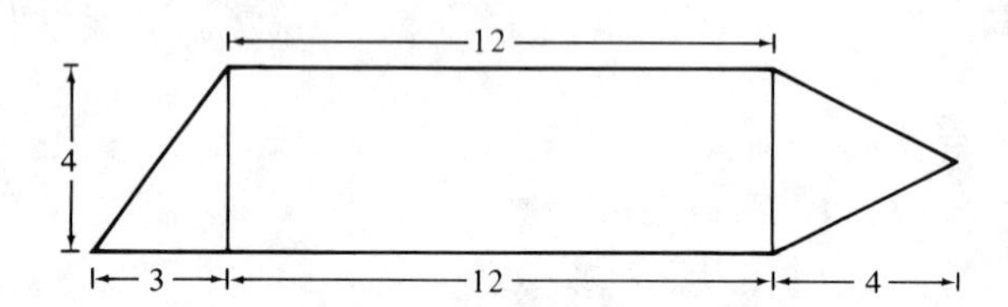

The areas of the triangles are $\frac{1}{2}(3 \times 4) = 6$ and $\frac{1}{2}(4 \times 4) = 8$, and the area of the rectangle is $12 \times 4 = 48$, so the total area is $6 + 8 + 48 = 62$ square units.

3 Find the area of the trapezoid[1] shown.

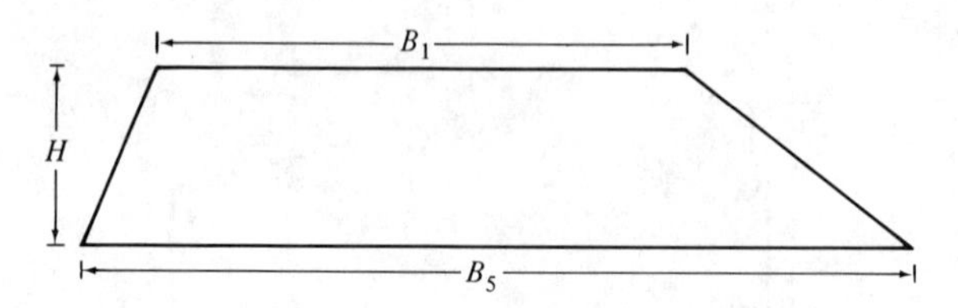

[1]A trapezoid is a four-sided figure, two of whose sides are parallel.

► Break the trapezoid up into two triangles and a rectangle as shown.

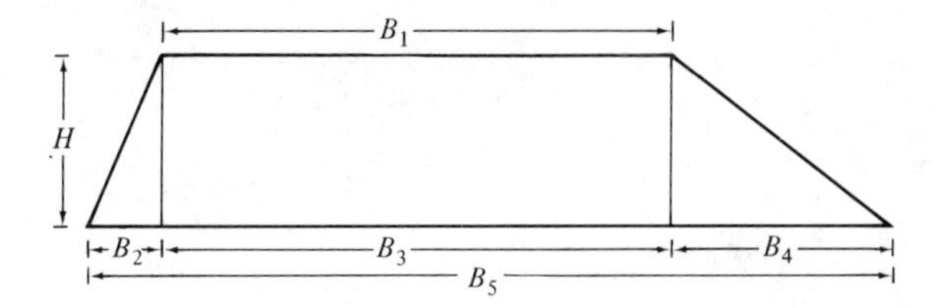

The areas of the triangles are $\frac{1}{2}B_2H$ and $\frac{1}{2}B_4H$, and the area of the rectangle is B_3H, so the total area is $\frac{1}{2}B_2H + \frac{1}{2}B_4H + B_3H$, which may be written

$$\begin{aligned}\tfrac{1}{2}(B_2 + B_4)H + \tfrac{1}{2}(2B_3)H &= \tfrac{1}{2}(B_2 + B_4)H + \tfrac{1}{2}(B_3 + B_1)H \\ &= \tfrac{1}{2}(B_2 + B_4 + B_3 + B_1)H = \tfrac{1}{2}(B_5 + B_1)H\end{aligned}$$

The area of the trapezoid is therefore given by

$$A = \tfrac{1}{2}(B_1 + B_5)H$$

In words, this says that the area of a trapezoid is one half the sum of its bases[1] times its height.

We look next at a figure whose boundaries are not straight line segments (see Figure 6.14). We will derive the formula for the area of a circle. A circle may be defined as a curve all of whose points are the same distance from a fixed point called its center. The center is not part of the circle. The distance from the center to the curve is called the *radius* and is denoted by R, whereas the distance around

figure 6.14

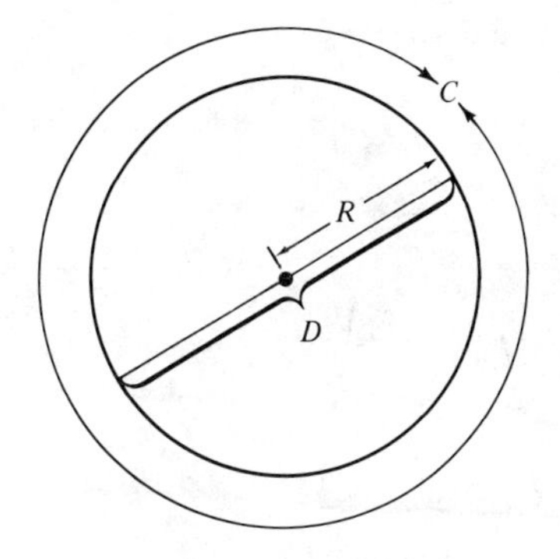

[1]The bases are the two sides that are parallel.

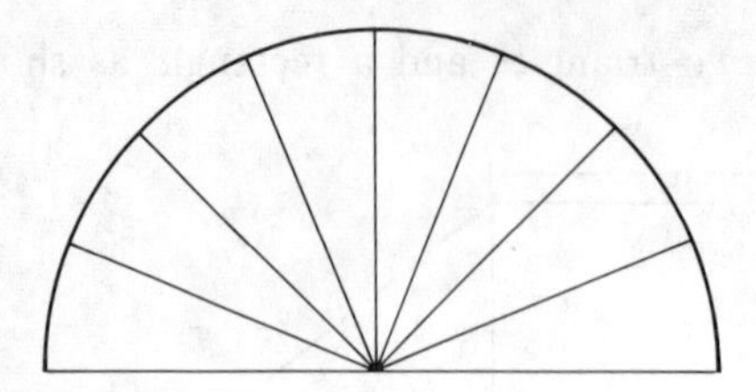

figure 6.15

the curve is called the *circumference* and is denoted by C. The diameter, denoted by D, is the distance across the circle through the center and is twice the radius. A *chord* is a line segment from one point of the circle to another, so every diameter is a chord but not vice versa. The ratio between the circumference and the diameter is constant, regardless of the size of the circle, and is an irrational number, denoted by the Greek letter π, which is approximately equal to 3.14 or $3\frac{1}{7}$. This approximation is sufficiently accurate for most computations and we shall use this value in the Examples and Exercises below. We have $C = \pi D$, and since $D = 2R$, we may write $C = 2\pi R$.

To find the area enclosed by a circle, take the top half[1] of the area and break it up as shown in Figure 6.15. Arrange these pieces as shown in Figure 6.16. The bottom half may be broken up and arranged similarly, and the pieces will fit into the top half as shown in Figure 6.17 to give a figure that is almost a rectangle.

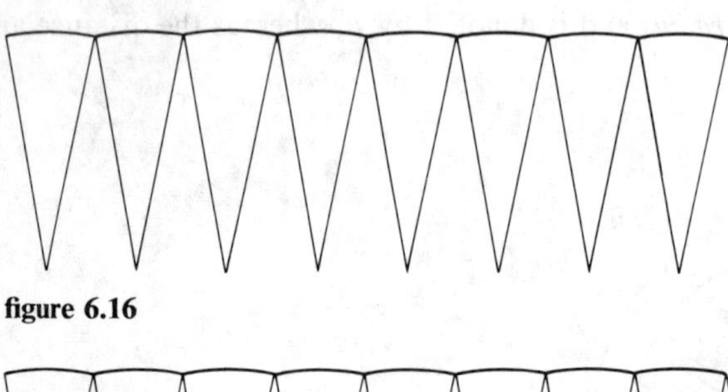

figure 6.16

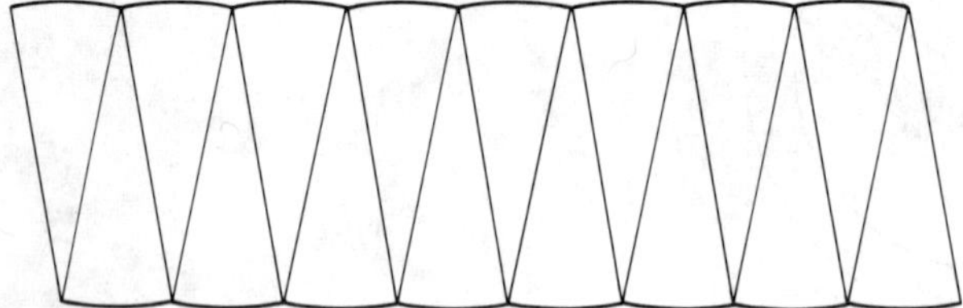

figure 6.17

[1]Half of a circle is called a *semicircle*, so we are talking about the area enclosed by the semicircle and the diameter.

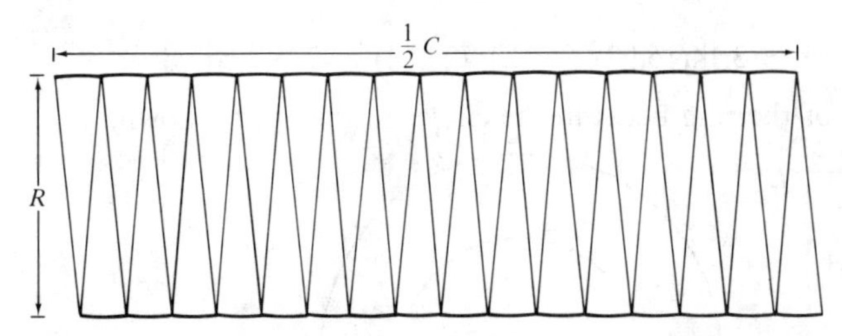

figure 6.18

This figure is not a rectangle because the top and bottom are not straight lines and the angles are not 90°. This shortcoming may be remedied by increasing the number of pieces into which we cut each half of the circle; the more pieces we make, the more the resulting figure approaches being a rectangle (see Figure 6.18). Since the area of the resulting figure is the same as the area of the circle, no matter how many pieces it consists of, and we can make it as close to a rectangle as we please, we say that in the limit, this figure becomes a rectangle whose area is the same as that of the circle. The length of this rectangle is $\frac{1}{2}C$ and its width is R, so its area is $\frac{1}{2}CR = \frac{1}{2} \cdot 2\pi R \cdot R = \pi R^2$. We have, therefore, for the area of a circle the formula

$$A = \pi R^2$$

examples

4 Find the area of a circle whose radius is 10 ft.

$$A = \pi R^2$$

We use $\pi = 3.14$ and $R = 10$ and obtain

$$A = 3.14 \cdot 10 \cdot 10 = 314 \text{ sq ft}$$

5 Find the area of a circle whose diameter is 80 yd.

The radius is half the diameter, so $R = 40$ yd and, therefore,

$$A = 3.14 \cdot 40 \cdot 40 = 5024 \text{ sq yd}$$

6 Find the area of a circle whose circumference is 100 yd.

$$C = 2\pi R$$

Therefore,

$$R = \frac{C}{2\pi} = \frac{100}{2 \cdot 3.14} = \frac{100}{6.28} \approx 15.92$$

and

$$A \approx 3.14(15.92)^2 \approx 3.14(253.4) \approx 795.7 \text{ sq yd}$$

7 Find the area of the ring between the circles in the figure shown.

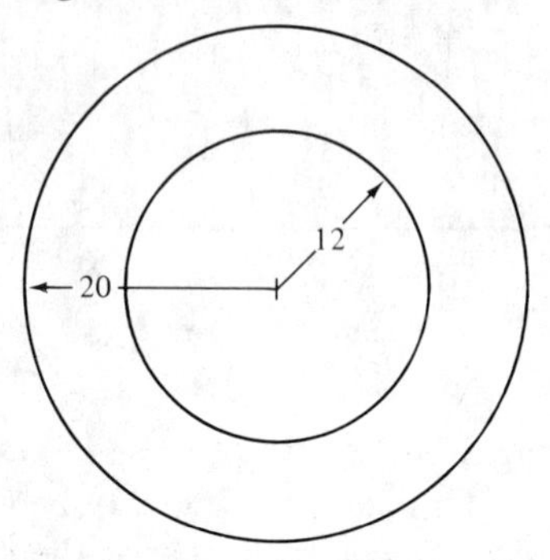

► The area of the outer circle is $3.14(20^2) = 1256$, and the area of the inner circle is $3.14(12^2) = 452.16$, so the area between the circles is $1256 - 452.16 = 803.84$ square units.

8 Find the area of the figure shown, where the top portion is a semicircle.

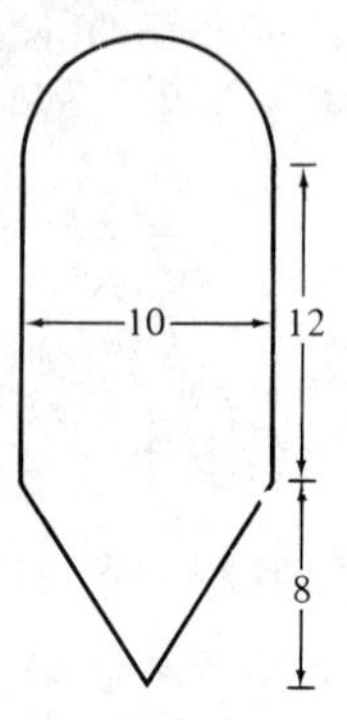

► The radius of the semicircle is 5 units, so its area is $\frac{1}{2} \cdot 3.14 \cdot 5^2 = 39.25$. The area of the rectangle is $12 \times 10 = 120$, and the area of the triangle is $\frac{1}{2} \cdot 10 \cdot 8 = 40$, so the entire area is $39.25 + 120 + 40 = 199.25$ square units.

9 What would it cost to put two coats of paint on the wall that is pictured if paint costs \$1.10 per quart and 1 qt covers 50 sq ft with one coat?

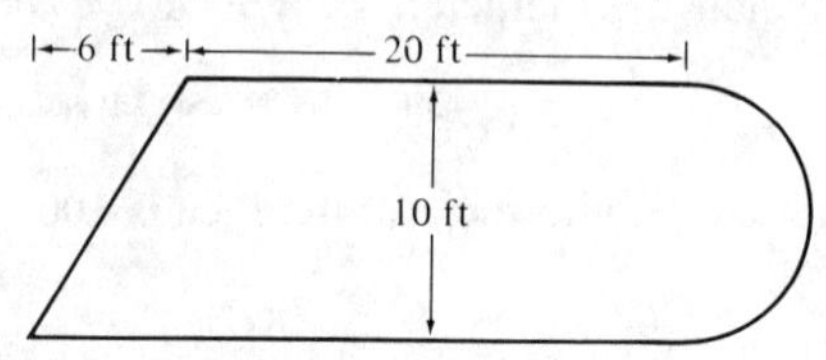

► The area of the semicircle is $\frac{1}{2} \cdot 3.14 \cdot 5^2 = 39.25$, the area of the rectangle is $20 \cdot 10 = 200$, and the area of the triangle is $\frac{1}{2} \cdot 10 \cdot 6 = 30$, so the total area is 269.25 sq ft. Two coats of paint requires covering 538.5 sq ft, so we need $538.5 \div 50 = 10.77$ qt of paint. At \$1.10 per quart, this would cost \$11.85, assuming that we pay only for the amount of paint that we actually use. If we had to buy 11 qt, it would cost \$12.10.

exercises 6.1

1 Determine the areas of the following figures:

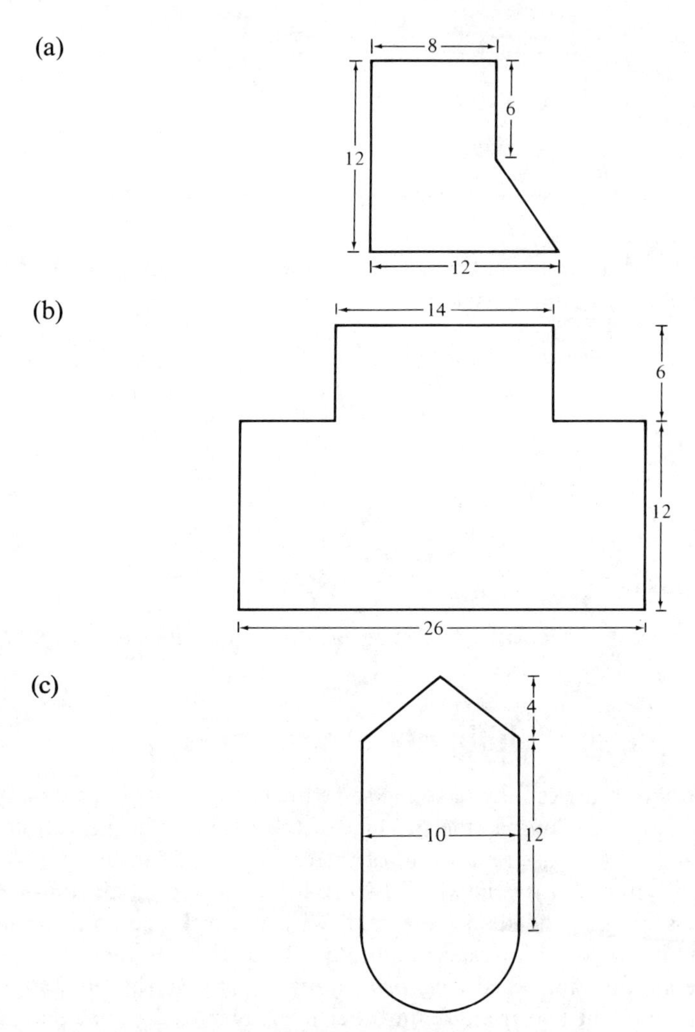

(The bottom portion is a semicircle.)

(d)

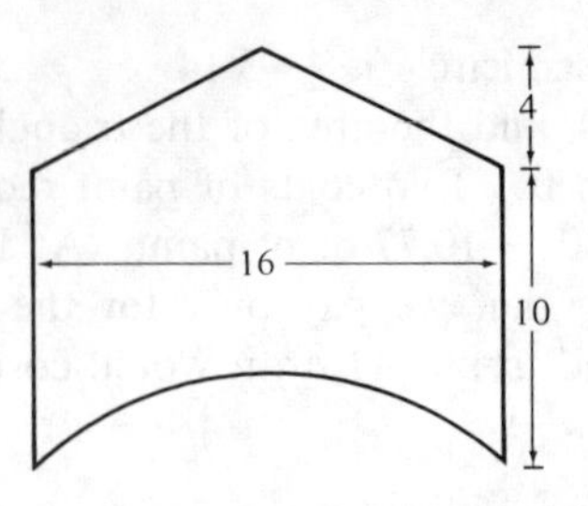

2 What would it cost to cover the surface shown with two coats of paint if 1 qt covers 60 sq ft of surface area and costs $1.20?

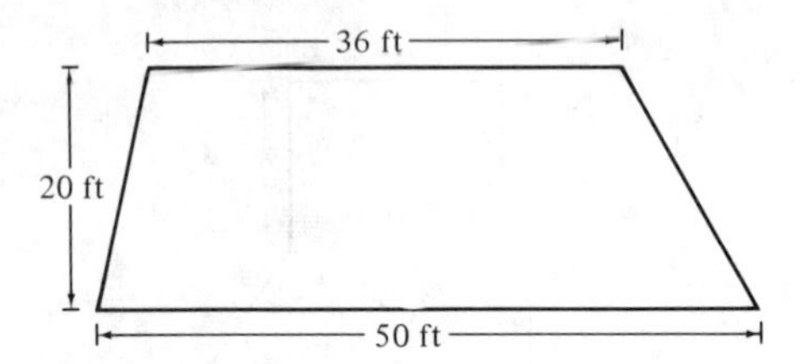

3 What would it cost to construct the circular cement sidewalk shown if the cost of cement is $0.60 per square foot?

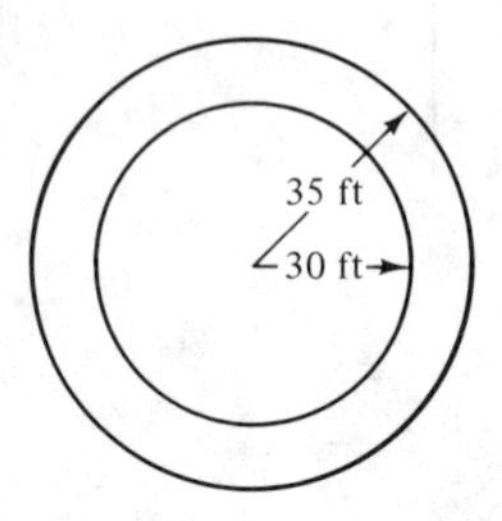

4 Determine the total surface area of a cube each of whose sides is 6 ft long.

6.2/congruence and similarity of triangles

It is obvious that two triangles may have the same area, even though their shapes are completely different, as long as the product of the base and height of one is equal to the product of the base and height of the other. In this section we shall be interested in the circumstances in which two triangles not only have the same area but have the same size and shape; that is, if we could pick one up and place it atop the other, it would be an exact duplicate. Two triangles that are exact duplicates of one another are called *congruent* triangles, and we shall investigate what we must know about two triangles in order to be able to say that they are congruent.

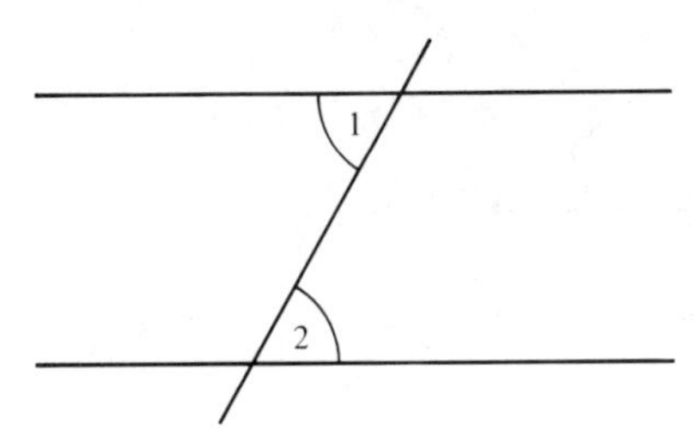

figure 6.19

First we need a fact from geometry, which states that if two parallel lines are cut by a third line, then the alternate interior angles (these are the angles in the interior of the *z*-shaped figure such as angles 1 and 2 of Figure 6.19) are equal.

Suppose that we have a triangle ABC as in Figure 6.20, and through the vertex we draw a line parallel to the opposite side. Since angles 1 and 2 are alternate interior angles, they are equal, and since angles 3 and 4 are alternate interior angles, they are equal. We know that the sum of angles 1, 5, and 3 is 180°, and so, by the axiom that states that any quantity may be substituted for its equal, we see that the sum of angles 2, 5, and 4 is 180°. This proves that the sum of the three angles of any triangle is equal to 180°; in other words, if two angles of one triangle are equal, respectively, to two angles of another triangle, then the third angle of the first triangle is equal to the third angle of the second triangle.

Assume now that we have two triangles such that a side of one is equal to a side of the other and the angles adjacent to the one side are equal, respectively, to the angles adjacent to the other side, as shown in Figure 6.21.[1] The shape and size of the first triangle is completely determined by the side AB and the angles 1 and 2, because these angles determine the direction of the other two sides uniquely (except for whether they are above AB or below AB) and hence determine their point of intersection, which is the third vertex of the triangle, uniquely (except for

figure 6.20

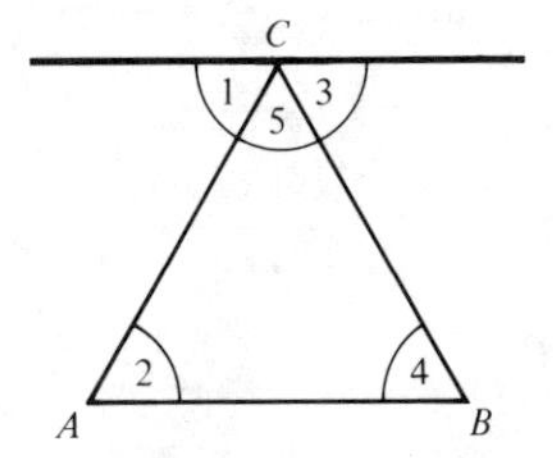

[1]The symbol ∠ stands for "angle," so ∠ 1 is read "angle 1." A line segment is denoted by the letters at its extremities; thus AB indicates the line segment between A and B.

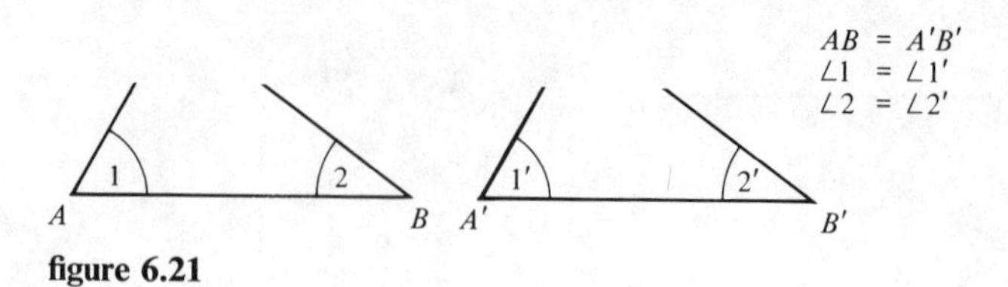

figure 6.21

whether it is above or below AB). The second triangle having the same side and angles must, therefore, be an exact duplicate of the first; that is, the triangles must be congruent. (If the third vertex of one triangle is above AB and the other third vertex is below $A'B'$, we can match the triangles by flipping the second one over so that its third vertex is above $A'B'$.) By what has been shown, if two triangles have a side of one equal to a side of the other and any two angles of one equal to the corresponding two angles of the other, then the three angles of the one must equal the corresponding three angles of the other, and the triangles are therefore congruent. We indicate the fact that two triangles (for which two angles and a side of one are equal to two angles and the corresponding side of the other) are congruent by writing *AAS*.

Next assume that two sides of one triangle and the angle included between them are equal, respectively, to the two sides and the included angle of another triangle, as in Figure 6.22. There is no choice left for the third side of the first triangle; it must connect A and C, and so the shape of the first triangle is completely determined. Since the same is true of the second triangle, it must be congruent to the first. We summarize this by saying that if two sides and the included angle of one triangle are equal to two sides and the included angle of another triangle, then the triangles are congruent; we abbreviate this by writing *SAS*.

There is one other criterion for the congruence of arbitrary triangles. Suppose that we are given three line segments of arbitrary lengths and asked to form a triangle with them. If we can form a triangle with them at all, then this can be done in only one way; that is, any two triangles that we might form would be congruent. This implies that if the three sides of one triangle are equal to the three sides of another triangle, then the triangles are congruent, and this is abbreviated by writing *SSS*.

figure 6.22

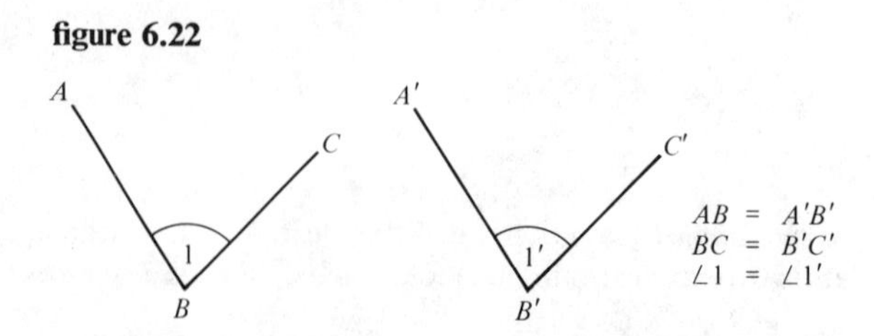

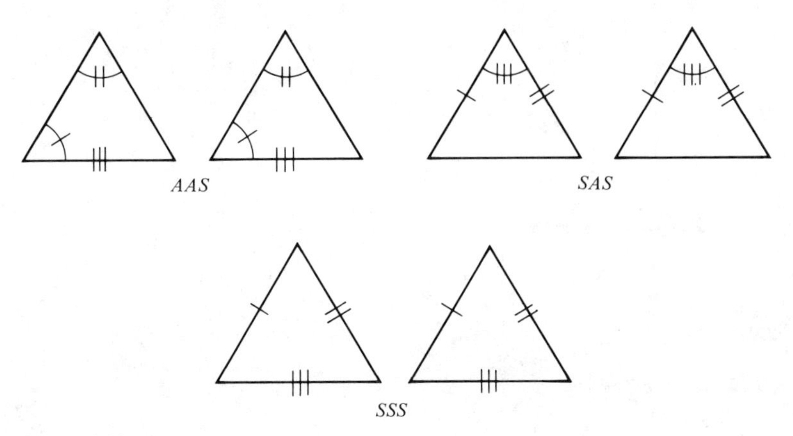

figure 6.23

We now have three ways at our disposal of proving that two triangles are congruent:

1. If two angles and a side of one are equal to two angles and the corresponding side of the other (*AAS*).
2. If two sides and the included angle of one are equal to two sides and the included angle of the other (*SAS*).
3. If the three sides of one are equal to the three sides of the other (*SSS*).

These situations are illustrated in Figure 6.23, where equal parts are indicated by the same number of dashes drawn through them. Since two triangles are congruent if and only if one is an exact duplicate of the other, it follows immediately that *corresponding parts of congruent triangles are equal*, and this fact will be abbreviated as *CPCT*.

The study of congruent triangles is a part of plane geometry, a discipline that illustrates the deductive method of proving certain facts or theorems on the basis of given facts or hypotheses and accepted postulates or axioms and other theorems. We will illustrate this method by working out several examples in which certain things are given and certain other things must be proved. The work is shown in compact tabular form giving a reason for each statement, and certain symbols are used to make the proofs even more compact. These symbols and their meanings are as follows:

$\triangle$ *triangle* A triangle is denoted by giving it three vertices; thus $\triangle ABC$ means the triangle whose vertices are A, B, and C.

$\angle$ *angle* An angle may be denoted by giving its vertex, or if there is more than one angle at a point, as in Figure 6.24, the angle may be denoted by giving the three points that trace the angle with its vertex as the point in the middle. Thus the angle indicated in Figure 6.24 is denoted by $\angle ABC$. Angles

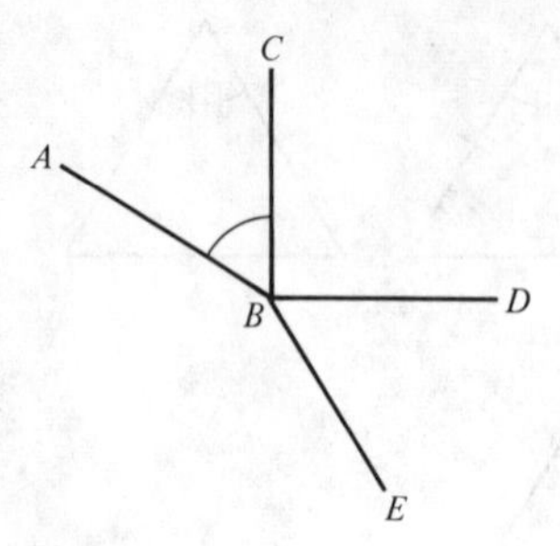

figure 6.24

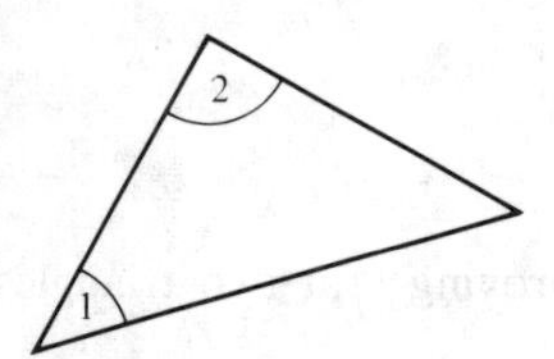

figure 6.25

may also be indicated by numbers such as $\angle 1$, $\angle 2$, etc., as shown in Figure 6.25.

$\therefore$ *therefore*
$\cong$ *is congruent to*
$\parallel$ *is parallel to*
$\perp$ *is perpendicular to*

These examples are intended more to illustrate the methods used rather than to be completely rigorous, so some of the steps and reasons have been shortened beyond what would be required for a rigorous proof.

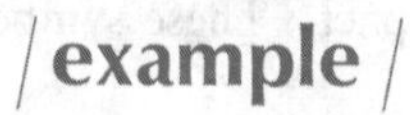

1

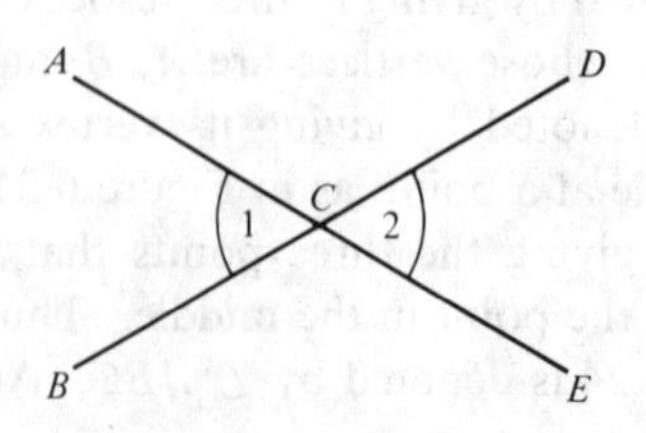

Given: Intersecting lines *AE* and *BD*
Prove: $\angle 1 = \angle 2$

Statement	Reason
1. $\angle ACE = \angle BCD$.	1. Both are equal to 180°.
2. $\angle ACD = \angle ACD$.	2. Any angle is equal to itself.
3. $\therefore \angle 1 = \angle 2$.	3. If equals are subtracted from equals, the results are equal (axiom): $\angle 1 = \angle BCD - \angle ACD$ $\angle 2 = \angle ACE - \angle ACD$

Angles such as $\angle 1$ and $\angle 2$ in Example 1 are known as *vertical* angles, and Example 1 shows that *vertical angles are equal.* We will use this theorem in working out Example 3.

/examples/

2

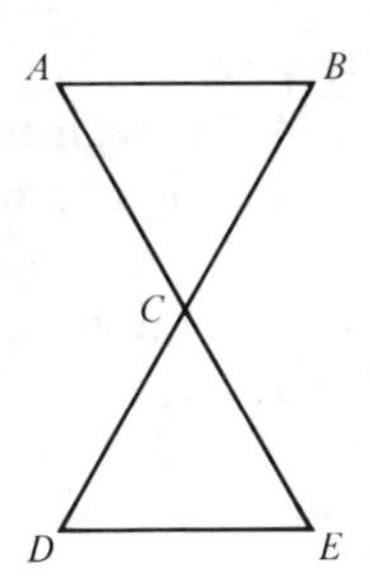

Given: $AB \parallel DE$
$AB = DE$
Prove: $AC = CE$

Statement	Reason
1. $AB \parallel DE$.	1. Given.
2. $\therefore \angle CAB = \angle CED$.	2. Alternate interior angles are equal.
3. $\angle ABC = \angle CDE$.	3. Same as 2.
4. $AB = DE$.	4. Given.
5. $\therefore \triangle ACB \cong \triangle CDE$.	5. *AAS.*
6. $\therefore AC = CE$.	6. *CPCT.*

3

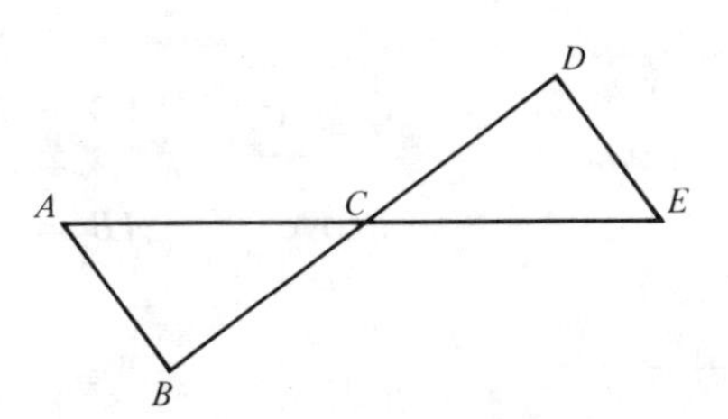

Given: $AC = CD$
$BC = CE$
Prove: $AB = DE$

Statement	Reason
1. $AC = CD$.	1. Given.
2. $BC = CE$.	2. Given.
3. $\angle ACB = \angle DCE$.	3. Vertical angles are equal.
4. $\therefore \triangle ACB \cong \triangle DCE$.	4. *SAS*.
5. $\therefore AB = DE$.	5. *CPCT*.

4

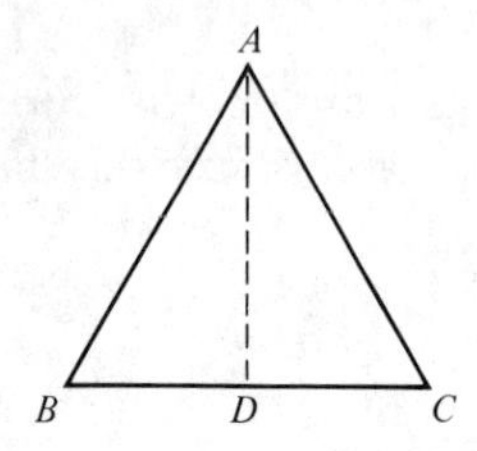

Given: $AB = AC$
Prove: $\angle B = \angle C$

Statement	Reason
1. $AB = AC$.	1. Given.
2. Draw line segment AD from A to the midpoint D of BC.	2. A line segment may be drawn connecting two points.
3. $AD = AD$.	3. Any line segment is equal to itself.
4. $BD = DC$.	4. Definition of midpoint.
5. $\therefore \triangle ABD \cong \triangle ADC$.	5. *SSS*.
6. $\therefore \angle B = \angle C$.	6. *CPCT*.

A triangle such as $\triangle ABC$ above with two equal sides is called an *isosceles* triangle. Example 4 shows that the angles opposite the equal sides of an isosceles triangle (these are usually called the *base* angles) are equal.

/ examples /

5

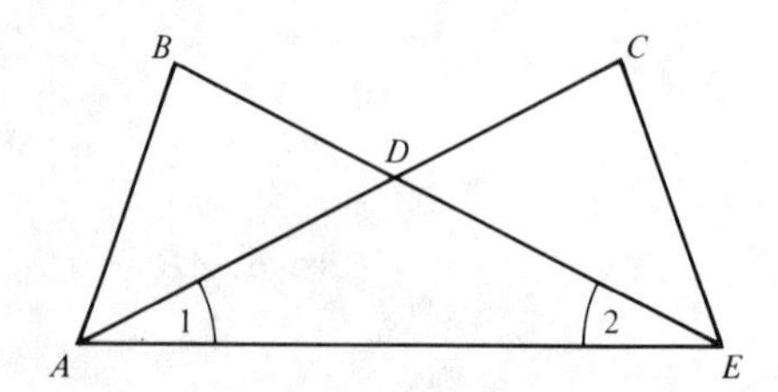

Given: $AD = DE$
$\angle BAD = \angle DEC$
Prove: $AB = CE$

Statement	Reason
1. $AD = DE$.	1. Given.
2. $\therefore \angle 1 = \angle 2$.	2. Base angles of an isosceles triangle are equal.
3. $\angle BAD = \angle DEC$.	3. Given.
4. $\therefore \angle BAE = \angle CEA$.	4. If equals are added to equals, the results are equal.
5. $AE = AE$.	5. Any line segment is equal to itself.
6. $\therefore \triangle ABE \cong \triangle CAE$.	6. *AAS.*
7. $\therefore AB = CE$.	7. *CPCT.*

6

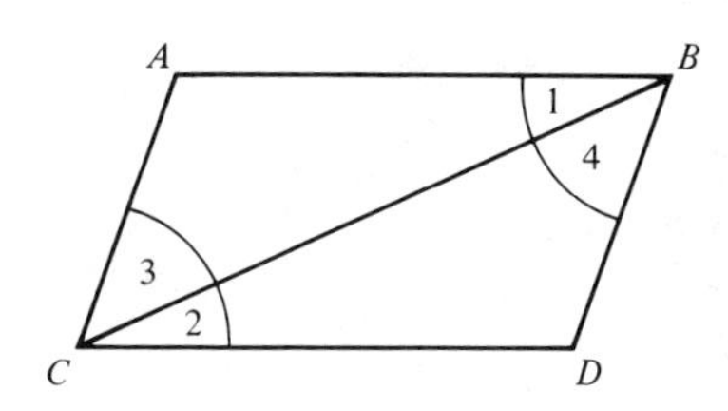

Given: $AB \parallel CD$
$AC \parallel BD$
Prove: $AB = CD$ and
$AC = BD$

Statement	Reason
1. Draw line segment BC.	1. A line segment may be drawn connecting two points.
2. $AB \parallel CD$.	2. Given.
3. $\therefore \angle 1 = \angle 2$.	3. Alternate interior angles are equal.
4. $AC \parallel BD$.	4. Given.
5. $\therefore \angle 3 = \angle 4$.	5. Alternate interior angles are equal.
6. $BC = BC$.	6. Any line segment is equal to itself.
7. $\therefore \triangle ABC \cong \triangle BCD$.	7. *AAS.*
8. $\therefore AB = CD$ and $AC = BD$.	8. *CPCT.*

Thus far we have derived three criteria for congruence of two triangles: *SAS*, *AAS*, and *SSS*. What about *SSA*, two sides and a nonincluded angle of one triangle being equal to two sides and the corresponding angle of another triangle? Figure 6.26 shows that this condition is not sufficient for congruence. $AB = BC$ in the figure, $BD = BD$, and $\angle A = \angle C$, so we have *SSA* for triangles ABD and BCD, but they are obviously not congruent.

There is only one other possibility for three parts of one triangle being equal to the corresponding three parts of another, that is, *AAA*, the case where the three angles of one are equal to the three angles of the other, as shown in Figure

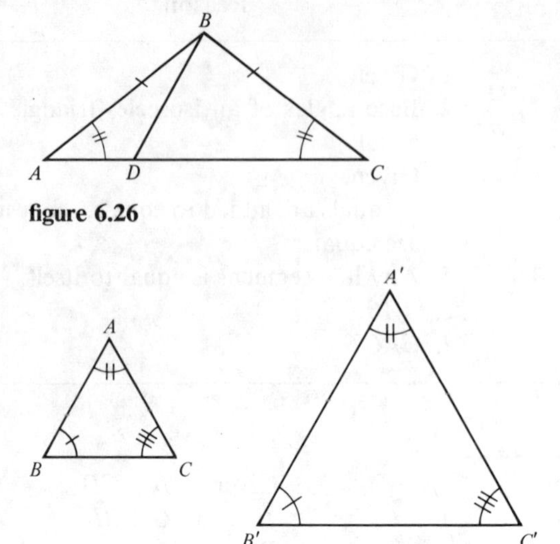

figure 6.26

figure 6.27

6.27. It is obvious from this figure that AAA is also not sufficient for congruence of triangles. When the three angles of one triangle are equal to the three angles of another triangle, we say that the two triangles are *similar*. The symbol $\sim$ is used to denote similarity of triangles. Thus in Figure 6.27, we would write $\triangle ABC \sim \triangle A'B'C'$. Note that if two triangles are congruent they are similar; the converse, however, is not true.

When two triangles are similar, it turns out that their sides are proportional. In Figure 6.27, for example, we have $AB/A'B' = BC/B'C' = CA/C'A'$. Since the three angles of any triangle add up to 180°, we can say that two triangles are similar if two angles of one are equal to two angles of the other, because the third angle of the first will then have to equal the third angle of the other.

examples

7

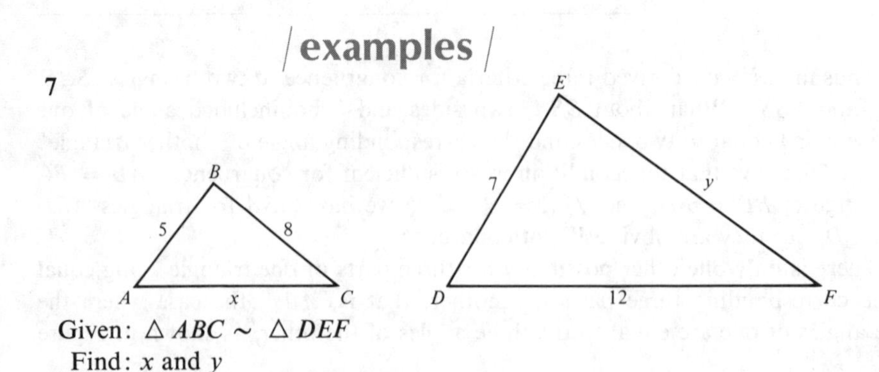

Given: $\triangle ABC \sim \triangle DEF$

Find: x and y

► Since the triangles are similar, their sides are proportional, so we have

$$\frac{5}{7} = \frac{x}{12} \qquad 7x = 60 \qquad \text{and} \qquad x = \frac{60}{7} = 8\,\frac{4}{7}$$

Also,

$$\frac{5}{7} = \frac{8}{y} \qquad 5y = 56 \qquad \text{and} \qquad y = \frac{56}{5} = 11\,\frac{1}{5}$$

8

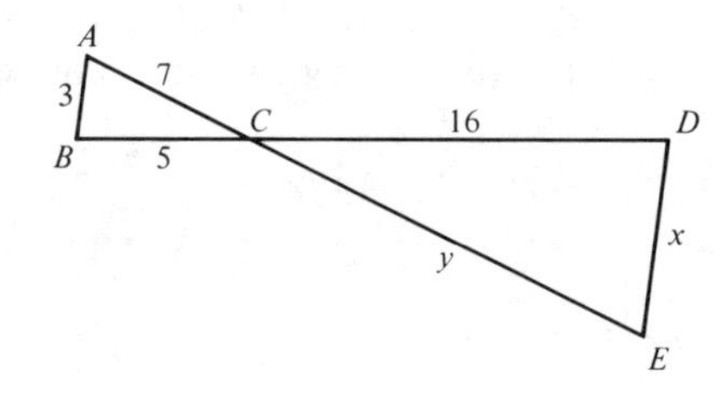

Given: $AB \parallel DE$
Find: x and y

► We have $AB \parallel DE$. Therefore $\angle ABC = \angle CDE$ and $\angle BAC = \angle CED$ because they are alternate interior angles. Therefore, $\triangle ABC \sim \triangle CDE$. This implies that

$$\frac{3}{x} = \frac{5}{16} \qquad 5x = 48 \qquad \text{and} \qquad x = 9\,\frac{3}{5}$$

and

$$\frac{7}{y} = \frac{5}{16} \qquad 5y = 112 \qquad \text{and} \qquad y = 22\,\frac{2}{5}$$

We will close this section by proving the famous *theorem of Pythagoras*, which says that in a right triangle (one that has a right angle), the sum of the squares of the legs (the two shorter sides) is equal to the square of the hypotenuse (the side that is opposite the right angle and is always the longest side).

If we start with the right triangle ABC in Figure 6.28 with right angle at C

figure 6.28

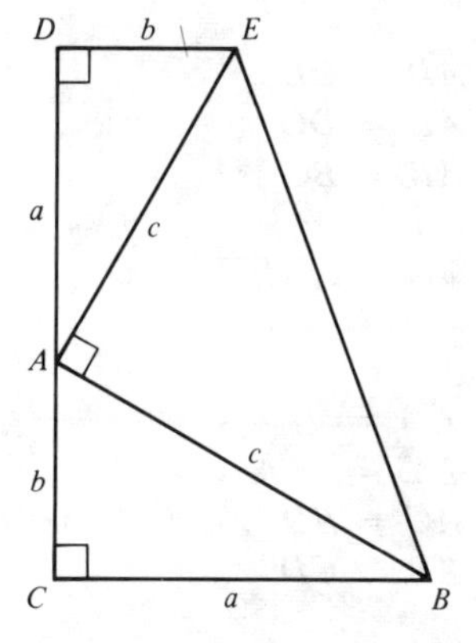

and extend side CA to D so that $AD = CB$, draw DE perpendicular to AD so that $DE = AC$, and then draw a line from E to A, we have constructed a right triangle ADE that is congruent to triangle ABC, because we have SAS. This implies that $EA = AB = c$. If we now draw a line from E to B, the figure $BCDE$ is a trapezoid, since $DE \parallel CB$, and its area is equal to the sum of the areas of triangles ABC, AEB, and ADE. $\angle DEA + \angle DAE = 90°$, and since $\angle DEA = \angle CAB$, we have $\angle CAB + \angle DAE = 90°$. This implies that $\angle EAB = 180° - (\angle CAB + \angle DAE) = 180° - 90° = 90°$; therefore, $\triangle EAB$ is a right triangle. The area of the trapezoid $BCDE = \frac{1}{2}(a + b)(a + b)$ by Example 3 of Section 6.1, and so we have

$$\tfrac{1}{2}(a + b)^2 = \tfrac{1}{2}ab + \tfrac{1}{2}c^2 + \tfrac{1}{2}ab$$

Multiplying through by 2 yields

$$a^2 + 2ab + b^2 = 2ab + c^2$$

or

$$\boxed{a^2 + b^2 = c^2}$$

This last formula is the form in which the theorem of Pythagoras is usually expressed.

exercises 6.2

1

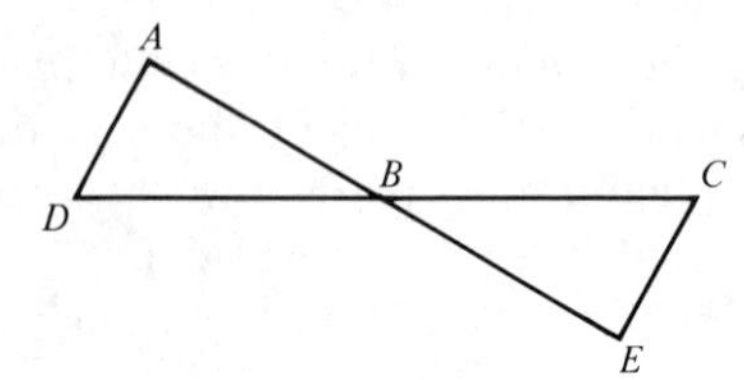

Given: $AB = BE$
$DB = BC$
Prove: $\angle DAB = \angle BEC$

2

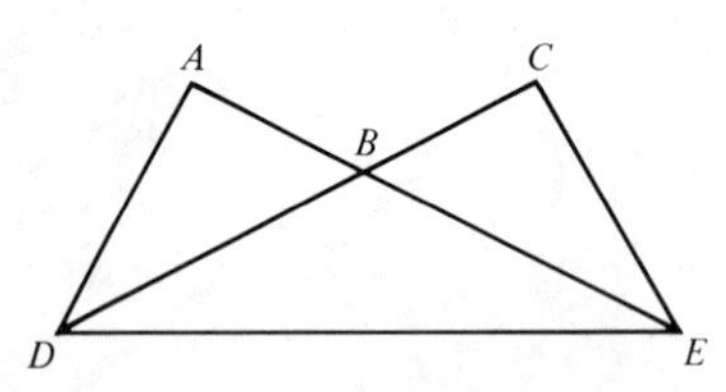

Given: $AD = CE$
$AE = DC$
Prove: $AB = BC$

3

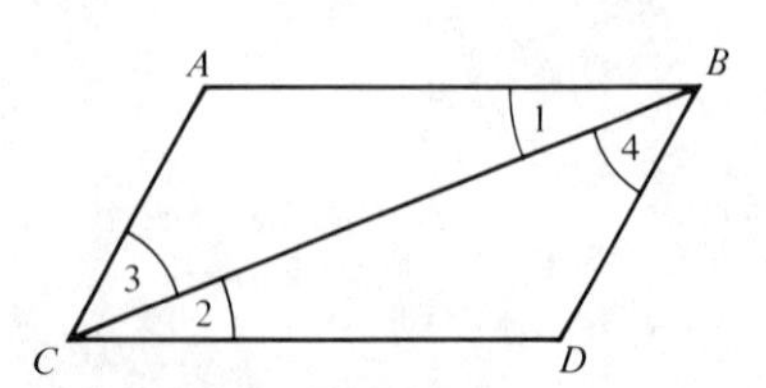

Given: $\angle 1 = \angle 2$
$\angle 3 = \angle 4$
Prove: $AC = BD$
$AB = CD$

4

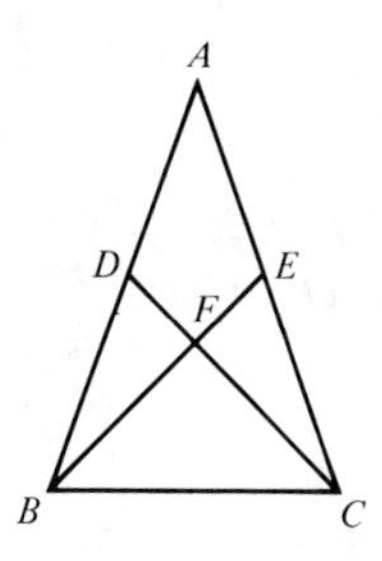

Given: $BD = EC$
$DC = BE$
Prove: $AB = AC$

5

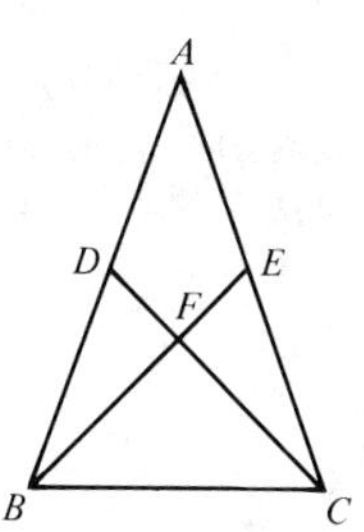

Given: $AB = AC$
$DB = EC$
Prove: $DC = BE$

6

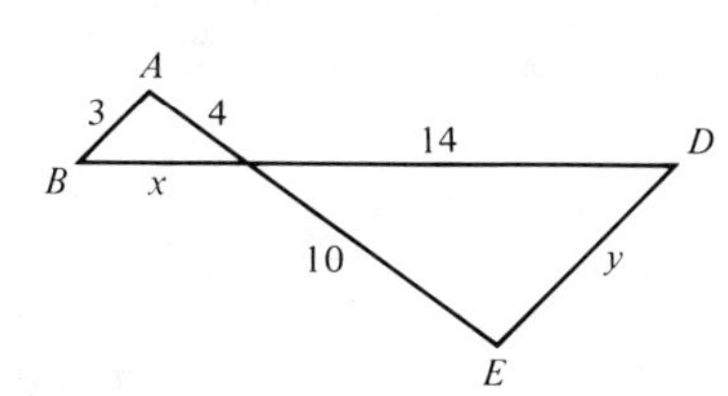

Given: $AB \parallel DE$
Find: x and y

7

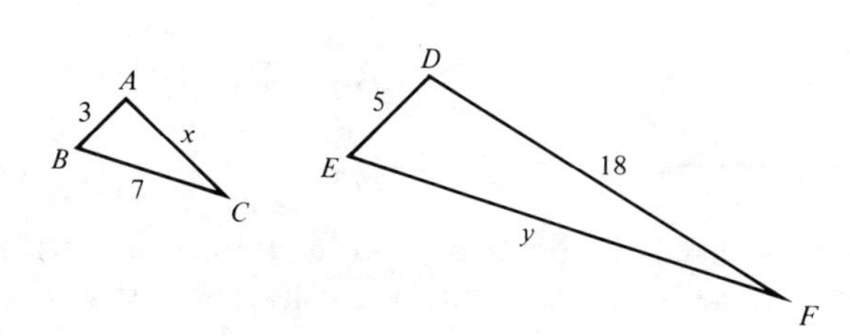

Given: $\triangle ABC \sim \triangle DEF$
Find: x and y

8

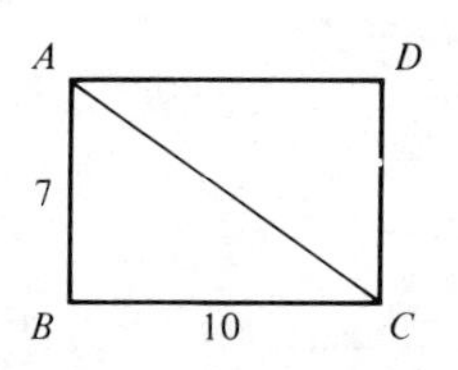

Given: Rectangle $ABCD$
Find: AC

9

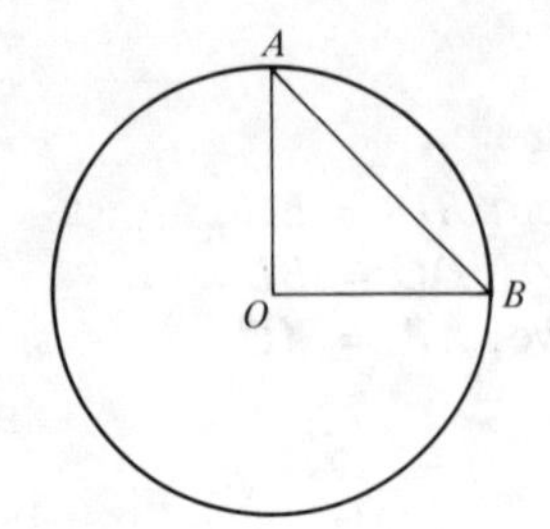

Given: $\angle AOB = 90°$, $OA = 10$
Find: AB

10

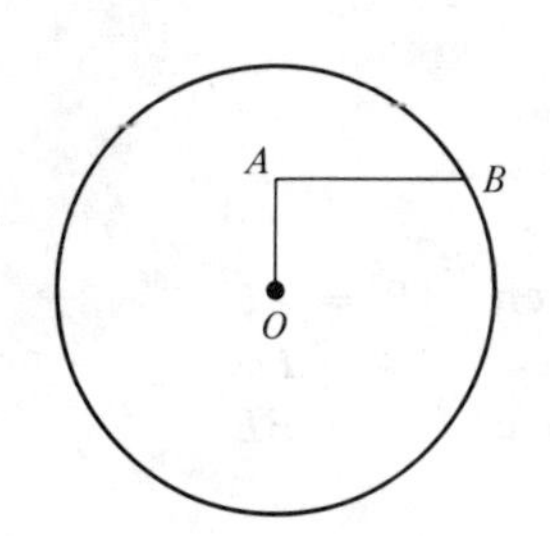

Given: $\angle OAB = 90°$
Radius of circle $= 10$
$AB = 8$
Find: OA

11

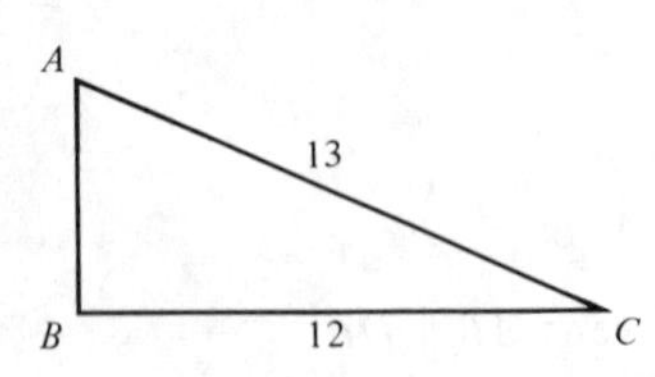

Given: $\angle ABC = 90°$
Find: Area of $\triangle ABC$

12

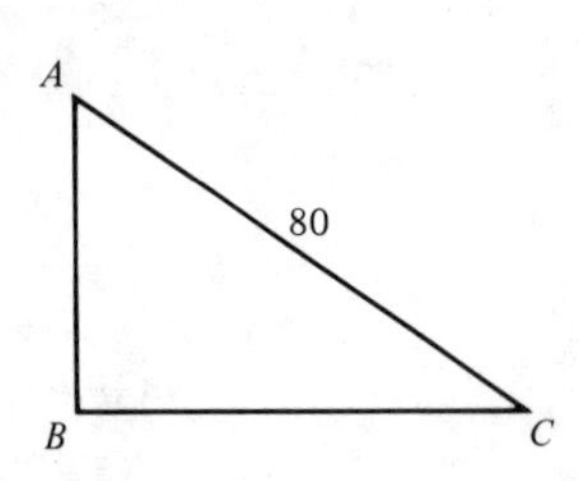

Given: $\angle ABC = 90°$
$BC = 2(AB)$
Find: Area of $\triangle ABC$

13 Prove that the shortest distance from a point to a straight line is along the perpendicular. (*Hint:* Use the fact that the hypotenuse of a right triangle is longer than either of the other two sides.)

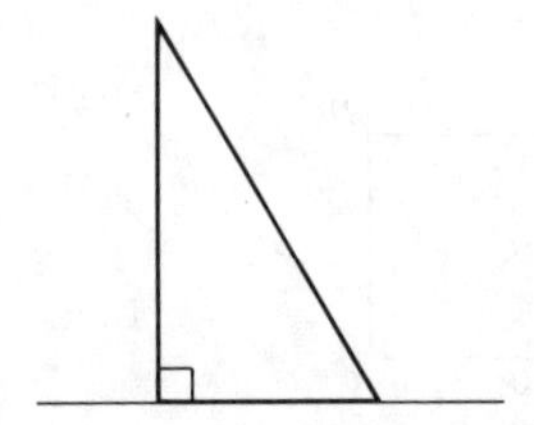

14 Prove that a tangent to a circle is perpendicular to the radius drawn to the point of tangency. A tangent to a circle is a line that intersects the circle in one and only one point and the rest of which lies completely exterior to the circle. (*Hint:* Use problem 13.)

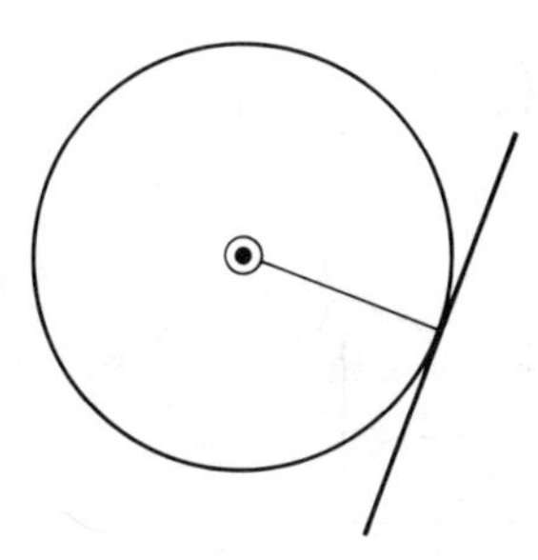

6.3 / solution of right triangles

Trigonometry means the measurement of three-sided figures, that is, the study of triangles. By means of similar triangles, we can find the length or size of certain parts of triangles if we know certain other parts. Suppose that we wish to find the distance across the river from point A to point B in Figure 6.29 without crossing it. Starting at point A we could mark off 100 yd perpendicular to AB along the shoreline to point C, and then sighting along a protractor or using some more sophisticated surveying instrument, we could sight from C to B and measure the angle ACB. Suppose that this angle turns out to be 37°. Take a sheet of paper and draw a right triangle with a 37° angle as in Figure 6.30. This triangle will be similar to the one in Figure 6.29, since two angles of one are equal to two angles of the other. Measure the sides of the triangle that you drew, and let us say that they come out to be 3 in. and 4 in. as indicated in Figure 6.30. We may then write

$$\frac{x \text{ yd}}{100 \text{ yd}} = \frac{3 \text{ in.}}{4 \text{ in.}}$$

figure 6.29

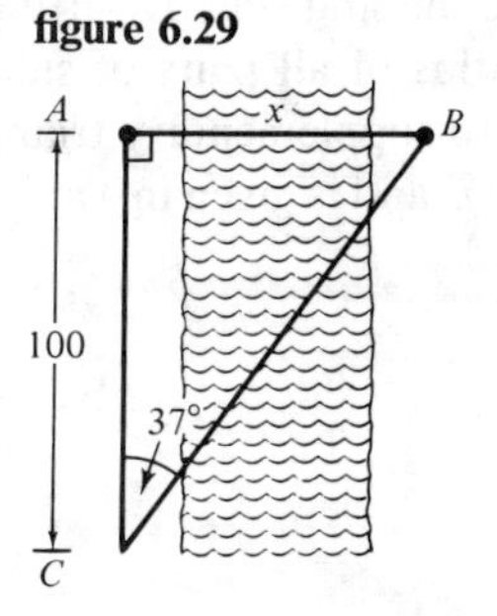

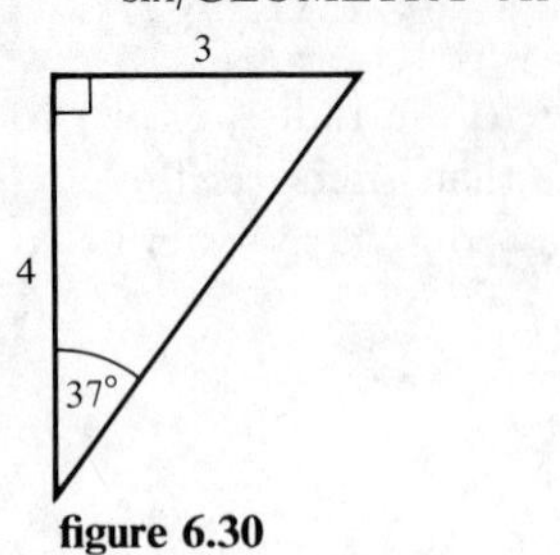

figure 6.30

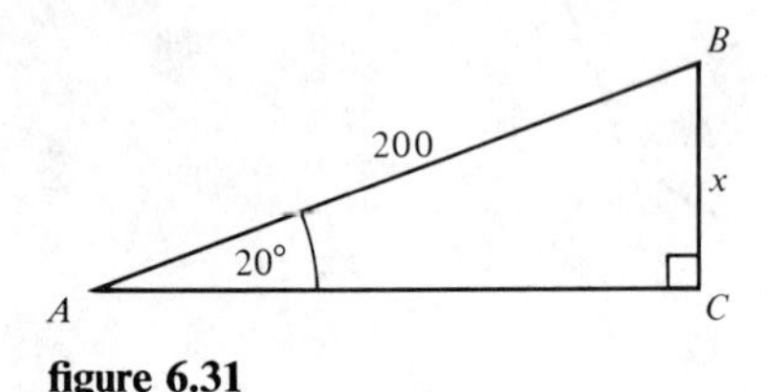

figure 6.31

Therefore, $4x = 300$, $x = 75$ yd, and we have determined the distance across the river without actually measuring it.

Let us consider another example. Suppose that a boy flying a kite has let out 200 ft of string at an angle of 20° as in Figure 6.31, and he wishes to determine the height, x, of his kite above the ground. If he were to draw a right triangle with a 20° angle on a sheet of paper as in Figure 6.32 and measure the sides shown and find that they were 1 in. and 3 in. long, respectively, he could write, since the triangles are similar,

$$\frac{x \text{ ft}}{200 \text{ ft}} = \frac{1 \text{ in.}}{3 \text{ in.}}$$

$$3x = 200$$

$$x = \tfrac{200}{3} = 66\tfrac{2}{3} \text{ ft}$$

which is the height of the kite.

Notice in each of these problems that to find an unknown side of some right triangle, one side of which was known, we drew a similar right triangle, determined the ratio of the two corresponding sides, and found the missing side by setting up a proportion. Now if we had a table that gave the ratios of all pairs of sides of all right triangles, then we would not have to draw these supplementary triangles.

Such a table is called a table of *trigonometric functions* and is given in Table 6.1.

figure 6.32

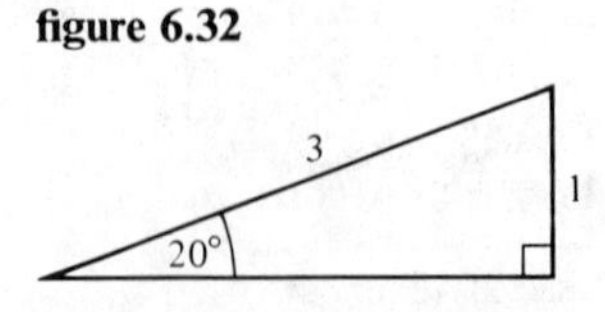

table 6.1 trigonometric functions

A	sin A	cos A	tan A	cot A	sec A	csc A	A
0°	0.0000	1.0000	0.0000	∞	1.000	∞	90°
1	0.0175	0.9998	0.0175	57.290	1.000	57.30	89
2	0.0349	0.9994	0.0349	28.636	1.001	28.65	88
3	0.0523	0.9986	0.0524	19.081	1.001	19.11	87
4	0.0698	0.9976	0.0699	14.301	1.002	14.34	86
5°	0.0872	0.9962	0.0875	11.430	1.004	11.47	85°
6	0.1045	0.9945	0.1051	9.5144	1.006	9.567	84
7	0.1219	0.9925	0.1228	8.1443	1.008	8.206	83
8	0.1392	0.9903	0.1405	7.1154	1.010	7.185	82
9	0.1564	0.9877	0.1584	6.3138	1.012	6.392	81
10°	0.1736	0.9848	0.1763	5.6713	1.015	5.759	80°
11	0.1908	0.9816	0.1944	5.1446	1.019	5.241	79
12	0.2079	0.9781	0.2126	4.7046	1.022	4.810	78
13	0.2250	0.9744	0.2309	4.3315	1.026	4.445	77
14	0.2419	0.9703	0.2493	4.0108	1.031	4.134	76
15°	0.2588	0.9659	0.2679	3.7321	1.035	3.864	75°
16	0.2756	0.9613	0.2867	3.4874	1.040	3.628	74
17	0.2924	0.9563	0.3057	3.2709	1.046	3.420	73
18	0.3090	0.9511	0.3249	3.0777	1.051	3.236	72
19	0.3256	0.9455	0.3443	2.9042	1.058	3.072	71
20°	0.3420	0.9397	0.3640	2.7475	1.064	2.924	70°
21	0.3584	0.9336	0.3839	2.6051	1.071	2.790	69
22	0.3746	0.9272	0.4040	2.4751	1.079	2.669	68
23	0.3907	0.9205	0.4245	2.3559	1.086	2.559	67
24	0.4067	0.9135	0.4452	2.2460	1.095	2.459	66
25°	0.4226	0.9063	0.4663	2.1445	1.103	2.366	65°
26	0.4384	0.8988	0.4877	2.0503	1.113	2.281	64
27	0.4540	0.8910	0.5095	1.9626	1.122	2.203	63
28	0.4695	0.8829	0.5317	1.8807	1.133	2.130	62
29	0.4848	0.8746	0.5543	1.8040	1.143	2.063	61
30°	0.5000	0.8660	0.5774	1.7321	1.155	2.000	60°
31	0.5150	0.8572	0.6009	1.6643	1.167	1.942	59
32	0.5299	0.8480	0.6249	1.6003	1.179	1.887	58
33	0.5446	0.8387	0.6494	1.5399	1.192	1.836	57
34	0.5592	0.8290	0.6745	1.4826	1.206	1.788	56
35°	0.5736	0.8192	0.7002	1.4281	1.221	1.743	55°
36	0.5878	0.8090	0.7265	1.3764	1.236	1.701	54
37	0.6018	0.7986	0.7536	1.3270	1.252	1.662	53
38	0.6157	0.7880	0.7813	1.2799	1.269	1.624	52
39	0.6293	0.7771	0.8098	1.2349	1.287	1.589	51
40°	0.6428	0.7660	0.8391	1.1918	1.305	1.556	50°
41	0.6561	0.7547	0.8693	1.1504	1.325	1.524	49
42	0.6691	0.7431	0.9004	1.1106	1.346	1.494	48
43	0.6820	0.7314	0.9325	1.0724	1.367	1.466	47
44	0.6947	0.7193	0.9657	1.0355	1.390	1.440	46
45°	0.7071	0.7071	1.0000	1.0000	1.414	1.414	45°
A	cos A	sin A	cot A	tan A	csc A	sec A	A

To explain how a table of trigonometric functions is used, we consider the right triangle ABC of Figure 6.33 with right angle C and fix our attention on angle A. The long side AB of the triangle, opposite the right angle, is called the *hypotenuse*; the side AC, which is part of angle A, is called the *adjacent* side; and the side, BC, opposite angle A, is called the *opposite* side. There are six ratios that may be formed with these three sides. Each of these ratios has a name, as indicated in the following list:

$$\text{sine } A = \frac{\text{opposite side}}{\text{hypotenuse}}$$

$$\text{cosine } A = \frac{\text{adjacent side}}{\text{hypotenusc}}$$

$$\text{tangent } A = \frac{\text{opposite side}}{\text{adjacent side}}$$

$$\text{cotangent } A = \frac{\text{adjacent side}}{\text{opposite side}}$$

$$\text{secant } A = \frac{\text{hypotenuse}}{\text{adjacent side}}$$

$$\text{cosecant } A = \frac{\text{hypotenuse}}{\text{opposite side}}$$

These six ratios are referred to as the six trigonometric functions, and are usually abbreviated as follows:

$$\sin A = \frac{\text{opp}}{\text{hyp}}$$

$$\cos A = \frac{\text{adj}}{\text{hyp}}$$

$$\tan A = \frac{\text{opp}}{\text{adj}}$$

$$\cot A = \frac{\text{adj}}{\text{opp}}$$

$$\sec A = \frac{\text{hyp}}{\text{adj}}$$

$$\csc A = \frac{\text{hyp}}{\text{opp}}$$

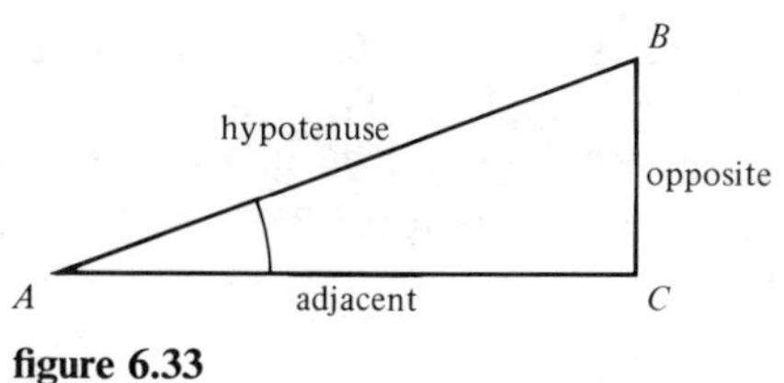

figure 6.33

If we were interested in the six trigonometric functions of angle B, the roles of the adjacent and opposite sides would be interchanged.

Since the two acute angles of a right triangle add up to 90°, we see that for any acute angle A, we have $\sin A = \cos(90° - A)$, $\tan A = \cot(90° - A)$, and $\sec A = \csc(90° - A)$[1]. Because of this the table need only include angles from 0 to 45°. For an angle between 45 and 90°, we look up 90° minus the angle and use the complementary function. Thus to find tan 72° we would look up cot 18°, for cos 64° we would use sin 26°, etc. In other words, for angles between 0 and 45°, find the angle in the left-hand column and the function on top; for angles between 45 and 90°, find the angle in the right-hand column and the function on the bottom.

/examples/

1 Find sin 23°.

► We look for 23° in the left-hand column and under sin at the top, we find that sin 23° = 0.3907.

2 Find cos 68°.

► We look for 68° in the right-hand column and above cos at the bottom, we find that cos 68° = 0.3746.

[1]The cosine, cotangent, and cosecant are sometimes referred to as the complementary sine, complementary tangent, and complementary secant, respectively. This is because two angles whose sum is 90° are called *complementary* angles. Similarly, two angles whose sum is 180° are called *supplementary* angles. Individual angles may also be classified according to the following chart:

value of angle	*name of angle*
0–90°	acute
90°	right
90–180°	obtuse
180°	straight
180–360°	reflex

Let us work the two examples of Figures 6.29 and 6.31 using the table of trigonometric functions instead of drawing the supplementary triangles.

In the first example, we find by referring to Figure 6.29 that $\tan C = x/100$ or $\tan 37° = x/100$. Referring to Table 6.1, we see that $\tan 37° = 0.7536$, so we have the equation

$$\frac{x}{100} = 0.7536$$

which we solve to obtain $x = 75.36$.

For the second example we see, by referring to Figure 6.31, that $x/200 = \sin 20°$, and, from Table 6.1, $\sin 20° = 0.3420$. This results in the equation $x/200 = 0.3420$, and so $x = 68.4$. This is the more accurate answer, because it is possible to be a bit off in measuring the sides of a triangle with a ruler.

The procedure for finding an unknown side of a right triangle in which one side and one acute angle are known may be summarized as follows:

1. Write the unknown side over the known side and determine which of the six functions this represents for the known acute angle.
2. Look up the value of this function for the known angle in Table 6.1.
3. This yields an equation of the form $x/a = b$ where a and b are known numbers and which may then be solved for the unknown side x.

/ examples /

3 A pole casts a shadow of 12 ft and the angle of elevation up to the sun from the tip of the shadow is 68°. How tall is the pole?

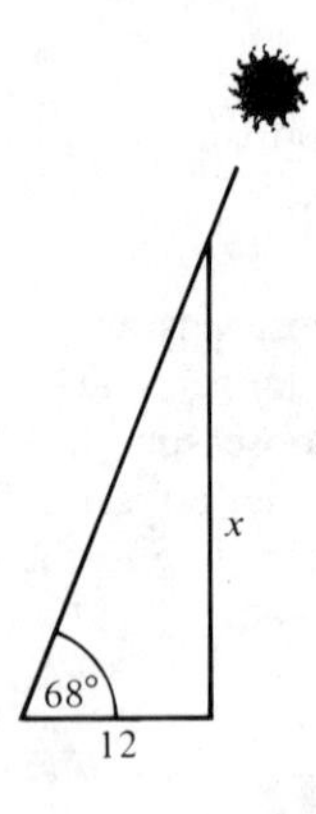

► We write $x/12 = \tan 68°$. From Table 6.1, we see that $\tan 68° = 2.4751$. This gives the equation $x/12 = 2.4751$, which gives $x = 29.7$ ft, the height of the pole.

4 A hill ascends at an angle of 26°. The top of the hill is 28 ft above the ground. How far must one walk in going from the bottom of the hill to the top?

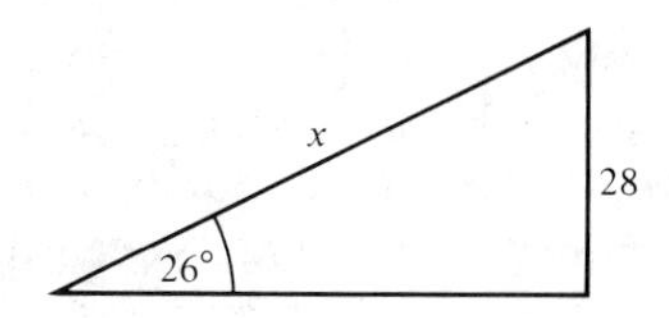

► From the diagram we see that $x/28 = \csc 26°$. From Table 6.1 we find that $\csc 26° = 2.2810$. This gives the equation $x/28 = 2.2810$, which gives $x = 63.87$ ft.

If two of the three sides of a right triangle are known, we may determine the third side by the theorem of Pythagoras. One of the acute angles may then be found using Table 6.1, and the other acute angle will be 90° minus the known one. When we find all the missing parts of a right triangle, we say that we are *solving the right triangle.* This procedure is illustrated in the two examples that follow.

examples

5 Solve the right triangle for which $C = 90°$, $b = 80$, and $c = 100$.

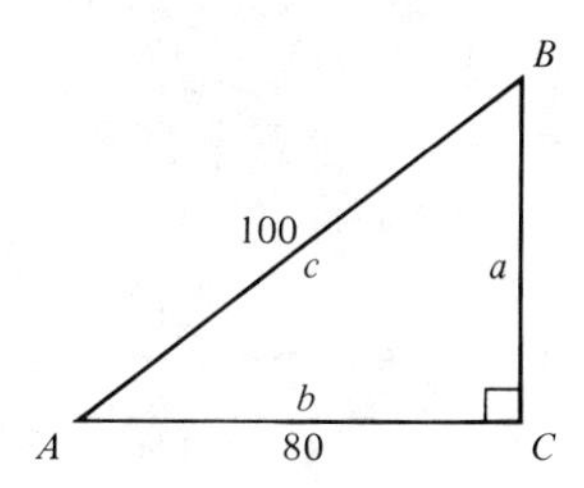

► By the theorem of Pythagoras, we have $a^2 = c^2 - b^2 = (100)^2 - (80)^2 = 10000 - 6400 = 3600$; therefore $a = 60$. Then $a/c = \sin A = \frac{60}{100} = 0.6000$. Looking in Table 6.1 in the sin column, we see that $\sin 36° = 0.5878$ and $\sin 37° = 0.6018$; because A is closer to 37°, we would write the answer as $A = 37°$. Then $B = 90° - 37° = 53°$.

6 Solve the right triangle for which $C = 90°$, $c = 15$, and $A = 75°$.

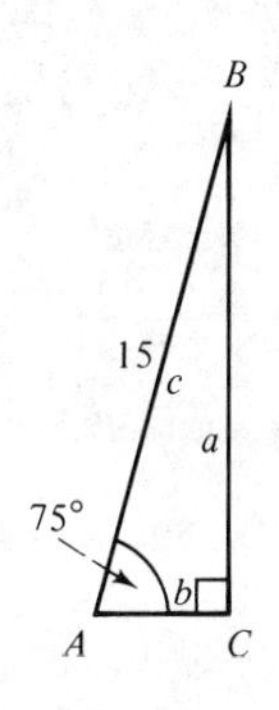

► We have immediately $B = 90° - 75° = 15°$. To find side a, we write $a/15 = \sin 75°$, $a/15 = 0.9659$, $a = 14.5$, $b/15 = \cos 75°$, $b/15 = 0.2588$, $b = 3.9$. Side b could also have been found by the theorem of Pythagoras, but this would involve more computation. We can check our answer, however, by substituting in $a^2 + b^2 = c^2$, which gives $(14.5)^2 + (3.9)^2 = 15^2$.

Some tables of the trigonometric functions do not give all six of them. They may merely give the values for sin, cos, and tan of an angle. This need not stop us from solving trigonometric problems, however, since there are certain identities that relate the trigonometric functions to one another. If we examine Figure 6.34, we see that

$$\sin A = \frac{a}{c} \qquad \cot A = \frac{b}{a}$$

$$\cos A = \frac{b}{c} \qquad \sec A = \frac{c}{b}$$

$$\tan A = \frac{a}{b} \qquad \csc A = \frac{c}{a}$$

It is easy to see that

$$\cot A = \frac{b}{a} = \frac{1}{\frac{a}{b}} = \frac{1}{\tan A}$$

$$\sec A = \frac{c}{b} = \frac{1}{\frac{b}{c}} = \frac{1}{\cos A}$$

$$\csc A = \frac{c}{a} = \frac{1}{\frac{a}{c}} = \frac{1}{\sin A}$$

$$\frac{\sin A}{\cos A} = \frac{\frac{a}{c}}{\frac{b}{c}} = \frac{a}{c} \cdot \frac{c}{b} = \frac{a}{b} = \tan A$$

figure 6.34

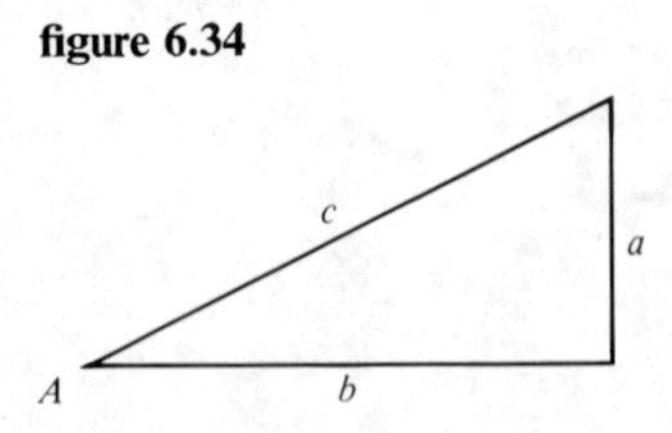

and

$$\frac{\cos A}{\sin A} = \frac{\frac{b}{c}}{\frac{a}{c}} = \frac{b}{c} \cdot \frac{c}{a} = \frac{b}{a} = \cot A$$

By making use of the theorem of Pythagoras, which tells us that $a^2 + b^2 = c^2$ in Figure 6.34, it is also easy to see that

$$\sin^2 A^1 + \cos^2 A = \frac{a^2}{c^2} + \frac{b^2}{c^2} = \frac{a^2 + b^2}{c^2} = \frac{c^2}{c^2} = 1$$

$$\tan^2 A + 1 = \frac{a^2}{b^2} + 1 = \frac{a^2 + b^2}{b^2} = \frac{c^2}{b^2} = \sec^2 A$$

and

$$\cot^2 A + 1 = \frac{b^2}{a^2} + 1 = \frac{b^2 + a^2}{a^2} = \frac{c^2}{a^2} = \csc^2 A$$

These eight formulas, which are summarized below for reference, are known as the *basic trigonometric identities* and are true for any angle, A[2] (see Exercise 4 of Section 6.4).

$$\cot A = \frac{1}{\tan A} \qquad \cot A = \frac{\cos A}{\sin A}$$

$$\sec A = \frac{1}{\cos A} \qquad \sin^2 A + \cos^2 A = 1$$

$$\csc A = \frac{1}{\sin A} \qquad \tan^2 A + 1 = \sec^2 A$$

$$\tan A = \frac{\sin A}{\cos A} \qquad \cot^2 A + 1 = \csc^2 A$$

By means of these eight basic identities, other trigonometric identities may be proved.

[1]This is the notation used for $(\sin A)^2$. Similarly, $\cos^2 A$ stands for $(\cos A)^2$, etc.
[2]We do have to exclude, however, those values of the angle that result in undefined expressions. For example, $\tan^2 A + 1 = \sec^2 A$ does not make sense when $A = 90°$.

examples

7 Prove the identity $\sec A \cot A = \csc A$.

► We substitute $1/\cos A$ for $\sec A$ and $\cos A/\sin A$ for $\cot A$ on the left-hand side and get

$$\frac{1}{\cos A} \cdot \frac{\cos A}{\sin A}$$

which reduces to $1/\sin A$ or $\csc A$; this agrees with the right-hand side, so we have proved the identity.

8 Prove the identity $\sin A = \csc A - \cot A \cos A$.

► We substitute $1/\sin A$ for $\csc A$ and $\cos A/\sin A$ for $\cot A$ on the right-hand side and get

$$\frac{1}{\sin A} - \left(\frac{\cos A}{\sin A}\right)(\cos A)$$

We can reduce this to $(1 - \cos^2 A)/\sin A$, which is equal to $\sin^2 A/\sin A$, and then we can cancel $\sin A$ from the top and bottom. This gives $\sin A$, which agrees with the left-hand side, so we have proved the identity.

9 Prove the identity

$$\frac{1 - \sin A}{\cos A} = \frac{\cos A}{1 + \sin A}$$

► If we cross-multiply, we get $1 - \sin^2 A = \cos^2 A$, which is one of our basic identities. This proves that the original identity is true.

10 Prove the identity

$$\frac{\sec A + \csc A}{\tan A + \cot A} = \sin A + \cos A$$

► If we replace $\sec A$ by $1/\cos A$, $\csc A$ by $1/\sin A$, $\tan A$ by $\sin A/\cos A$, and $\cot A$ by $\cos A/\sin A$, we get on the left-hand side

$$\frac{\dfrac{1}{\cos A} + \dfrac{1}{\sin A}}{\dfrac{\sin A}{\cos A} + \dfrac{\cos A}{\sin A}}$$

If we now multiply numerator and denominator by $\cos A \sin A$, we get $(\sin A + \cos A)/(\sin^2 A + \cos^2 A)$. Since $\sin^2 A + \cos^2 A = 1$, this reduces to $\sin A + \cos A$, which agrees with what was on the right-hand side in the original identity, so we have proved the identity.

A trigonometric identity may be proved by working on the left side and showing that it is equal to the right side, by working on the right side and showing that it is equal to the left side, or by working on both sides simultaneously and getting them to come out equal.

exercises 6.3

1 If you walk 50 ft away from the base of a building, the angle of elevation when you look up at the top is 62°. How tall is the building?

2 A pole casts a shadow 10 ft long, and the angle of elevation of the sun from the tip of the shadow is 42°. How tall is the pole?

3 A boy flying a kite has let out 150 ft of string at an angle of 34°. How far above the ground is the kite?

4 A building is 200 ft tall. If you stand at a distance from its base and look up at the top, the angle of elevation is 55°. How far from the base are you?

5 A boy is flying a kite that is 120 ft above the ground. The string ascends at an angle of 40°. How much string has he let out?

6 A side of a square is 8 ft long. How long is the diagonal?

7 Solve the right triangle for which $C = 90°$, $b = 12$, and $B = 34°$.

8 Solve the right triangle for which $C = 90°$, $a = 5$, and $b = 12$.

9 Solve the right triangle for which $C = 90°$, $A = 67°$, and $a = 5$.

10 Prove the following identities:

(a) $\cos A + \sin A \tan A = \sec A$

(b) $\sin A \sec A \cot A = 1$

(c) $\sec^2 A \csc^2 a = \sec^2 A + \csc^2 A$

(d) $\dfrac{\sin A}{\csc A} + \dfrac{\cos A}{\sec A} = 1$

(e) $\dfrac{1}{\sec A + \tan A} = \sec A - \tan A$

(f) $\dfrac{\sin A + \tan A}{\cot A + \csc A} = \sin A \tan A$

11 Plot the graph of the equation $y = \sin x$. On the x axis use the values 0, 30, 60, 90, 120, up to 360°, and also -30, -60, -90, -120, up to $-360°$, and on the y axis use the values -1, $-\frac{3}{4}$, $-\frac{1}{2}$, $-\frac{1}{4}$, 0, $\frac{1}{4}$, $\frac{1}{2}$, $\frac{3}{4}$, and 1.

12 Plot the graph of the equation $y = \cos x$.

13 Plot the graph of the equation $y = \tan x$.

6.4 / solutions of arbitrary triangles

Our definitions of the trigonometric functions of an angle required that the angle be part of a right triangle, and so these functions were defined only for angles between 0 and 90°. An arbitrary triangle, however, could have an angle larger

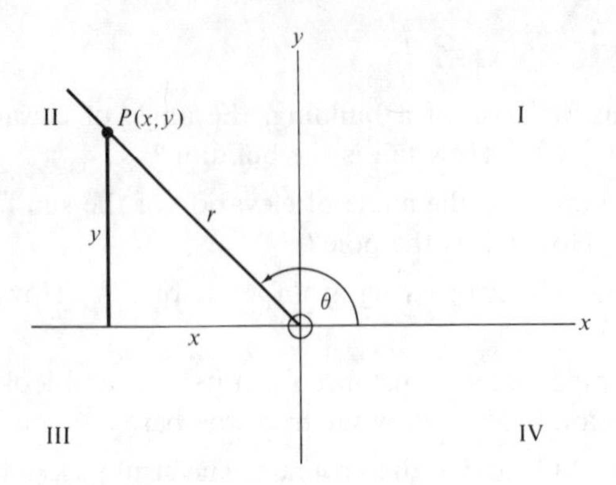

figure 6.35

than 90°, and in solving arbitrary triangles, we will see that at times we will have to determine the trigonometric functions of such angles. Our first task in this section will therefore be to extend the definitions of the trigonometric functions to arbitrary angles.

To do this we set up a rectangular coordinate system, as in Figure 6.35, take the positive x axis as one side of the angle for which we wish to define the trigonometric functions, measure the angle in a counterclockwise direction from the positive x axis, and then draw the other side of the angle starting from the origin. We pick any point, P, with coordinates (x, y) on this second line, and we let r be the distance (always considered positive) from the origin to P. If we call the angle θ, then the six trigonometric functions of θ are defined as follows:

$$\sin\theta = \frac{y}{r} \qquad \cot\theta = \frac{x}{y}$$

$$\cos\theta = \frac{x}{r} \qquad \sec\theta = \frac{r}{x}$$

$$\tan\theta = \frac{y}{x} \qquad \csc\theta = \frac{r}{y}$$

It does not matter where we pick the point P on the second side of the angle; the values of the six trigonometric functions are always the same, because corresponding sides of similar triangles are proportional.

The x and y axes divide the plane into four portions called *quadrants*. We label these quadrants I, II, III, and IV as in Figure 6.35. Although r is always positive, x is negative in some quadrants and y is negative in some quadrants, so the trigonometric functions are not always positive. The signs of these functions in each quadrant may be deduced from their definitions and are given in Table 6.2.

table 6.2

	quadrant			
function	I (0–90°)	II (90–180°)	III (180–270°)	IV (270–360°)
sine	+	+	−	−
cosine	+	−	−	+
tangent	+	−	+	−
cotangent	+	−	+	−
secant	+	−	−	+
cosecant	+	+	−	−

Notice that when θ is in the first quadrant, the new definition of the trigonometric functions agrees with the old one.

Since the trigonometric tables only give us the values of the trigonometric functions for angles up to 90°, we must devise a rule for determining the values of these functions for angles larger than 90°.

We can see from Figure 6.36 that when θ is in the second quadrant, the angle $180° - \theta$ will have the same values for x, y, and r as θ except that x will have the opposite sign. The numerical values of the trigonometric functions for $180° - \theta$ will therefore be the same as those for θ, but their signs may not be the same.

When θ is in the third quadrant, we see from Figure 6.37 that $\theta - 180°$ has the same numerical values for its trigonometric functions as θ except possibly for the signs, and when θ is in the fourth quadrant, we see from Figure 6.38 that $360° - \theta$ has the same numerical values for its trigonometric functions as θ except possibly for the signs.

figure 6.36

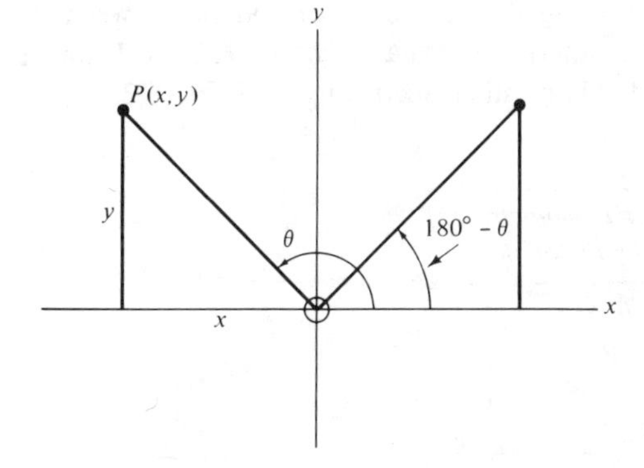

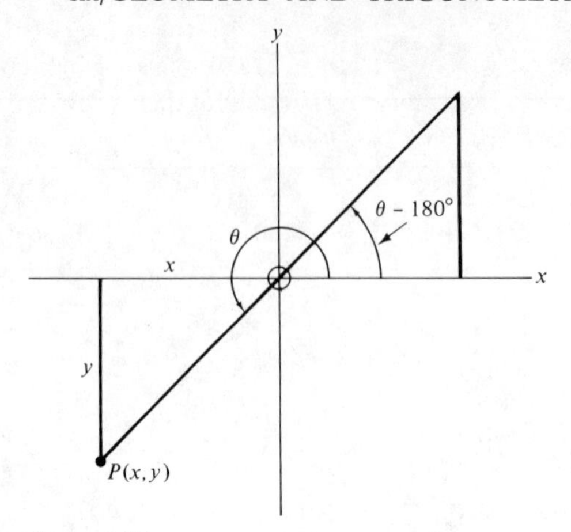

figure 6.37

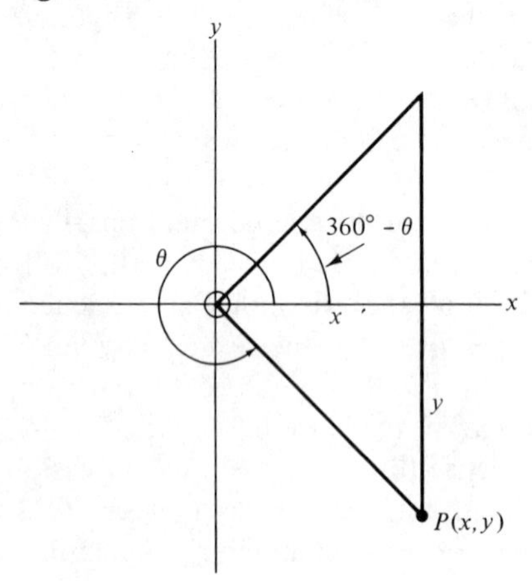

figure 6.38

Thus we can determine the value of any trigonometric function for any angle between 0 and 360° by first determining the corresponding angle between 0 and 90° from Table 6.3, looking up the function for that angle in Table 6.1, and then putting the proper sign in front of this number according to Table 6.2.

table 6.3

when θ is in quadrant	*look up the trigonometric function for the angle*
I	θ
II	$180°-\theta$
III	$\theta-180°$
IV	$360°-\theta$

If it should ever become necessary to determine a trigonometric function of an angle larger than 360°, we divide that angle by 360° and then look up the trigonometric function of the remainder.

/examples/

1 Determine sin 238°.

► 238° is in the third quadrant, so we use 238° − 180° = 58°. From Table 6.1, sin 58° = 0.8480. From Table 6.2 sin is negative in the third quadrant, so sin 238° = −0.8480.

2 Determine cot 314°.

► 314° is in the fourth quadrant, so we use 360° − 314° = 46°. From Table 6.1, cot 46° = 0.9657. From Table 6.2, cot is negative in the fourth quadrant, so cot 314° = −0.9657.

3 Determine cos 1060°.

► Dividing 1060° by 360° yields a quotient of 2 and a remainder of 340°. Since 340° is in the fourth quadrant, we use 360° − 340° = 20°. Cos 20° = 0.9397, and cos is positive in the fourth quadrant, so cos 1060° = 0.9397.

As we saw in Section 6.3, two triangles are congruent if any one of the conditions *SSS*, *AAS*, or *SAS* are satisfied. This means that if we wish to give enough information about an arbitrary triangle to determine it completely, we must give its three sides, two of its angles and a side, or two of its sides and the included angle. Once one of these sets of information is given, the remaining parts of the triangle may be found by utilizing one or both of two trigonometric formulas, which we now prove.

The Sine Law

In triangle ABC (Figure 6.39), we have $h/c = \sin A$, so $h = c \sin A$. Also, $h/a = \sin C$, so $h = a \sin C$. Therefore, $c \sin A = a \sin C$ or $a/\sin A = c/\sin C$. Stated in words, this says that *the ratio of any side of a triangle to the sine of its*

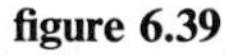

figure 6.39

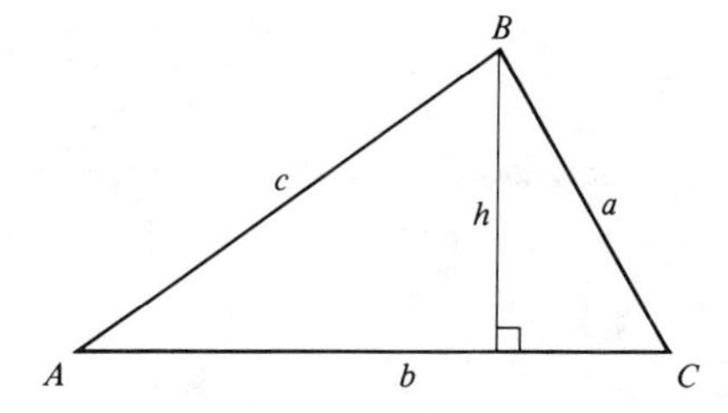

opposite angle is equal to the ratio of any other side to the sine of its opposite angle. The sine law may be written

$$\frac{a}{\sin A} = \frac{b}{\sin B} = \frac{c}{\sin C}$$

The Cosine Law

In triangle ABC (Figure 6.40), we have $d/c = \cos A$, so $d = c \cos A$. Also, $a^2 = h^2 + (b - d)^2 = h^2 + b^2 - 2bd + d^2 = c^2 - d^2 + b^2 - 2bd + d^2 = c^2 + b^2 - 2bd = b^2 + c^2 - 2b(c \cos A)$. This cosine law may thus be written, $a^2 = b^2 + c^2 - 2bc \cos A$. Stated in words, it says that *the square of any side of a triangle is equal to the sum of the squares of the other two sides minus twice the product of these sides times the cosine of their included angle.* Since this is true for any side, the cosine law may be written in any of the three forms

$$\begin{aligned} a^2 &= b^2 + c^2 - 2bc \cos A \\ b^2 &= a^2 + c^2 - 2ac \cos B \\ c^2 &= a^2 + b^2 - 2ab \cos C \end{aligned}$$

Any of these equations may be solved for the cosine of the angle to give the following three alternative forms of the cosine law:

$$\cos A = \frac{b^2 + c^2 - a^2}{2bc}$$

$$\cos B = \frac{a^2 + c^2 - b^2}{2ac}$$

$$\cos C = \frac{a^2 + b^2 - c^2}{2ab}$$

These formulas are sometimes referred to as the alternative form of the cosine law.

figure 6.40

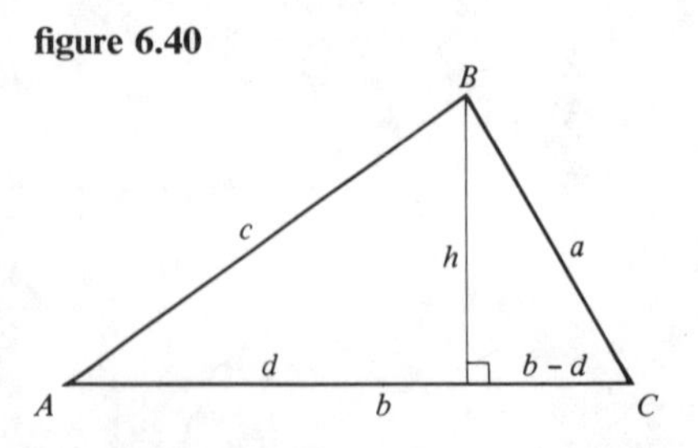

A specific series of steps may now be described for solving a triangle, given any one of the three sets of information *SSS*, *AAS*, or *SAS*:

1. Given *SSS*.

Procedure: Find cos A and cos B using the alternative form of the cosine law. Using Table 6.1, find A and B. (Remember that cosine is positive in the first quadrant and negative in the second, so if the cosine is negative, the value in the table must be subtracted from 180°.) The formula $C = 180° - (A + B)$ then gives the value of C.

2. Given *AAS* (assume b, B, and C are given).

Procedure: Find A from $A = 180° - (B + C)$. Use the sine law $a/\sin A = b/\sin B$ to find a and the sine law $c/\sin C = b/\sin B$ to find c.

3. Given *SAS* (assume A, b, and c are given).

Procedure: Use the cosine law $a^2 = b^2 + c^2 - 2bc \cos A$ to find a. Then use the alternative form of the cosine law to find cos B, and then B. Then C is determined from the formula $C = 180° - (A + B)$.

examples

4 Solve the triangle for which $A = 120°$, $B = 40°$, and $c = 10$.

A
120°
10
40°
C
B

$$C = 180° - (120° + 40°) = 20°$$

$$\frac{a}{\sin A} = \frac{c}{\sin C} \qquad \frac{a}{\sin 120°} = \frac{10}{\sin 20°}$$

$$\frac{a}{0.8660} = \frac{10}{0.3420} \qquad a = \frac{10 \times 0.8660}{0.3420} = \frac{8.66}{0.342} = 25.3$$

$$\frac{b}{\sin B} = \frac{c}{\sin C} \qquad \frac{b}{\sin 40°} = \frac{10}{\sin 20°} \qquad \frac{b}{0.6428} = \frac{10}{0.3420}$$

$$b = \frac{10 \times 0.6428}{0.3420} = \frac{6.428}{0.342} = 18.8$$

5 Solve the triangle for which $a = 6$, $b = 9$, and $c = 12$.

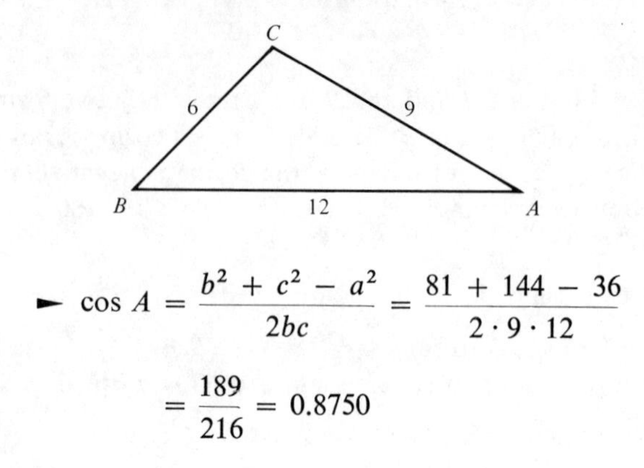

$$\blacktriangleright \cos A = \frac{b^2 + c^2 - a^2}{2bc} = \frac{81 + 144 - 36}{2 \cdot 9 \cdot 12}$$

$$= \frac{189}{216} = 0.8750$$

From Table 6.1, $A \approx 29°$.

$$\cos B = \frac{a^2 + c^2 - b^2}{2ac} = \frac{36 + 144 - 81}{2 \cdot 6 \cdot 12} = \frac{99}{144} = 0.6875$$

From Table 6.1, $B \approx 47°$.

$$C \approx 180° - (29° + 47°) = 104°$$

6 Solve the triangle for which $a = 8$, $b = 5$, and $C = 105°$.

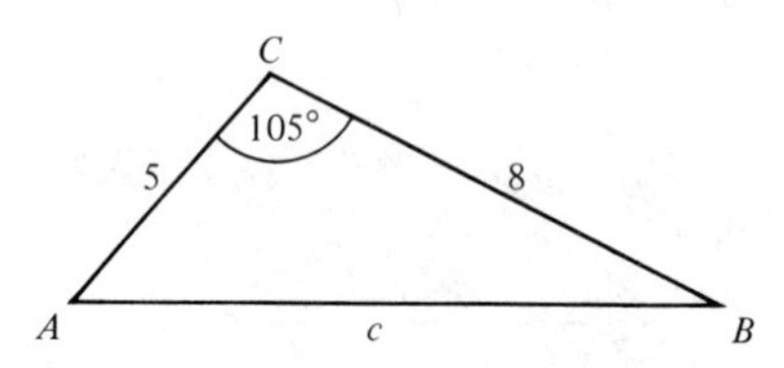

► By the cosine law,

$$c^2 = a^2 + b^2 - 2ab \cos C$$

$$= 64 + 25 - 2 \cdot 8 \cdot 5 \cdot \cos 105°$$

$$= 89 - 80(-0.2588) = 89 + 20.7 = 109.70$$

$$c = \sqrt{109.70} = 10.5$$

$$\cos A = \frac{b^2 + c^2 - a^2}{2bc} = \frac{25 + 109.7 - 64}{2 \cdot 5 \cdot 10.5} = \frac{70.7}{105} = 0.6733$$

$$A \approx 48° \qquad B \approx 180° - (A + B) = 180° - (48° + 105°) = 27°$$

exercises 6.4

1 Evaluate the following trigonometric functions:

(a) sin 290° (b) cot 145° (c) sec 312° (d) tan 138°
(e) csc 215° (f) cos 300° (g) sin 800° (h) tan 950°

2 Solve the triangle that is described by the following data:

(a) $a = 5, b = 8, c = 12$ (b) $A = 30°, B = 80°, b = 10$
(c) $a = 8, b = 6, C = 85°$ (d) $A = 112°, C = 32°, a = 25$
(e) $a = 15, c = 8, B = 140°$ (f) $a = 12, b = 10, c = 17$

3 Three towns, A, B, and C, are joined by roads that run in a straight line. The road from A to B makes an angle of 110° with the road from B to C. A is 12 miles from B, and B is 15 miles from C. What is the distance from A to C?

4 Prove that the eight basic trigonometric identities derived in Section 6.3 are still valid when the angle A is in any of the four quadrants. (*Hint:* Use the general definitions of the six trigonometric functions given in this section and the fact that in all four quadrants, $x^2 + y^2 = r^2$.)

review test 6

NAME ____________

DATE ____________

1 Determine the areas of the following figures:

(a)

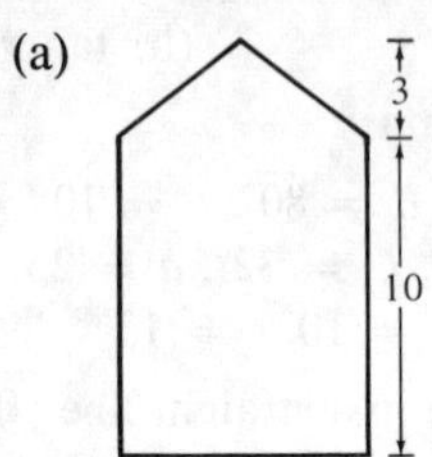

(b)

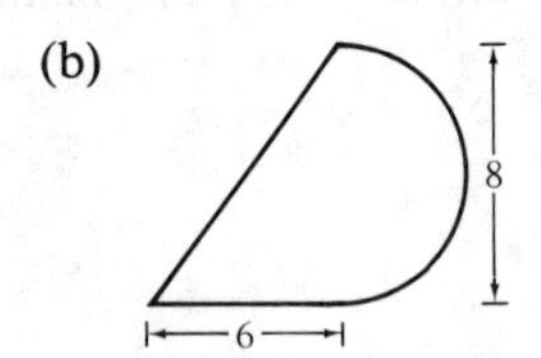

(c)

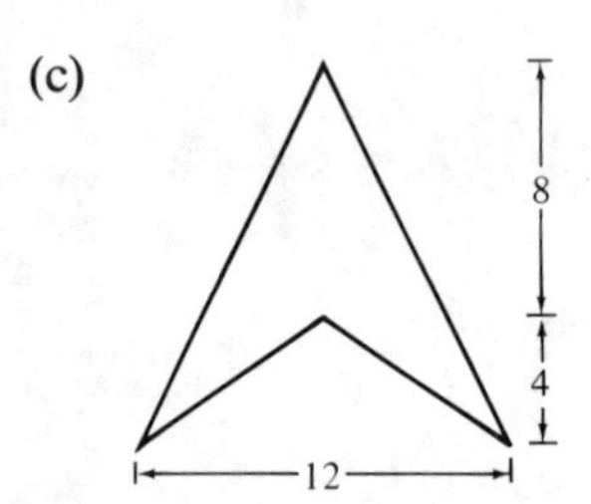

2 What would it cost to paint a semicircular wall of diameter 10 ft if it costs \$0.12 to cover each square foot?

3 Which has the larger area, a square whose side is 9 ft or a circle whose diameter is 10 ft?

4

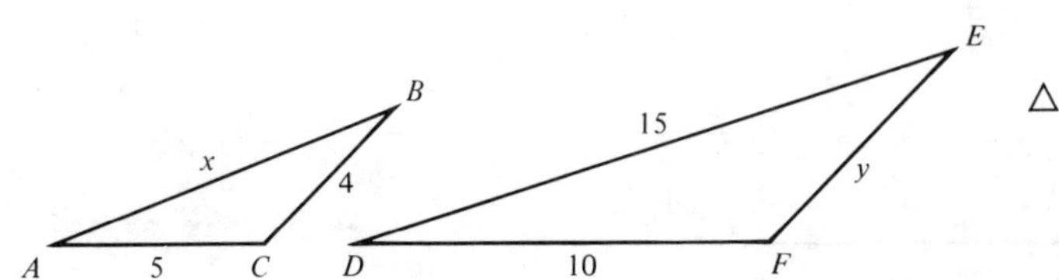

$\triangle$ Given: $\triangle ABC \sim \triangle DEF$
Find: x and y

5

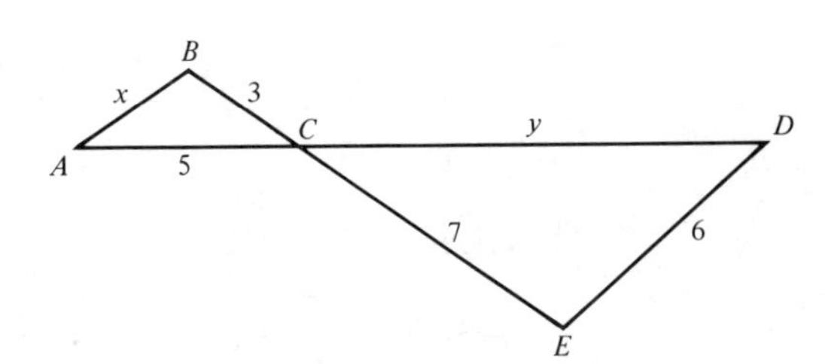

Given: $\angle A = \angle D$
Find: x and y

6

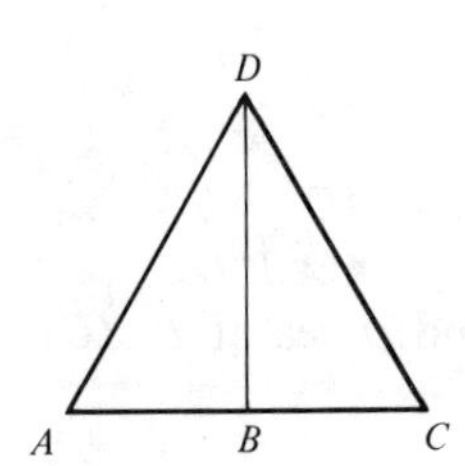

Given: $AD = DC$
$DB \perp AC$
Prove: $AB = BC$

7

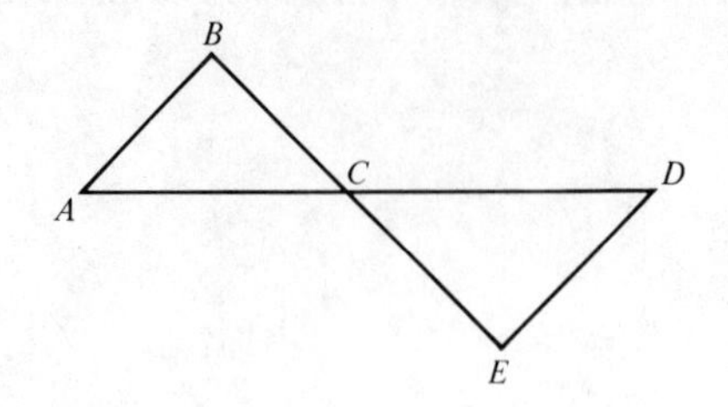

Given: $AB = DE$

$AB \parallel DE$

Prove: $AC = CD$

8

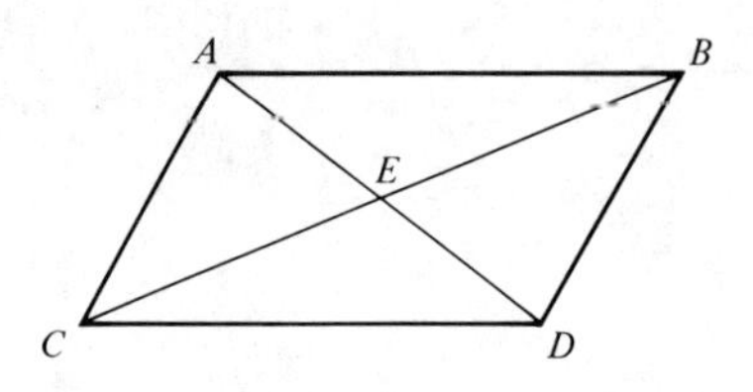

Given: $ABCD$ is a parallelogram

Prove: $AE = ED$

9

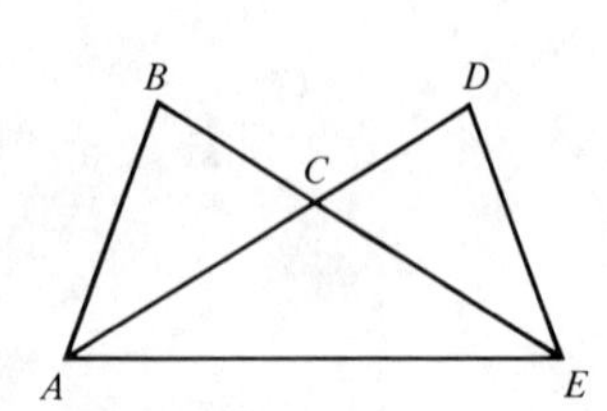

Given: $BC = CD$

$AC = CE$

Prove: $AB = DE$

10

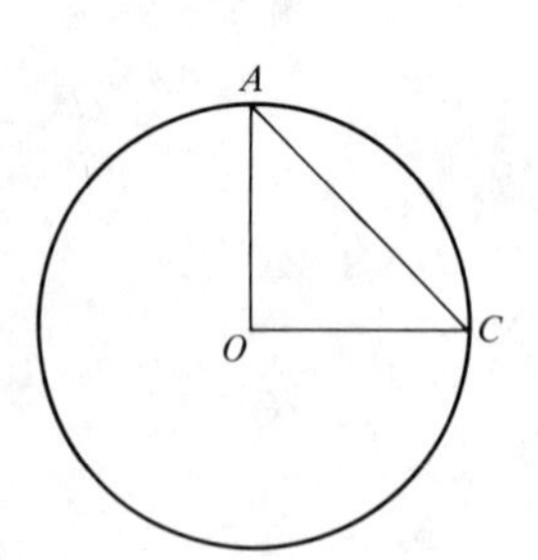

Given: $AO = 6$

$AO \perp OC$

Find: Area of $\triangle AOC$

11

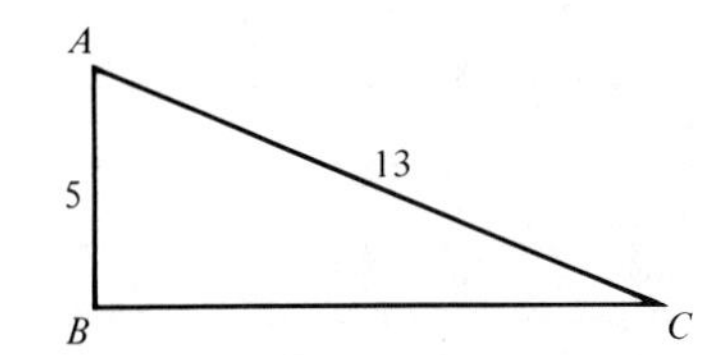

Given: $AB \perp BC$
Find: Area of $\triangle ABC$

12 Solve the right triangle for which $C = 90°$, $A = 22°$, and $a = 12$.

13 Solve the right triangle for which $C = 90°$, $a = 10$, and $b = 24$.

14 If you walk 80 ft away from the base of a building and then look up at the top, the angle of elevation is 75°. How tall is the building?

15 At a certain time a pole 60 ft tall casts a shadow of 42 ft. What is the angle of elevation of the sun?

16 A hill ascends at an angle of 34°. The top of the hill is 60 ft above the ground. What is the distance from the foot of the hill to the top?

17 Evaluate the following trigonometric functions:

(a) $\sin 165°$ __________

(b) $\cos 95°$ __________

(c) $\cot 200°$ __________

(d) $\sin 175°$ __________

(e) $\tan 310°$ __________

(f) $\cos 1200°$ __________

18 Solve the triangle for which $a = 10$, $b = 6$, and $c = 15$.

19 Solve the triangle for which $A = 70°$, $B = 30°$, and $a = 6$.

20 Solve the triangle for which $a = 8$, $b = 6$, and $C = 100°$.

21 Two adjacent sides of a parallelogram are 7 ft and 10 ft long, and they meet at an angle of 40°. How long are the diagonals of the parallelogram?

22 Prove the following identities:

(a) $\sin \theta \cot \theta = \cos \theta$

(b) $\cos \theta(1 + \sec \theta) = 1 + \cos \theta$

(c) $\cos\theta(1 + \tan\theta) = \sin\theta + \cos\theta$

(d) $\tan\theta + \cot\theta = \sec\theta\csc\theta$

(e) $\dfrac{1 + \tan\theta}{1 - \tan\theta} = \dfrac{\cos\theta + \sin\theta}{\cos\theta - \sin\theta}$

(f) $\dfrac{1 + \sec\theta}{\tan\theta} = \dfrac{\cos\theta + 1}{\sin\theta}$

(g) $\dfrac{\sec\theta + 1}{\tan\theta} = \dfrac{\tan\theta}{\sec\theta - 1}$

23 (a) If $\sin\theta = a$, show that $\tan\theta = a/\sqrt{1 - a^2}$.

(b) Find similar expressions for $\cos\theta$, $\cot\theta$, $\sec\theta$, and $\csc\theta$. (*Hint:* Construct a right triangle whose hypotenuse has length 1 and whose side opposite the angle θ has length a.)

24 Find all the values of θ between 0 and 360° for which $2 \sin^2 \theta - 3 \sin \theta + 1 = 0$. (*Hint:* Factor the left-hand side of the equation and then use the table of trigonometric functions to find the values of θ that satisfy the resulting linear equations.)

chapter seven

ANALYTIC GEOMETRY

7.1 / straight lines

We have seen in Chapter Five that a linear equation could be plotted on the coordinate axes and that the resulting curve turned out to be a straight line. Similarly, other equations may be plotted on these coordinate axes. The study of the relationship between equations and their curves is known as *analytic geometry*. We shall concern ourselves in this chapter only with equations in two variables, x and y, and of degrees one and two.

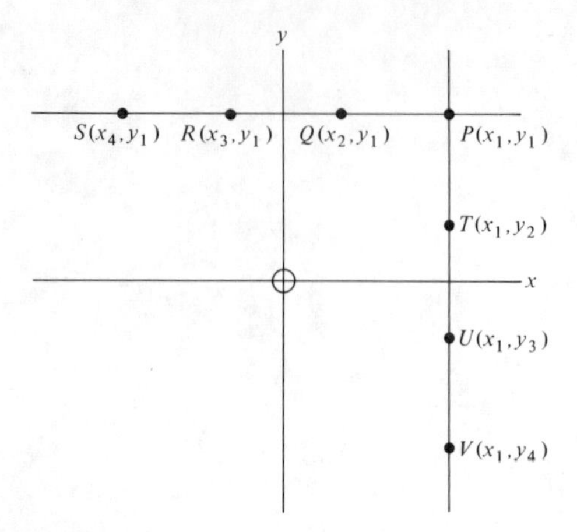

figure 7.1

The portion of a line that goes from one point in the plane to another is called a *line segment*, and our first task will be to derive a formula for the length of a line segment connecting two points. Notice first (Figure 7.1) that all points on a horizontal line have the same y coordinate and that all points on a vertical line have the same x coordinate. The distance between any two points on a horizontal line is the difference of their x coordinates, and the distance between any two points on a vertical line is the difference in their y coordinates. In Figure 7.1, for example, the distance from P to T is $y_2 - y_1$, the distance from V to U is $y_3 - y_4$, the distance from U to V is $y_4 - y_3$, and the distance from R to Q is $x_2 - x_3$. Notice that the distance from V to U is the negative of the distance from U to V.

For two arbitrary points P and Q with coordinates (x_1, y_1) and (x_2, y_2), respectively, as in Figure 7.2, we draw a vertical line through one point and a horizontal line through the other and assume that these meet at the point R. Since R is on the same horizontal line as P, its y coordinate is y_1, and since R is on the same vertical line as Q, its x coordinate is x_2. We make use of the theorem of Pythagoras and have

$$PQ = \sqrt{(PR)^2 + (RQ)^2} = \sqrt{(x_2 - x_1)^2 + (y_2 - y_1)^2}$$

That is, the distance d between the points with coordinates (x_1, y_1) and (x_2, y_2) is given by

$$\boxed{d = \sqrt{(x_2 - x_1)^2 + (y_2 - y_1)^2}}$$

This is known as the *distance formula*.

Next let us determine in Figure 7.3 the coordinates (x, y) of the midpoint M of the line segment PQ halfway between P and Q. We draw a vertical line through

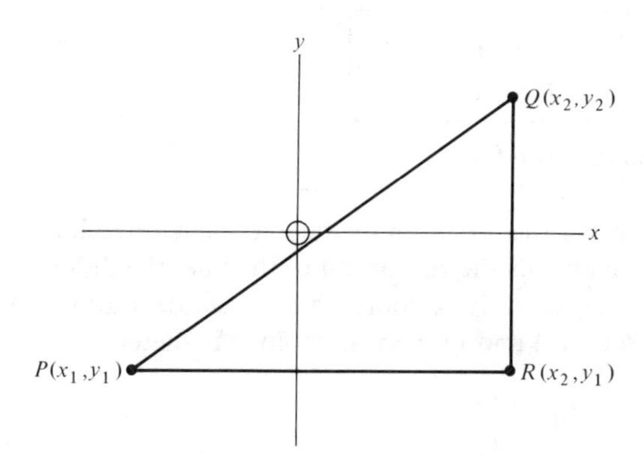

figure 7.2

M meeting PR at N and a horizontal line through M meeting QR at O. Since triangles PMN and PQR are similar, we have

$$\frac{PN}{PR} = \frac{PM}{PQ} = \frac{1}{2}$$

The coordinate x of N must therefore satisfy the equation $x - x_1 = \frac{1}{2}(x_2 - x_1)$, which may be solved $2(x - x_1) = x_2 - x_1$, $2x = x_1 + x_2$, $x = \frac{1}{2}(x_1 + x_2)$.

figure 7.3

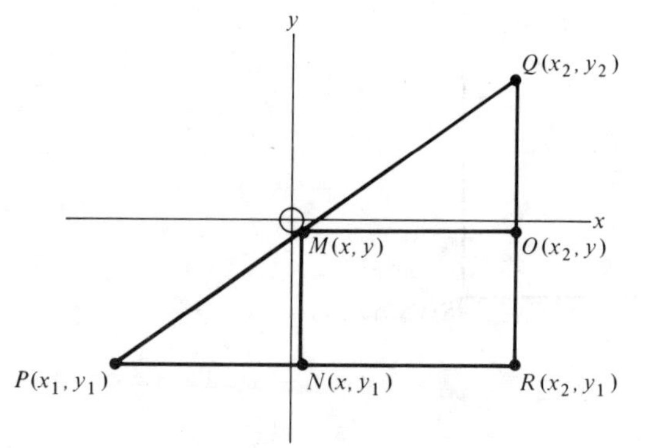

Similarly the coordinate y of O is determined as $y = \frac{1}{2}(y_1 + y_2)$. The coordinates (x, y) of the midpoint M are therefore given by

$$x = \tfrac{1}{2}(x_1 + x_2)$$
$$y = \tfrac{1}{2}(y_1 + y_2)$$

These are known as the *midpoint formulas.*

Two points determine a line, so if we are given the coordinates of two points on a line, we should be able to tell the direction of the line. The direction of the line is described by its *slope*, which tells the rise or fall of the line (the difference in y coordinates) for each unit increase in the x coordinate.[1] If m denotes the slope of a line through the points $P(x_1, y_1)$ and $Q(x_2, y_2)$, its slope is therefore given by

$$m = \frac{y_2 - y_1}{x_2 - x_1}$$

Another way of defining the slope is to say that it is the tangent of the angle that the line makes with a horizontal line measured from the right half of the horizontal line counterclockwise to the given line. Since corresponding angles are equal, it does not matter which horizontal line we use. In Figure 7.4, the slope, m, of the line through P and Q is the tangent of the angle θ, and this is RQ/PR, so once again we have the formula for the slope,

$$m = \frac{y_2 - y_1}{x_2 - x_1}$$

figure 7.4

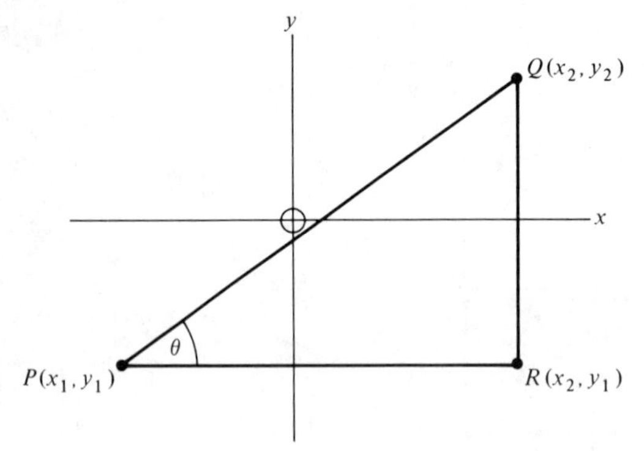

[1]Some people prefer to say that the slope measures the steepness of a line.

As we move from left to right, a line that is rising has a positive slope, and a line that is falling has a negative slope. A horizontal line has slope 0, whereas a vertical line has an infinite slope.

We have already discussed in Chapter Five the problem of plotting a line, given its equation. We turn now to the problem of determining the equation of a line, given sufficient information to fully describe the line.

Remember that the equation of a line is a linear equation in x and y such that the x and y coordinates of any point on the line will satisfy that equation. To determine the equation of a line we therefore pick an arbitrary point on the line, denote its coordinates by (x, y) and try to write a true equation about x and y.

Point-slope form

A line is completely determined if we know one of its points and its slope. In Figure 7.5, the point (x_1, y_1) is a fixed point on the line, the slope of the line is denoted by m, and (x, y) is any point on the line. It is clear that, regardless of the position of the point (x, y), the slope of the line is always given by

$$m = \frac{y - y_1}{x - x_1}$$

Therefore, the equation of the line is given by

$$\boxed{y - y_1 = m(x - x_1)}$$

This is known as the *point-slope form* of the equation of a straight line.

Slope-intercept form

A special case of the point-slope form occurs when the specified point on the line is its y *intercept* (the point where it intersects the y axis). If the y intercept has coordinates $(0, b)$, then the previous equation becomes

$$y - b = m(x - 0)$$

figure 7.5

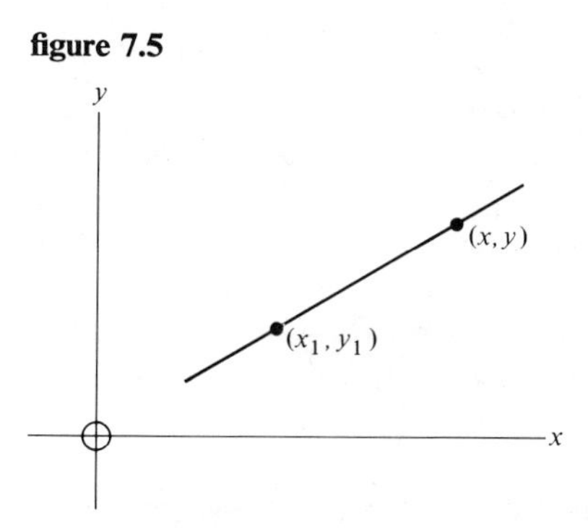

which may be reduced to

$$y = mx + b$$

This is the *slope-intercept form* of the equation of a straight line.

The general equation of a straight line may be written $Ax + By = C$ for real numbers A, B, and C. If this is solved for y, we get

$$y = -\frac{A}{B}x + \frac{C}{B}$$

Thus the slope of the line $Ax + By = C$ is $-(A/B)$, and it intersects the y axis at the point $(0, C/B)$.

Two-point form

A straight line is also completely determined if we are given two of its points. Let (x_1, y_1) and (x_2, y_2) be two fixed points on the straight line in Figure 7.6, and let (x, y) be any point on the line. It is clear that the slope m of the line is given by $m = (y - y_1)/(x - x_1)$ and is also given by $m = (y_1 - y_2)/(x_1 - x_2)$.

We have, therefore, $(y - y_1)/(x - x_1) = (y_1 - y_2)/(x_1 - x_2)$, which may be reduced to

$$y - y_1 = \frac{y_1 - y_2}{x_1 - x_2}(x - x_1)$$

This is known as the *two-point form* of the equation of a straight line. Notice that this reduces to the point-slope form if we replace $(y_1 - y_2)/(x_1 - x_2)$ by m.

figure 7.6

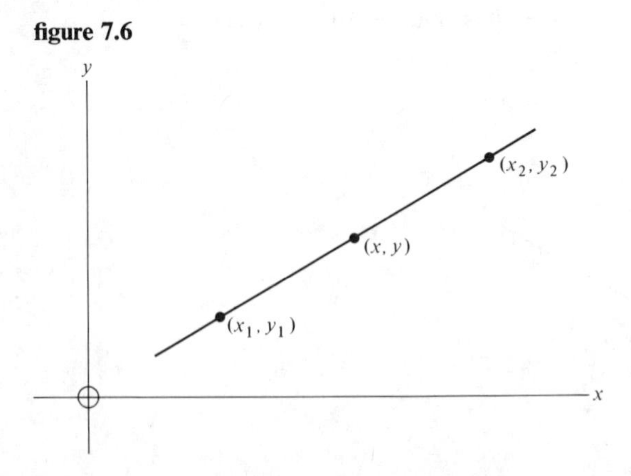

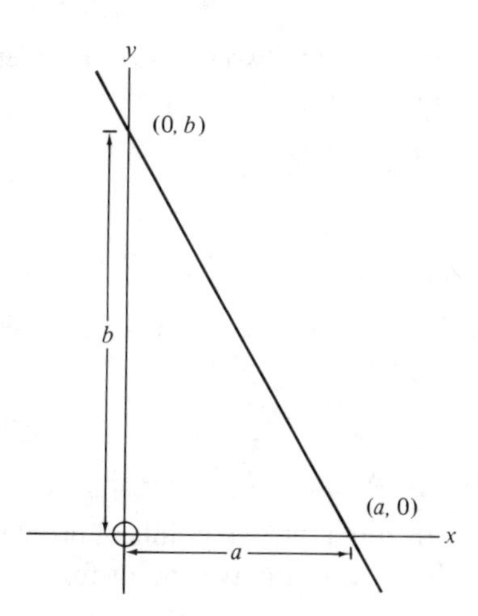

figure 7.7

Intercept form

When the two given points are the *intercepts* of the line (the points where it intersects the x and y axes), we have, calling the x intercept $(a, 0)$ and the y intercept $(0, b)$, $x_1 = a$, $y_1 = 0$, $x_2 = 0$, and $y_2 = b$. Substituting this in the equation above yields $y - 0 = [(0 - b)/(a - 0)](x - a)$, which may be reduced to

$$\boxed{\frac{x}{a} + \frac{y}{b} = 1}$$

This is known as the *intercept form* of the equation of a straight line and is illustrated in Figure 7.7. The equation $Ax + By = C$ may be transformed into

$$\frac{A}{C}x + \frac{B}{C}y = 1$$

so we see that its x and y intercepts are $(C/A, 0)$ and $(0, C/B)$, respectively.

/examples/

1 Find the distance between the points $(-4, 3)$ and $(-1, -6)$.

► Substituting the values $x_1 = -4$, $y_1 = 3$, $x_2 = -1$, and $y_2 = -6$ into the distance formula yields

$$\begin{aligned} d &= \sqrt{[-1 - (-4)]^2 + (-6 - 3)^2} \\ &= \sqrt{3^2 + (-9)^2} = \sqrt{90} \approx 9.5 \end{aligned}$$

2 Find the coordinates of the point halfway between the two points of problem 1.
► Substituting the above values in the midpoint formulas yields

$$x = \tfrac{1}{2}[-4 + (-1)] = -\tfrac{5}{2}$$

$$y = \tfrac{1}{2}[3 + (-6)] = -\tfrac{3}{2}$$

so the coordinates of the midpoint may be written $(-\frac{5}{2}, -\frac{3}{2})$.

3 Determine the slope of the line passing through the two points of problem 1.
► Substituting the coordinates in the slope formula yields

$$m = \frac{-6 - 3}{-1 - (-4)} = \frac{-9}{3} = -3$$

4 Determine the equation of the line passing through the two points of problem 1.
► Substituting the coordinates of problem 1 into the two-point form of the equation of a straight line yields

$$y - 3 = \frac{3 - (-6)}{-4 - (-1)}[x - (-4)]$$

$$y - 3 = \frac{9}{-3}(x + 4)$$

$$-3y + 9 = 9x + 36$$

$$9x + 3y = -27$$

which may be reduced to

$$3x + y = -9$$

5 Determine the equation of the line whose intercepts are (4, 0) and (0, −3).
► Substituting $a = 4$ and $b = -3$ into the intercept form of the equation of a straight line yields

$$\frac{x}{4} + \frac{y}{-3} = 1$$

which reduces to

$$3x - 4y = 12$$

6 Determine the slope and y intercept of the line

$$3x - 2y = -5$$

► Comparing this equation with the general equation $Ax + By = C$ yields $A = 3$, $B = -2$, and $C = -5$, so the slope is $-(A/B)$ or $\frac{3}{2}$, and its y intercept is C/B or $\frac{5}{2}$.

7 Determine the x and y intercepts of the equation $x + 5y = 2$.
► We have $A = 1$, $B = 5$, and $C = 2$, so the x intercept is (C/A) or 2, and the y intercept is C/B or $\frac{2}{5}$.

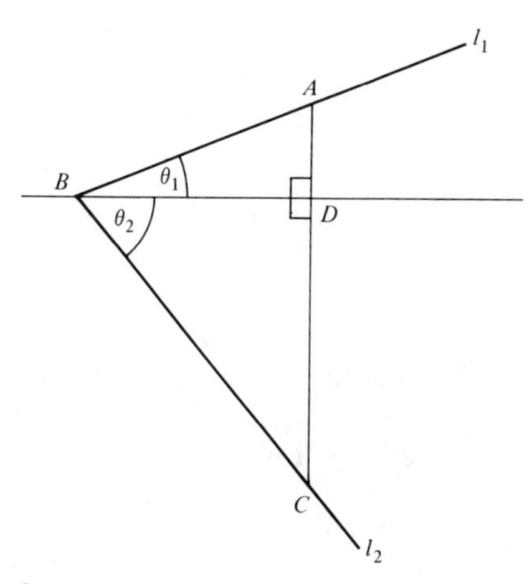

figure 7.8

Note in this last example that we could just as well have found the x intercept by setting y equal to 0 and finding that $x = 2$, and we could have found the y intercept by setting x equal to 0 and solving $5y = 2$ to get $y = \frac{2}{5}$.

It is obvious that two parallel lines have the same slope. When two lines, l_1 and l_2, are perpendicular as in Figure 7.8, the angles θ_1 and θ_2, which determine their slope, must lie in two successive quadrants, and so their tangents must have opposite signs. The angle θ_1 is equal to $\angle\, BCD$, since each is $90° - \theta_2$. The triangles ABD and BCD are therefore similar, so that

$$\frac{AD}{BD} = \frac{BD}{DC}$$

$$\tan \theta_1 = -\cot \theta_2 = -\frac{1}{\tan \theta_2}$$

Thus if the slopes of l_1 and l_2 are denoted by m_1 and m_2, respectively, we have

$$\boxed{m_1 = -\frac{1}{m_2}}$$

This is stated in words by saying that the slopes of perpendicular lines are the negative reciprocals of each other.

examples

8 Write the equation of the line parallel to the line $2x + 5y = 1$ and passing through the point $(-1, 2)$.

► The slope of the line $2x + 5y = 1$ is $-(A/B)$ or $-\frac{2}{5}$. Using the point-slope form gives

$$y - 2 = -\tfrac{2}{5}(x + 1)$$

or

$$2x + 5y = 8$$

9 Write the equation of the line perpendicular to the line $2x + 5y = 1$ and passing through the point $(-1, 2)$.

► The slope of the desired line is $-1/(-\frac{2}{5})$ or $\frac{5}{2}$. Using the point-slope form gives

$$y - 2 = \tfrac{5}{2}(x + 1)$$

or

$$5x - 2y = -9$$

10 Find the equation of the line that passes through the point $(1, -2)$ and also through the point of intersection of the lines $2x - y = 3$ and $x + 3y = 5$.

► To find the point of intersection of the lines $2x - y = 3$ and $x + 3y = 5$, we solve their equations simultaneously:

$$2x - y = 3$$
$$x + 3y = 5$$

Multiplying the second equation by 2 yields

$$2x - y = 3$$
$$2x + 6y = 10$$
$$7y = 7 \qquad y = 1$$
$$x + 3y = 5 \qquad x = 5 - 3y = 5 - 3 \cdot 1 = 2$$

The coordinates of the point of intersection of the lines are therefore $(2, 1)$.

Substituting $x_1 = 1$, $y_1 = -2$, $x_2 = 2$, and $y_2 = 1$ into the two-point form of the equation of a straight line yields

$$y - (-2) = \frac{-2 - 1}{1 - 2}(x - 1)$$

which reduces to

$$3x - y = 5$$

exercises 7.1

1 (a) Find the distance between the points $(-1, 3)$ and $(5, -2)$.
(b) What are the coordinates of the midpoint of the line segment joining these points?

(c) What is the slope of the line through these points?
(d) What is the equation of the line through these points?

2 Determine the perimeter of the triangle whose vertices are at $(1, 4)$, $(-4, 3)$, and $(5, 0)$.

3 Given the parallelogram with vertices at $(-1, 4)$, $(3, 4)$, $(-6, -1)$, and $(-2, -1)$, show that the coordinates of the midpoint of one diagonal are the same as those of the midpoint of the other diagonal.

4 Show that the line through the points $(1, 5)$ and $(-3, 1)$ is parallel to the line through the points $(2, 0)$ and $(-2, -4)$ and is perpendicular to the line through the points $(-3, 0)$ and $(1, -4)$.

5 Determine the equation of the line whose slope is -2 and whose y intercept is 3.

6 Determine the equation of the line passing through the point $(4, -1)$ and whose slope is -2.

7 Determine the equation of the line passing through the points $(1, -3)$ and $(0, 2)$.

8 Find the equation of the line whose intercepts are $(-2, 0)$ and $(0, 5)$.

9 Determine the equation of the line passing through the point $(-1, 3)$ and parallel to the line $2x - 5y = 1$.

10 Find the equation of the line passing through the point $(3, 7)$ and perpendicular to the line $3x + 2y = 6$.

11 Determine the slope and the y intercept of the line $5x + 3y = 1$.

12 Determine the intercepts of the line $4x - 3y = 9$.

13 Plot the line described in problem 10.

14 Plot the line described in problem 11.

15 Plot the line described in problem 12.

7.2 / conic sections

In this section we shall discuss the curves corresponding to polynomial equations of degree 2 in x and y. These are usually referred to as *conic sections* because, as we shall see later, each one may be realized as the intersection of a plane with a double cone. There are three types of conic sections: the *parabola*, the *ellipse* (a circle is a special form of an ellipse), and the *hyperbola*.

The Parabola

A parabola may be described as a curve such that any point on it is the same distance from a fixed point, called the *focus*, as it is from a fixed line, called the *directrix*. If in Figure 7.9, we take the point $F(0, a)$ as the focus and the line

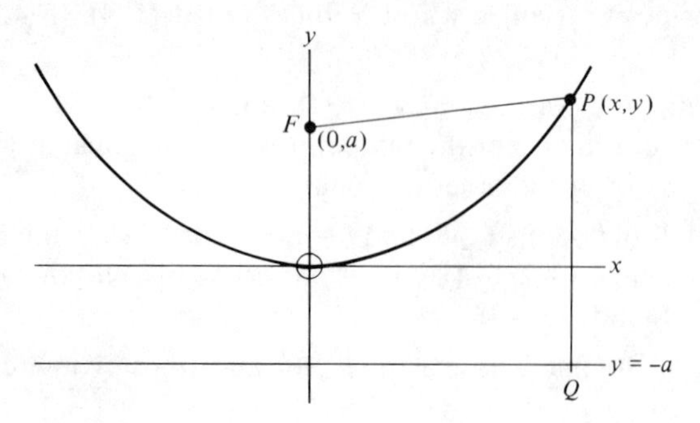

figure 7.9

$y = -a$ as the directrix, then the distance from the arbitrary point $P(x, y)$ on the parabola to the focus $F(0, a)$ is $\sqrt{x^2 + (y - a)^2}$, and the distance from P to the directrix is $y - (-a) = y + a$. The equation of the parabola is therefore

$$\sqrt{x^2 + (y - a)^2} = y + a$$

This may be simplified,

$$x^2 + (y - a)^2 = (y + a)^2$$

$$x^2 + y^2 - 2ya + a^2 = y^2 + 2ya + a^2$$

$$\boxed{x^2 = 4ay}$$

The point on the parabola halfway between the focus and the directrix is called the *vertex*, and so $x^2 = 4ay$ is the equation of a parabola whose vertex is at the origin and that opens upward. Similarly, a parabola with its vertex at the origin and that opens downward has equation $x^2 = -4ay$; a parabola with its

figure 7.10

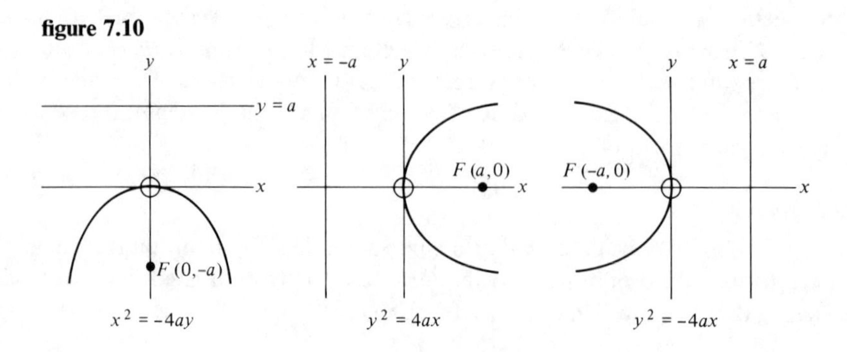

vertex at the origin and that opens to the right has equation $y^2 = 4ax$; and a parabola with its vertex at the origin and that opens to the left has equation $y^2 = -4ax$. This is illustrated in Figure 7.10. The number a is known as the *focal distance* and represents the distance between the vertex and focus.

/ examples /

1 Write the equation of the parabola whose focus is the point $(-3, 0)$ and whose directrix is the line $x = 3$.

► By referring to the relative positions of the focus, directrix, and parabola in Figure 7.9, we see that the desired parabola has its vertex at the origin and opens to the left. The focal distance is 3. Referring to Figure 7.10, we use the equation $y^2 = -4ax$ and substitute $a = 3$ to obtain the equation $y^2 = -12x$.

2 Determine the focus and directrix of the parabola $y^2 = 24x$.

► Referring to Figure 7.10 we see that this parabola has its vertex at the origin and opens to the right. Comparing the equations $y^2 = 24x$ and $y^2 = 4ax$, we see that $a = 6$, so the focus is at $(6, 0)$ and the equation of the directrix is $x = -6$.

The Ellipse

An ellipse may be described as a curve such that the sum of the distances from any of its points to two fixed points, called *foci*, is constant.

If, as in Figure 7.11, the foci F_1 and F_2 have coordinates $(-c, 0)$ and $(c, 0)$, respectively, and the sum of the distances PF_1 and PF_2 is denoted by $2a$, then the equation of the ellipse is

$$\sqrt{(x + c)^2 + y^2} + \sqrt{(x - c)^2 + y^2} = 2a$$

figure 7.11

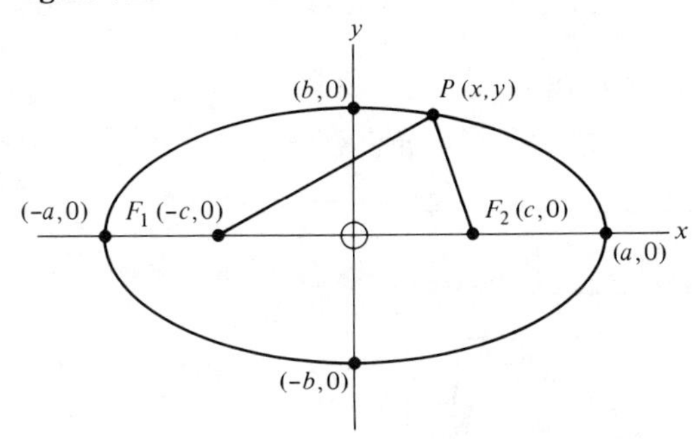

This may be simplified as follows:

$$\sqrt{(x + c)^2 + y^2} = 2a - \sqrt{(x - c)^2 + y^2}$$

$$(x + c)^2 + y^2 = 4a^2 - 4a\sqrt{(x - c)^2 + y^2} + (x - c)^2 + y^2$$

$$x^2 + 2xc + c^2 + y^2 = 4a^2 - 4a\sqrt{(x - c)^2 + y^2} + x^2 - 2xc + c^2 + y^2$$

$$4xc - 4a^2 = -4a\sqrt{(x - c)^2 + y^2}$$

$$xc - a^2 = -a\sqrt{(x - c)^2 + y^2}$$

$$(xc - a^2)^2 = a^2[(x - c)^2 + y^2]$$

$$x^2c^2 - 2xca^2 + a^4 = a^2x^2 - 2a^2xc + a^2c^2 + a^2y^2$$

$$a^2(a^2 - c^2) = (a^2 - c^2)x^2 + a^2y^2$$

Letting $a^2 - c^2 = b^2$ and interchanging the two sides of the equation yields

$$\boxed{b^2x^2 + a^2y^2 = a^2b^2}$$

which may also be written as

$$\boxed{\frac{x^2}{a^2} + \frac{y^2}{b^2} = 1}$$

The point midway between the two foci is called the *center* of the ellipse. Notice that the x intercepts of the ellipse (obtained by setting $y = 0$ in either of the above equations) are $(a, 0)$ and $(-a, 0)$, and the y intercepts (similarly obtained) are $(0, b)$ and $(0, -b)$. For the particular ellipse shown in Figure 7.11, the line segment between the x intercepts is called the *major axis* of the ellipse, and the line segment between the y intercepts is called the *minor axis* of the ellipse. The segment between the origin and either x intercept is called the *semimajor axis*, and the segment between the origin and either y intercept is called the *semiminor axis*. We see then that either of the two previous equations may be taken as the equation of an ellipse whose center is at the origin, whose semimajor and semiminor axes are of length a and b, respectively, and whose major axis lies along the x axis.

When the foci lie on the y axis, the major axis lies along the y axis, as in Figure 7.12, and the equation of the ellipse with center at the origin is written in either of the two forms

$$\boxed{a^2x^2 + b^2y^2 = a^2b^2}$$

or

$$\boxed{\frac{x^2}{b^2} + \frac{y^2}{a^2} = 1}$$

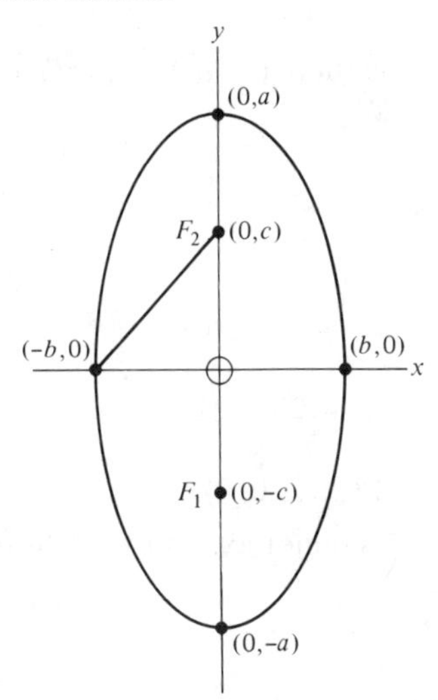

figure 7.12

The distance from the origin to either focus is called the *focal distance*, which we know is equal to c. By using the equation $a^2 = b^2 + c^2$ we may determine any of the quantities a, b, or c, provided that the other two are known.[1] Notice that a is always larger than b except in the case of a circle, where $a = b$, which becomes the radius of the circle. The equation of a circle with its center at the origin and whose radius is a thus becomes

$$x^2 + y^2 = a^2$$

In the case of a circle, the focal distance, c, is 0, and both foci are considered to be located at the center.

examples

3 Write the equation of the ellipse whose center is at the origin, whose semimajor axis is 6, whose semiminor axis is 4, and whose major axis lies along the x axis.

[1]The fact that $a^2 = b^2 + c^2$ for the ellipse may be remembered more easily by referring to Figure 7.12 and noting that the line from the origin to F_2 has length c, the line from the origin to the point $(-b, 0)$ has length b, and the line from F_2 to the point $(-b, 0)$ has length a. By the theorem of Pythagoras, we have $a^2 = b^2 + c^2$.

► Substituting $a = 6$ and $b = 4$ in the equation $(x^2/a^2) + (y^2/b^2) = 1$ yields the equation of the ellipse

$$\frac{x^2}{36} + \frac{y^2}{16} = 1$$

4 Locate the foci of the ellipse of problem 3.

► We have

$$c^2 = a^2 - b^2 = 36 - 16 = 20$$

Thus $c = \sqrt{20}$ and the foci are at $(\sqrt{20}, 0)$ and $(-\sqrt{20}, 0)$.

5 The foci of an ellipse are at (0, 5) and (0, −5). Its minor axis is 10. Determine its equation.

► We have $c = 5$, $2b = 10$, and $b = 5$, so

$$a^2 = b^2 + c^2 = 25 + 25 = 50$$

Substituting these values in $(x^2/b^2) + (y^2/a^2) = 1$ yields

$$\frac{x^2}{25} + \frac{y^2}{50} = 1$$

6 Determine the foci of the ellipse $16x^2 + 9y^2 = 144$.

► Dividing both sides by 144 yields $(x^2/9) + (y^2/16) = 1$. Thus $a = 4$, $b = 3$, and the major axis lies along the y axis.

$$c^2 = a^2 - b^2 = 16 - 9 = 7 \qquad c = \sqrt{7}$$

So the foci are at $(0, \sqrt{7})$ and $(0, -\sqrt{7})$.

7 Determine the equation of the line tangent to the circle $x^2 + y^2 = 25$ at the point (3, 4).

► The radius of the circle from the origin to (3, 4) has slope $\frac{4}{3}$. The tangent is perpendicular to this radius and therefore has slope $-\frac{3}{4}$. Using the point-slope form $y - y_1 = m(x - x_1)$ with $m = -\frac{3}{4}$, $x_1 = 3$, and $y_1 = 4$ yields $y - 4 = -\frac{3}{4}(x - 3)$ or $3x + 4y = 25$.

The Hyperbola

A hyperbola may be described as a curve such that the difference of the distances from any of its points to two points called foci is constant. If this difference is denoted by $2a$ and the foci are at $(c, 0)$ and $(-c, 0)$, the derivation of the equation of the hyperbola is quite similar to that for the ellipse, except that, as may be seen from Figure 7.13, the coordinates of the x intercept, x_1, must satisfy the equation $x_1 + c - (c - x_1) = 2a$. This means that $2x_1 = 2a$, so $x_1 = a$; it is obvious from the figure that $a < c$, so instead of making the substitution

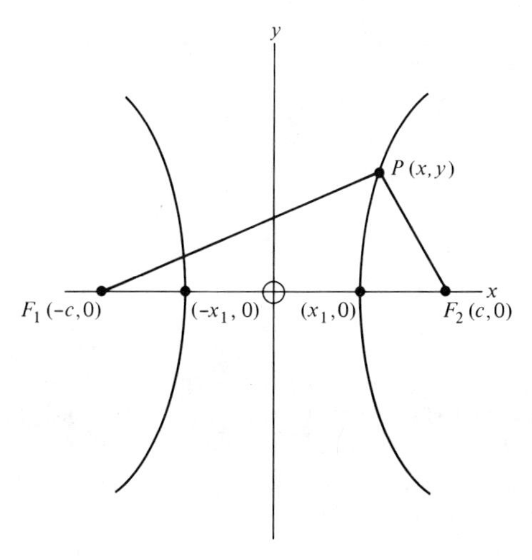

figure 7.13

$a^2 - c^2 = b^2$, as for the ellipse, we substitute $c^2 - a^2 = b^2$, because otherwise b^2 would be negative.[1]

The equation of the hyperbola may be written

$$\sqrt{(x + c)^2 + y^2} - \sqrt{(x - c)^2 + y^2} = 2a$$

This may then be simplified,

$$\sqrt{(x + c)^2 + y^2} = 2a + \sqrt{(x - c)^2 + y^2}$$

$$(x + c)^2 + y^2 = 4a^2 + 4a\sqrt{(x - c)^2 + y^2} + (x - c)^2 + y^2$$

$$x^2 + 2xc + c^2 + y^2 = 4a^2 + 4a\sqrt{(x - c^2) + y^2} + x^2 - 2xc + c^2 + y^2$$

$$4xc - 4a^2 = 4a\sqrt{(x - c)^2 + y^2}$$

$$xc - a^2 = a\sqrt{(x - c)^2 + y^2}$$

$$x^2c^2 - 2xca^2 + a^4 = a^2(x^2 - 2xc + c^2 + y^2)$$

$$x^2c^2 - 2xca^2 + a^4 = a^2x^2 - 2a^2xc + a^2c^2 + a^2y^2$$

$$(c^2 - a^2)x^2 - a^2y^2 = a^2(c^2 - a^2)$$

[1]The length b does not appear anywhere in Figure 7.13. In order to see the significance of b for the hyperbola and to see a visual demonstration that $c^2 = a^2 + b^2$ for the hyperbola, we would have to draw two lines called *asymptotes* of the hyperbola. We shall not do this here, but the student who is interested should consult a textbook on analytic geometry for more information on this subject.

Substituting $c^2 - a^2 = b^2$ yields

$$b^2x^2 - a^2y^2 = a^2b^2$$

which may also be written

$$\frac{x^2}{a^2} - \frac{y^2}{b^2} = 1$$

The point midway between the foci is called the *center* of the hyperbola, and the line segment between the two x intercepts is called the *transverse* axis of the hyperbola. The x intercepts of this hyperbola are $(a, 0)$ and $(-a, 0)$, and it does not intersect the y axis. We see that either of the two previous equations may be taken as the equation of a hyperbola whose center is at the origin, whose transverse axis is of length $2a$, and whose foci are at $(c, 0)$ and $(-c, 0)$.

As before, the distance from the center to either focus is called the *focal distance.* By using the equation $c^2 = a^2 + b^2$, we may determine any of the quantities a, b, or c provided that the other two are known. Notice that c is larger than a or b. When $a = b$, the hyperbola is known as an *equilateral* hyperbola.

When the foci of the hyperbola are at $(0, c)$ and $(0, -c)$, its equation may be written in either of the forms

$$b^2y^2 - a^2x^2 = a^2b^2$$

figure 7.14

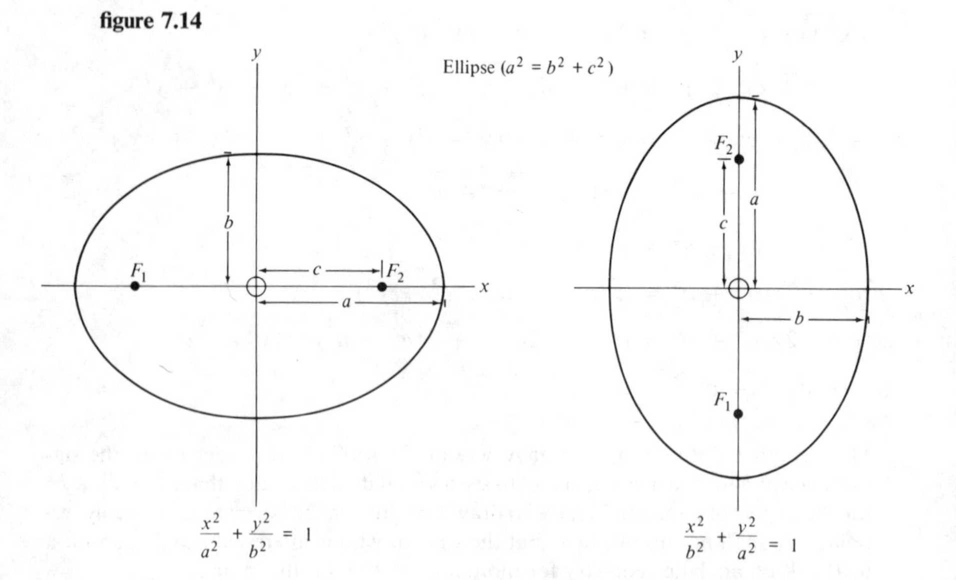

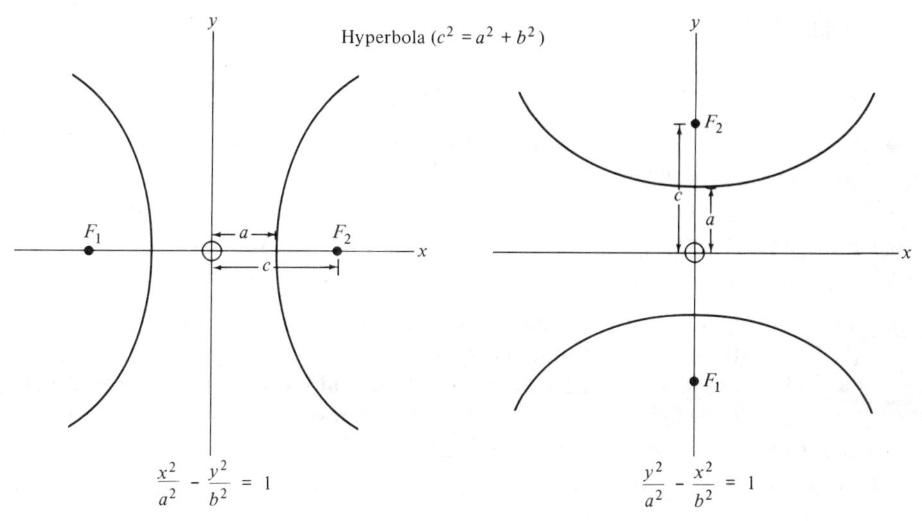

figure 7.15

or

$$\frac{y^2}{a^2} - \frac{x^2}{b^2} = 1$$

This hyperbola has y intercepts $(0, a)$ and $(0, -a)$ and has no x intercepts.

If we agree to reduce the equation of a hyperbola with its center at the origin to one of the four forms above, we note that when the x^2 term is positive, the foci are on the x axis, and when the y^2 term is positive, the foci are on the y axis. Note also that the parabola and ellipse are *connected* curves; that is, they may be drawn without lifting the pencil, whereas the hyperbola consists of two disjoint pieces.

Figures 7.14 and 7.15 summarize the positions and equations for ellipses and hyperbolas whose center is at the origin.

/ examples /

8 Write the equation of the hyperbola whose foci are at $(0, 5)$ and $(0, -5)$ and whose y intercepts are $(0, 3)$ and $(0, -3)$.

► According to Figure 7.15, the equation of this hyperbola has the form $(y^2/a^2) - (x^2/b^2) = 1$. We have $c = 5$, $a = 3$, and $b^2 = c^2 - a^2 = 25 - 9 = 16$, so $b = 4$. Substituting these values in the equation

$$\frac{y^2}{a^2} - \frac{x^2}{b^2} = 1$$

yields

$$\frac{y^2}{9} - \frac{x^2}{16} = 1$$

9 Determine the foci and intercepts of the hyperbola $4x^2 - y^2 = 4$.

► Dividing both sides by 4 yields $(x^2/1) - (y^2/4) = 1$, so $a^2 = 1$, $b^2 = 4$, $c^2 = a^2 + b^2 = 5$, and $c = \sqrt{5}$. Since the x^2 term is positive, the foci are on the x axis and are located at $(\sqrt{5}, 0)$ and $(-\sqrt{5}, 0)$. We have $a = 1$, so the x intercepts are $(1, 0)$ and $(-1, 0)$, and there are no y intercepts.

10 The intercepts of a hyperbola are at $(0, 2)$ and $(0, -2)$ and its foci are at $(0, 5)$ and $(0, -5)$. Determine its equation.

► The transverse axis of this hyperbola lies along the y axis. $a = 2$ and $c = 5$, so $b^2 = c^2 - a^2 = 25 - 4 = 21$. Substituting these values in the equation $b^2y^2 - a^2x^2 = a^2b^2$ yields $21y^2 - 4y^2 = 84$.

11 Determine the equation of the equilateral hyperbola whose foci are at $(6, 0)$ and $(-6, 0)$.

► We have $a = b$ and $a^2 + b^2 = c^2$, so $2a^2 = 36$ and $a^2 = 18$. Substituting this value in the equation $x^2 - y^2 = a^2$ yields $x^2 - y^2 = 18$.

The reason why curves whose equations are of the second degree are called conic sections is illustrated in Figure 7.16. It can be seen that the intersection of a plane with a double cone, except for a couple of unusual cases, is one of the curves discussed in this section. When the plane is parallel to a generator of the cone, the intersection is a parabola. When the plane intersects only one section of the cone, this intersection is an ellipse; when the plane intersects both sections of the cone, this intersection is a hyperbola. The unusual cases occur when the plane passes through the vertex of the cone. When it intersects the cone only at the vertex, we get the point circle whose equation is $x^2 + y^2 = 0$. The plane may also intersect the cone in two intersecting lines that are generators of the cone; their equations are $bx + ay = 0$ and $bx - ay = 0$, which may be combined into

figure 7.16

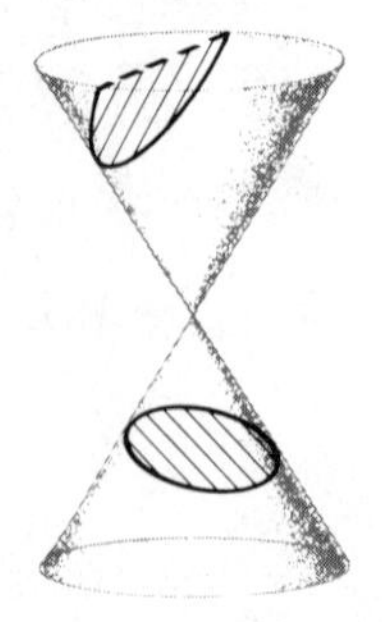

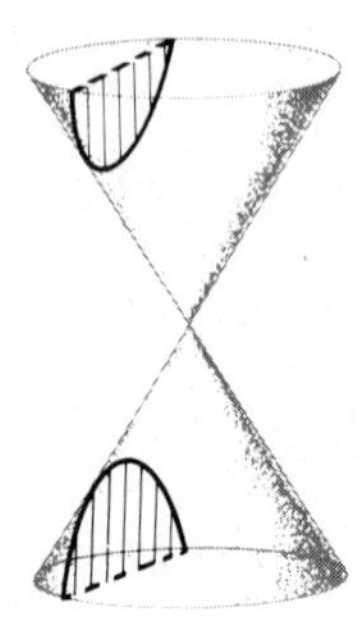

the single equation $b^2x^2 - a^2y^2 = 0$. When the plane is tangent to the cone, the intersection is one line considered twice, the equation of which may be taken as $(bx - ay)^2 = 0$.

exercises 7.2

1 Write the equation of the parabola whose focus is at the point $(0, -4)$ and whose vertex is at the origin.

2 Locate the focus and directrix of the parabola $x^2 = -12y$.

3 Write the equation of the ellipse whose center is at the origin, whose semimajor axis is 10, whose semiminor axis is 2, and whose major axis lies along the y axis.

4 Locate the foci of the ellipse of problem 3.

5 The foci of an ellipse are at $(5, 0)$ and $(-5, 0)$ and its major axis is 26. Determine its equation.

6 Locate the foci of the ellipse $25x^2 + 144y^2 = 3600$.

7 Write the equation of the hyperbola whose foci are at $(5, 0)$ and $(-5, 0)$ and whose x intercepts are $(4, 0)$ and $(-4, 0)$.

8 Locate the foci and intercepts of the hyperbola $9y^2 - 16x^2 = 144$.

9 Determine the equation of the equilateral hyperbola whose foci are at $(0, 4)$ and $(0, -4)$.

10 Write the equation of the hyperbola whose foci are at $(0, 13)$ and $(0, -13)$ and for which $b = 5$.

11 Explain what happens to the hyperbola $b^2x^2 - a^2y^2 = a^2b^2$ when a becomes 0. What happens when b becomes 0?

12 Describe the curve $b^2x^2 - a^2y^2 = 0$. (*Hint:* Factor the left side of the equation.)

7.3 / translation of axes

The conic sections that we discussed in Section 7.2 all had their vertex or center at the origin. It is apparent that when the vertex or center is moved away from the origin, the equation will be different; we will investigate here how the equation changes.

It is apparent from Figure 7.17 that if we take a conic whose center is at the origin and move it without rotating it so that its center is at the point R with coordinates (h, k), the result is the same as if we left the conic fixed and moved the axes so that the origin is at the point Q whose coordinates are $(-h, -k)$. We must determine what happens to the coordinates of a point $P(x, y)$ when the axes are translated so that the origin is at the point whose coordinates are $(-h, -k)$ with respect to the original axes. In Figure 7.18, in which the original axes and

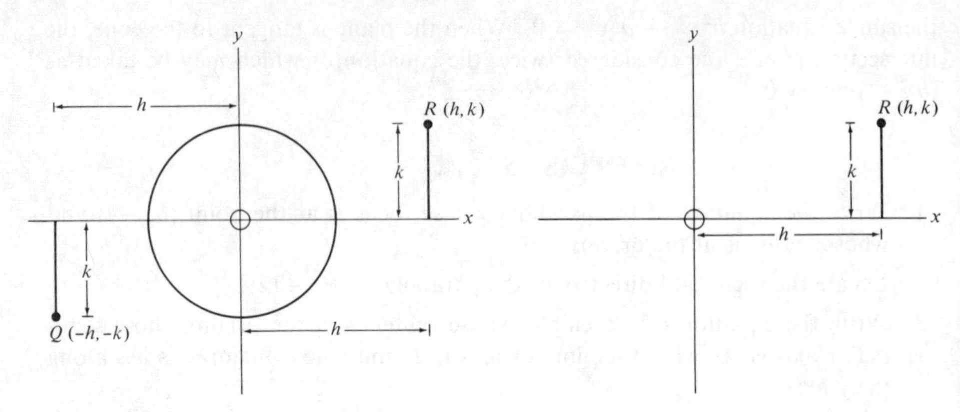

figure 7.17

origin are denoted by x, y, and O, respectively, the translated axes and origin are denoted by x', y', and O', respectively, and O' is at the point $(-h, -k)$ with respect to the original coordinate system, we see that the coordinates of the point P, which are (x, y) with respect to the original coordinate system, become $x' = x + h$ and $y' = y + k$ with respect to the translated coordinate system whose origin is at O'. This change in coordinates may be summarized by the equations

$$x = x' - h$$

$$y = y' - k$$

This means that if we know the equation of a conic whose vertex or center is at the origin, and we want to know how this equation is transformed when the vertex or center is moved to the point $R(h, k)$, we replace x by $x' - h$ and y by

figure 7.18

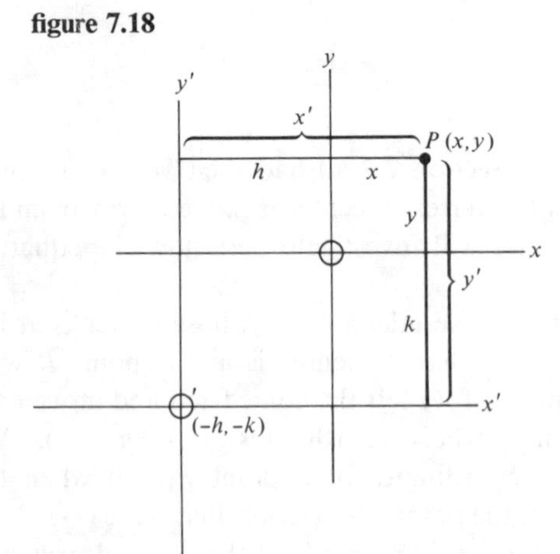

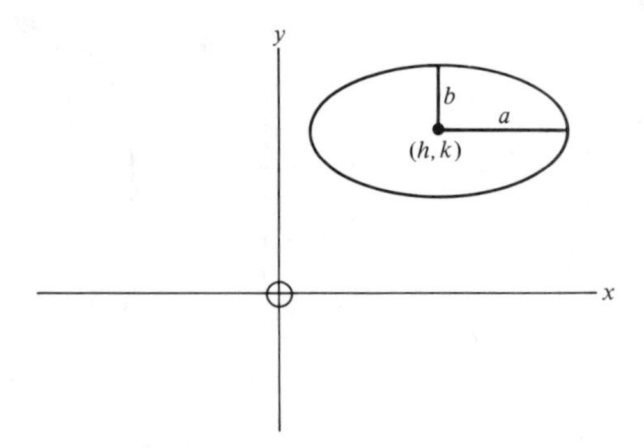

figure 7.19

$y' - k$ in the original equation to obtain the transformed equation. Thus the ellipse in Figure 7.19 would have as its equation

$$\frac{(x - h)^2}{a^2} + \frac{(y - k)^2}{b^2} = 1$$

where we have dropped the primes in writing the equation, because we have no old equation to be concerned with. Similarly, by referring back to Figures 7.9 through 7.15, we may summarize the equations of parabolas whose vertices are at (h, k) and ellipses and hyperbolas whose centers are at (h, k) by means of Figures 7.20, 7.21, and 7.22.

figure 7.20

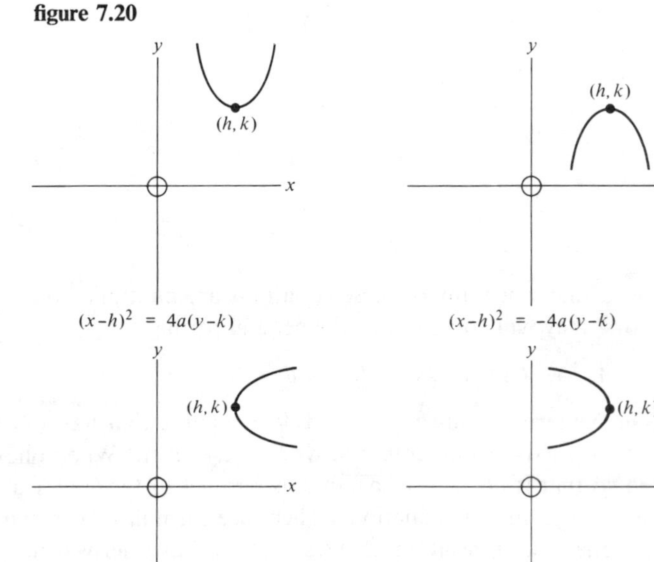

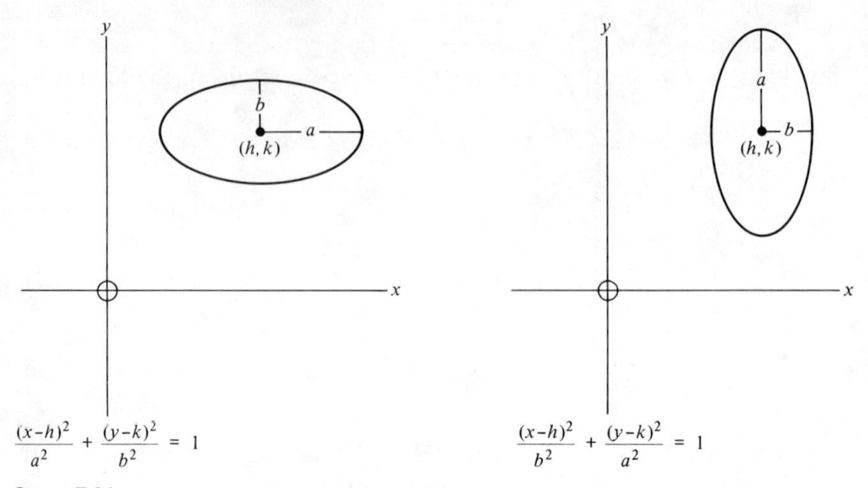

figure 7.21

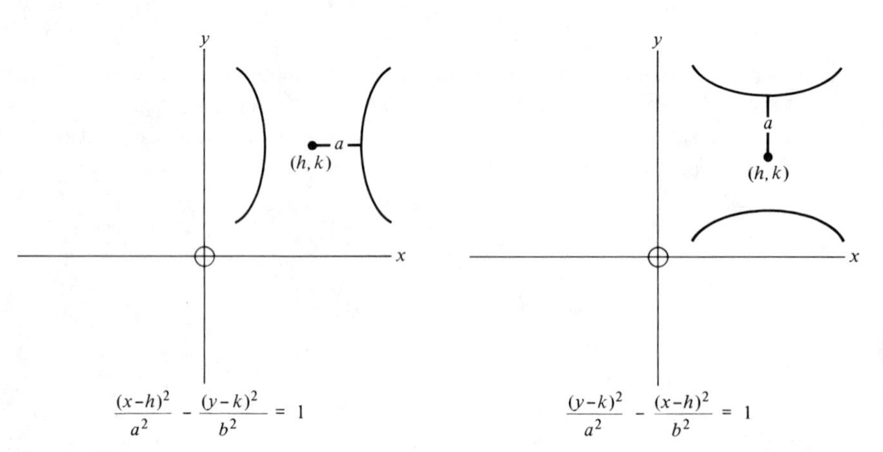

figure 7.22

It should be noticed that when any of these equations are multiplied out and cleared of denominators, they will come out in the general form

$$Ax^2 + Cy^2 + Dx + Ey + F = 0$$

that is, there will be no xy term. Thus any second-degree curve that has an axis parallel to either the x or y axis has an equation with no xy term. When one of these curves is rotated so that it no longer has an axis parallel to the x or y axis, then an xy term is introduced into its equation. There are formulas for rotation of axes similar to the ones for translation of axes that we have derived in this section, but we shall not discuss them in this book.

When the equation of a conic is expressed in the general form $Ax^2 + Cy^2 + Dx + Ey + F = 0$, we can identify the conic by means of the following rules:[1]

1. The equation of a parabola has either an x^2 term or a y^2 term but not both ($A = 0$ or $C = 0$ but not both).
2. The coefficients of x^2 and y^2 in the equation of an ellipse have the same sign ($AC > 0$).
3. The coefficients of x^2 and y^2 in the equation of a hyperbola have opposite signs ($AC < 0$).

/ examples /

1 Write the equation of the parabola whose vertex is at the point $(-1, 5)$, which opens to the right, and whose focal distance is 2.

► Referring to Figure 7.20 we see that the equation is of the form $(y - k)^2 = 4a(x - h)$ and that $h = -1$, $k = 5$, and $a = 2$. Making these substitutions yields the equation $(y - 5)^2 = 8(x + 1)$.

2 Write the equation of the ellipse whose foci are at $(1, -2)$ and $(9, -2)$ and whose semiminor axis is 2 units long.

► Referring to Figure 7.21, we see that the equation of the ellipse is of the form

$$\frac{(x - h)^2}{a^2} + \frac{(y - k)^2}{b^2} = 1$$

$a = \frac{1}{2}(9 - 1) = 4$, and $b = 2$. The center is halfway between $(1, -2)$ and $(9, -2)$, so $h = 5$ and $k = -2$. The equation is therefore

$$\frac{(x - 5)^2}{16} + \frac{(y + 2)^2}{4} = 1$$

3 Write the equation of the hyperbola whose foci are at $(3, 8)$ and $(3, -2)$ and for which $a = 2$.

► Referring to Figure 7.22, we see that the equation of the hyperbola is of the form

$$\frac{(y - k)^2}{a^2} - \frac{(x - h)^2}{b^2} = 1$$

The center is midway between $(3, 8)$ and $(3, -2)$, so $h = 3$ and $k = 3$. $c = \frac{1}{2}(8 + 2) = 5$ and $b^2 = c^2 - a^2 = 21$. The equation is therefore

$$\frac{(y - 3)^2}{4} - \frac{(x - 3)^3}{21} = 1$$

[1]We must restrict these rules to those equations that correspond to "real" conics. In this book we exclude "imaginary" conics, which result from equations such as $x^2 + y^2 + 10 = 0$, which are not satisfied by any real values of x and y.

4 Describe the curve whose equation is $9x^2 + 16y^2 - 36x + 32y = 92$. Locate its focus or foci and sketch the curve.

► Since $A = 9$ and $C = 16$, $AC > 0$, so the curve is an ellipse. We first separate the terms containing x and those containing y and factor out the coefficients of x^2 and y^2. This yields

$$9(x^2 - 4x \qquad) + 16(y^2 + 2y \qquad) = 92$$

We now complete the squares for the expressions inside the parentheses by adding the square of one half the coefficient of the linear term to obtain

$$9(x^2 - 4x + 4) + 16(y^2 + 2y + 1) = 92 + 36 + 16$$

Notice that adding 4 inside the parentheses means that we have added 9×4 to the left side of the equation; adding 1 inside the parentheses means that we have added 16×1 to the left side. This may be written

$$9(x - 2)^2 + 16(y + 1)^2 = 144$$

or

$$\frac{(x-2)^2}{16} + \frac{(y+1)^2}{9} = 1$$

This is an ellipse with its major axis horizontal and its center at $(2, -1)$, as shown. We have $a = 4$ and $b = 3$, so $c = \sqrt{16 - 9} = \sqrt{7}$. The foci are at $(2 + \sqrt{7}, -1)$ and $(2 - \sqrt{7}, -1)$.

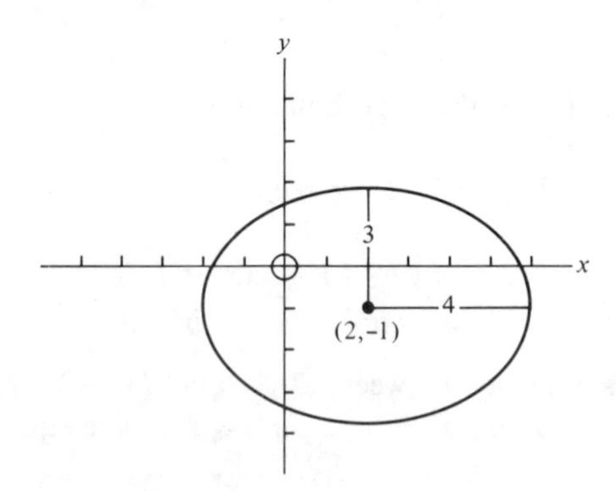

5 Describe the curve whose equation is $y^2 - 12x - 4y = 32$. Locate its focus and sketch the curve.

► Since there is a y^2 term but no x^2 term, this is a parabola. We collect the terms containing y on the left and those containing x and constant terms on the right. This gives

$$y^2 - 4y = 12x + 32$$

We next complete the square on the left by adding 4 to obtain

$$y^2 - 4y + 4 = 12x + 32 + 4$$

which may be written

$$(y - 2)^2 = 12(x + 3)$$

This is a parabola that opens to the right and whose vertex is at $(-3, 2)$, as shown. Since $a = 3$, the focus is at $(0, 2)$.

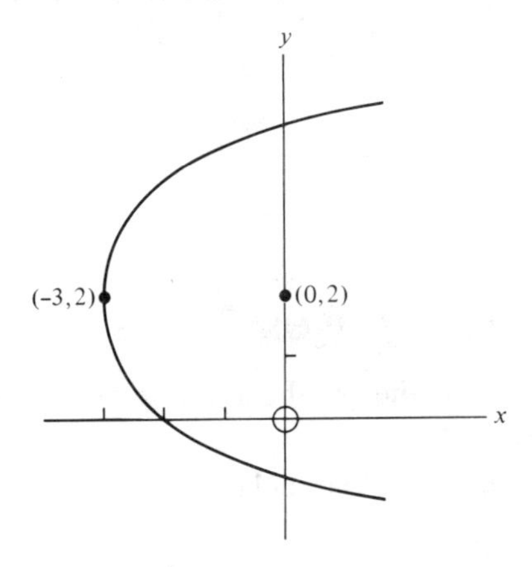

6 Describe the curve whose equation is $-4x^2 + 16x + 9y^2 - 18y = 43$. Locate its foci and sketch the curve.

► Since $A = -4$ and $C = 9$, $AC < 0$, so the curve is a hyperbola. We separate the terms containing x and those containing y and factor out the coefficients of x^2 and y^2 to obtain

$$-4(x^2 - 4x \qquad) + 9(y^2 - 2y \qquad) = 43$$

We complete the squares within the parentheses by adding the square of one half the coefficient of the linear term to obtain

$$-4(x^2 - 4x + 4) + 9(y^2 - 2y + 1) = 43 - 16 + 9$$

This may be written

$$9(y - 1)^2 - 4(x - 2)^2 = 36$$

or

$$\frac{(y - 1)^2}{4} - \frac{(x - 2)^2}{9} = 1$$

This is a hyperbola with its transverse axis vertical and its center at (2, 1), as shown. We have $a = 2$ and $b = 3$, so $c = \sqrt{4 + 9} = \sqrt{13}$. The foci are at $(2, 1 + \sqrt{13})$ and $(2, 1 - \sqrt{13})$.

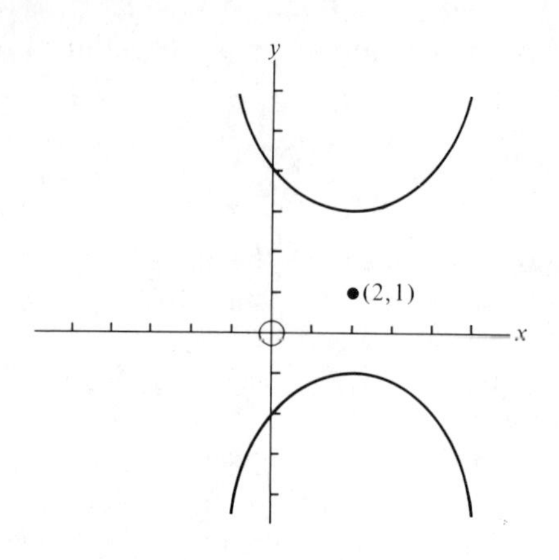

exercises 7.3

1 Write the equation of the parabola whose vertex is at (4, −1) and whose focus is at (4, −3).

2 Write the equation of the parabola whose vertex is at (−1, 3) and whose focus is at (2, 3).

3 Write the equation of the ellipse whose foci are at (2, −6) and (2, −2) and whose semimajor axis is 5 units long.

4 Write the equation of the ellipse for which the endpoints of the semimajor axis are at (−1, 4) and (9, 4) and for which $b = 3$.

5 Write the equation of the hyperbola whose foci are at (−3, −2) and (7, −2) and for which $a = 4$.

6 Write the equation of the equilateral hyperbola whose transverse axis is parallel to the x axis, whose center is at (−2, 4), and for which $a = 2$.

7 Describe the curve whose equation is $4x^2 + 25y^2 + 24x - 50y - 39 = 0$. Locate its foci and sketch the curve.

8 Describe the curve whose equation is $y^2 - 16x - 4y - 44 = 0$. Locate its focus and sketch the curve.

9 Describe the curve whose equation is $4x^2 - 9y^2 + 16x + 54y - 29 = 0$. Locate its foci and sketch the curve.

10 How is the equation of the line $5x - 2y = 6$ transformed when the origin is moved to the point (−1, 4)?

7.4 / polar coordinates

The rectangular coordinate system that we have been using is a convenient way for locating points in the plane by means of two coordinates. Starting from the origin, we determine from the x coordinate how far to go to the right or left and then from the y coordinate how far to go up or down. There are other coordinate systems that may be used to locate points in the plane. One of the most useful of these is the *polar coordinate system*, illustrated in Figure 7.23. Instead of locating points by starting at the origin and going a certain distance to the right or left and a certain distance up or down, we locate points by starting at the origin, facing in a certain direction (indicated by specifying an angle between 0 and 360°), and marking off a certain distance in that direction. In specifying the polar coordinates of a point, the distance (referred to as the *radius vector* or r coordinate) is given first, and the angle (referred to as the *vectorial angle* or θ coordinate and measured counterclockwise from the positive x axis) is given last. The positive x axis serves as the 0° line, the positive y axis serves as the 90° line, the negative x axis serves as the 180° line, and the negative y axis serves as the 270° line. Thus to locate the point (3, 210°), we would start at the origin and measure off 3 units along the 210° line to locate the point, as shown in Figure 7.24.

Several other points along with their polar coordinates are located in Figure 7.25. A negative radius vector means that we measure the distance in the opposite direction; that is, we either add 180° to θ or we subtract 180° from θ. Thus the point with coordinates $(-2, 120°)$, for example, would be the same as the point with coordinates (2, 300°), and the point with coordinates $(-5, 260°)$ would be the same as the point with coordinates (5, 80°). The origin has a radius vector of 0 and its vectorial angle is not unique, so $(0, \theta)$ would be the coordinates of the origin, where θ may be any angle whatsoever.

figure 7.23

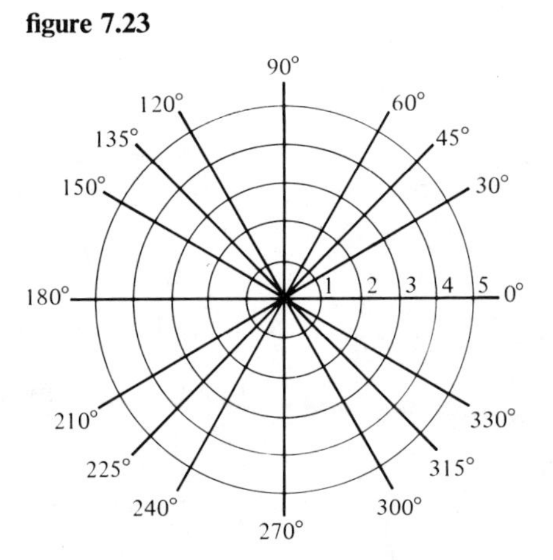

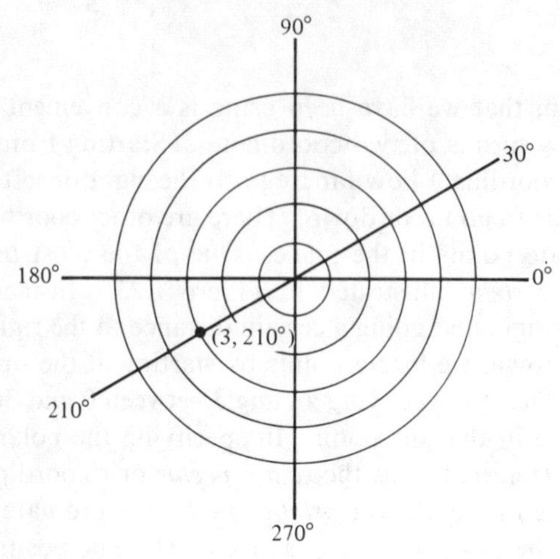

figure 7.24

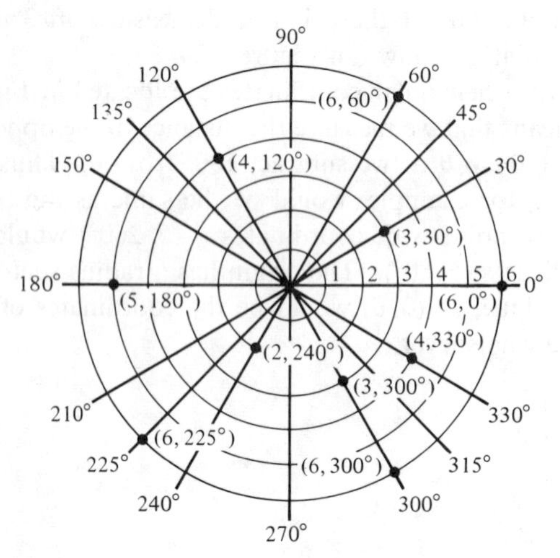

figure 7.25

Certain curves lend themselves much more to polar coordinates than to rectangular coordinates, and we shall give several examples of these. The equations of these curves are given in polar form rather than in rectangular form; that is, they are equations in r and θ rather than in x and y. To plot a curve in polar coordinates, we proceed in the same manner as for rectangular coordinates. We first make a table and list several pairs of values that satisfy the equation. The points corresponding to these coordinates are then plotted on a polar coordinate scale, and a smooth curve is drawn through these points.

examples

1 Plot the curve of the equation $r = \cos\theta$.

► It is usually easiest when setting up the table of pairs of values to choose several values of θ arbitrarily and then to determine the corresponding values of r. A table of values for the equation above would appear as follows:

θ (*degrees*)	r	θ (*degrees*)	r
0	1	180	−1
30	0.866	210	−0.866
45	0.707	225	−0.707
60	0.5	240	−0.5
90	0	270	0
120	−0.5	300	0.5
135	−0.707	315	0.707
150	−0.866	330	0.866
		360	1

Plotting these points and connecting them with a smooth curve results in the curve that is shown. Notice that the points from 210° on merely repeat the previously plotted points and add nothing new to the curve. This curve appears to be a circle, and we will verify that this is the case very shortly.

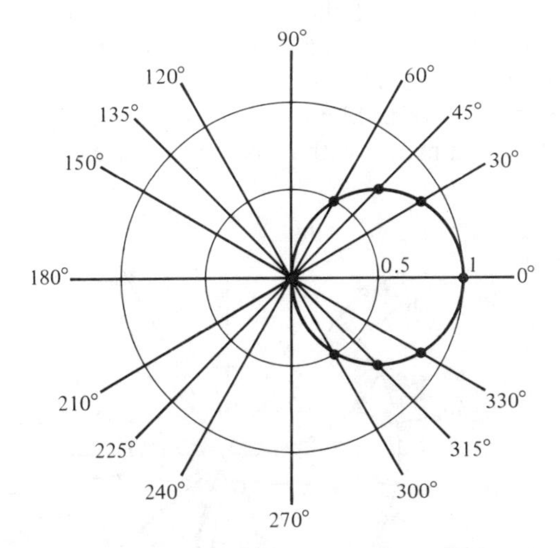

2 Plot the curve of the equation $r = 2 \sin 3\theta$.

► A table of values for this equation would be as follows:

θ (degrees)	r	θ (degrees)	r	θ (degrees)	r
0	0	120	0	240	0
10	1.0	130	1.0	250	1.0
20	1.73	140	1.73	260	1.73
30	2.0	150	2.0	270	2.0
40	1.73	160	1.73	280	1.73
50	1.0	170	1.0	290	1.0
60	0	180	0	300	0
70	−1.0	190	−1.0	310	−1.0
80	−1.73	200	−1.73	320	−1.73
90	−2.0	210	−2.0	330	−2.0
100	−1.73	220	−1.73	340	−1.73
110	−1.0	230	−1.0	350	−1.0
				360	0

We have listed many more values of θ here, since 3θ increases more rapidly than θ; furthermore, if we had used the same values as before, we might have missed some of the detail of the curve, which is plotted as shown. This curve is known as a *three-leaved rose*. An equation of the form $r = a \sin n\theta$ or $r = a \cos n\theta$ will lead to an n-leaved rose when n is an odd integer and to a $2n$-leaved rose when n is an even integer. When $n = 1$, the curve is a circle. When plotting equations of the form $r = a \cos n\theta$, we may save ourselves some work by noting that $\cos(-n\theta) = \cos n\theta$, so if we plot the portion of the curve between 0 and 180°, the remaining portion will be symmetric to this with respect to the x axis. Similarly, curves whose equations are of the form $r = a \sin n\theta$ are symmetric with respect to the y axis.

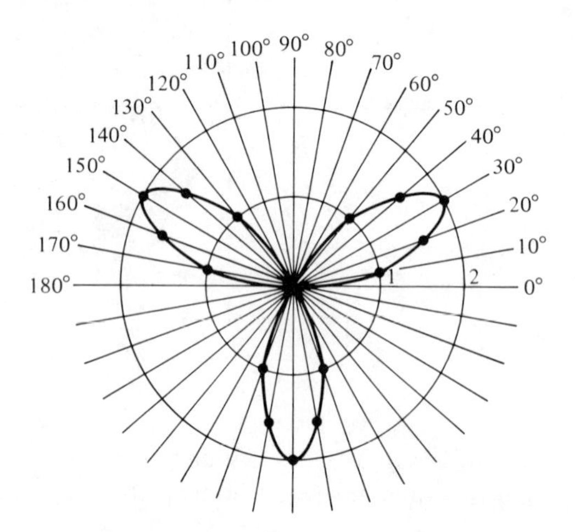

3 Plot the curve of the equation $r = 4(1 + \cos \theta)$.

► A table of values for this equation is as follows:

θ *(degrees)*	r	θ *(degrees)*	r
0	8	210	0.5
30	7.5	240	2
60	6	270	4
90	4	300	6
120	2	330	7.5
150	0.5	360	8
180	0		

This curve, plotted as shown, is known as a *cardioid.*

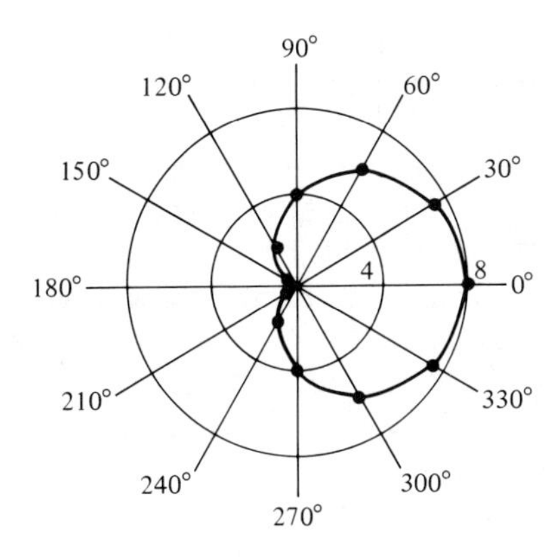

4 Plot the curve of the equation $r^2 = 4 \cos 2\theta$.

► A table of values is as follows:

θ *(degrees)*	r
0	± 2
15	± 1.8
30	± 1.4
45	0
45–135	(r is imaginary)
135	0
150	± 1.4
165	± 1.8
180	± 2

The reason why we say that r is imaginary between $\theta = 45°$ and $\theta = 135°$ is that r^2 is negative and no real-number value of r satisfies the equation.

Since $\cos(-2\theta) = \cos 2\theta$, the curve is symmetric with respect to the x axis. This curve, known as a *lemniscate*, is plotted as shown.

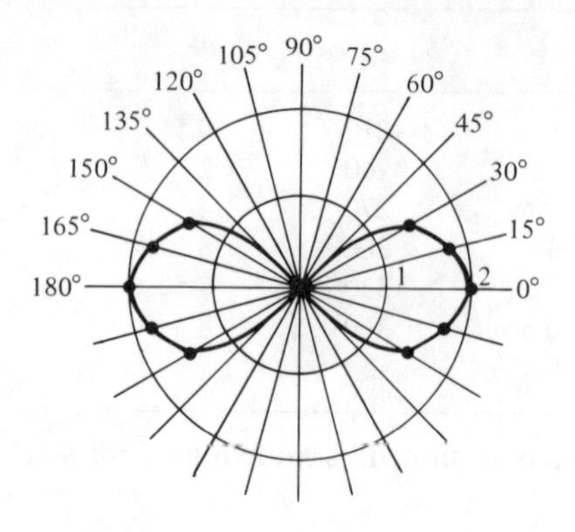

5 Plot the curve of the equation $r = a\theta$.

► A table of values is as follows:

θ *(degrees)*	θ *(radians)*	r	θ *(degrees)*	θ *(radians)*	r
0	0	0	180	π	$3.1a$
30	$\frac{1}{6}\pi$	$0.52a$	270	$\frac{3}{2}\pi$	$4.7a$
60	$\frac{1}{3}\pi$	$1.0a$	360	2π	$6.3a$
90	$\frac{1}{2}\pi$	$1.6a$	450	$2\frac{1}{2}\pi$	$7.9a$
120	$\frac{2}{3}\pi$	$2.1a$	540	3π	$9.4a$
150	$\frac{5}{6}\pi$	$2.6a$			

This curve, known as the *Spiral of Archimedes*, is plotted as shown.

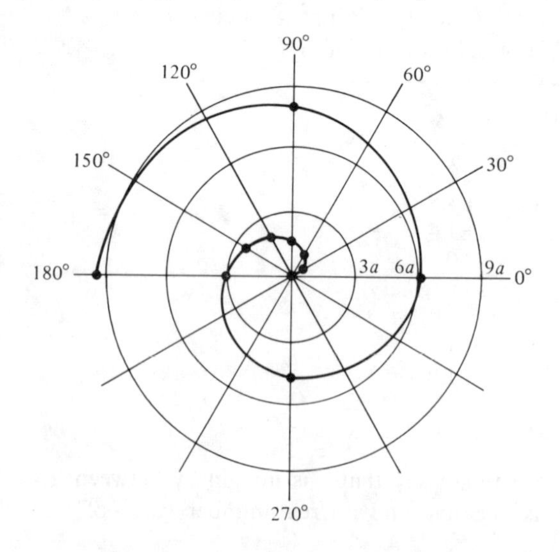

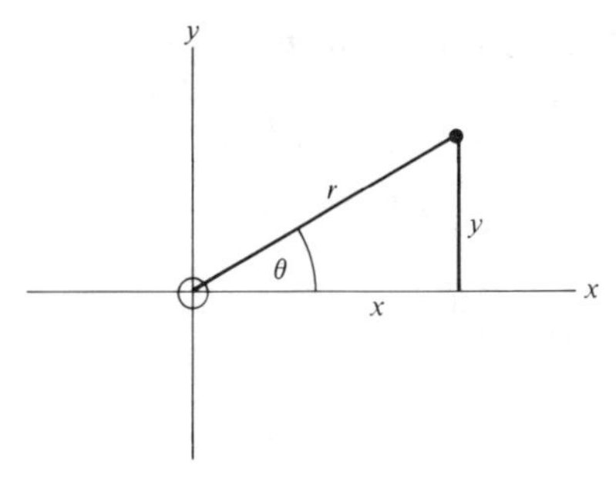

figure 7.26

The relationship between rectangular and polar coordinates may be seen from Figure 7.26. It is clear from the figure that

$$r = \sqrt{x^2 + y^2} \qquad \text{and} \qquad \theta = \arctan \frac{y}{x}^{1}$$

It is also clear that

$$x = r \cos \theta \qquad \text{and} \qquad y = r \sin \theta$$

Thus to change an equation in rectangular coordinates to one in polar coordinates we substitute $r \cos \theta$ for x and $r \sin \theta$ for y. To change an equation in polar coordinates to one in rectangular coordinates we substitute $\sqrt{x^2 + y^2}$ for r and $\arctan (y/x)$ for θ. Since most equations in polar coordinates involve $\sin \theta$ or $\cos \theta$ and $\sin \theta = y/r = y/\sqrt{x^2 + y^2}$ and $\cos \theta = x/r = x/\sqrt{x^2 + y^2}$, it is usually easier to substitute $y/\sqrt{x^2 + y^2}$ for $\sin \theta$ and $x/\sqrt{x^2 + y^2}$ for $\cos \theta$.

/examples/

6 Change the coordinates of the point (1, 3) to polar coordinates.

► We substitute $x = 1$ and $y = 3$ in the equations $r = \sqrt{x^2 + y^2}$ and $\theta = \arctan (y/x)$ and get $r = \sqrt{1^2 + 3^2} = \sqrt{10}$, $\theta = \arctan \frac{3}{1} = 72°$ approximately, according to Table 6.1, so the coordinates are $(\sqrt{10}, 72°)$.

7 Change the coordinates of the point (3, 120°) to rectangular coordinates.

► We substitute $r = 3$ and $\theta = 120°$ in the equations $x = r \cos \theta$ and $y = r \sin \theta$, and we get $x = 3(\cos 120°) = 3(-0.5000) = -1.5$ and $y = 3(\sin 120°) = 3(0.8660) = 2.598$, so the coordinates are $(-1.5, 2.598)$.

8 Change the equation $y^2 = 16x + 2$ to polar coordinates.

► Substituting $x = r \cos \theta$ and $y = r \sin \theta$ yields $r^2 \sin^2 \theta = 16r \cos \theta + 2$.

[1]This is read "θ is the angle whose tangent is y/x."

9 Change the equation $r = \cos\theta$ to rectangular coordinates.

► Substituting $r = \sqrt{x^2 + y^2}$ and $\cos\theta = x/\sqrt{x^2 + y^2}$ yields

$$\sqrt{x^2 + y^2} = \frac{x}{\sqrt{x^2 + y^2}}$$

which may be simplified to $x^2 + y^2 = x$. This curve is a circle with center at $(\frac{1}{2}, 0)$ and radius $\frac{1}{2}$.

exercises 7.4

1 Change to polar coordinates:

(a) $(4, -2)$ (b) $(3, 4)$ (c) $(5, -12)$
(d) $(-6, 8)$ (e) $(5, 0)$ (f) $(0, 1)$

2 Change to rectangular coordinates:

(a) $(1, 30°)$ (b) $(4, 150°)$ (c) $(3, 560°)$
(d) $(-1, 80°)$ (e) $(-3, 210°)$ (f) $(2, 320°)$

3 Plot the curve of the equation $r = 3 \sin\theta$.

4 Plot the curve of the equation $r = 3 \cos 4\theta$.

5 Plot the curve of the equation $r = 2(1 - \sin\theta)$.

6 Change the equation $3x^2 - 2y^2 + x = 5$ to polar coordinates.

7 Change the equation $r = 3 \sin\theta$ to rectangular coordinates.

8 Change the equation $r = a\theta$ to rectangular coordinates.

7.5 / parametric equations

The equations that we have dealt with thus far have involved two unknowns, x and y. Assigning a certain value to x determined a corresponding value or values of y and vice versa. It is sometimes desirable to express x and y in terms of a third variable t. Assigning a value t determines a value or values for each x and y. This third variable t is referred to as a *parameter*, and when two equations, one expressing x in terms of t and the other expressing y in terms of t, are given, we have what are called *parametric equations*. Consider, for example, the parametric equations

$$x = 2t$$

$$y = t + 1$$

We can plot the curve of these equations by making a table of values for t, x, and y by choosing several values of t and determining the corresponding values of x and y.

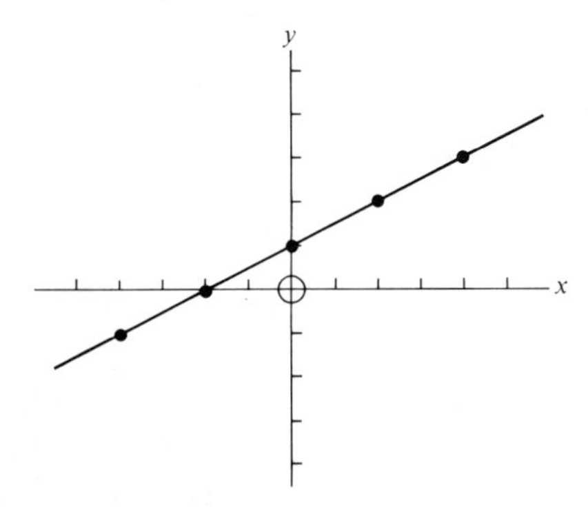

figure 7.27

t	x	y
0	0	1
1	2	2
2	4	3
3	6	4
−1	−2	0
−2	−4	−1

We then plot the curve by ignoring the values of t and plotting the points corresponding to the pairs of values of x and y. Thus the curve for the parametric equations above would appear as in Figure 7.27. It is easy to verify that this is a straight line, because if we solve the first equation for t in terms of x and then substitute this value of t in the second equation, we have $t = \frac{1}{2}x$, $y = \frac{1}{2}x + 1$ or $2y = x + 2$, which is the equation of a straight line.

In a similar manner, any set of two parametric equations may be transformed into a single equation in x and y by solving one of the equations for t in terms of x or y and then substituting this value of t in the other equation.

examples

1 Plot the curve corresponding to the parametric equations

$$x = t^2 - 2$$

$$y = 3t$$

Transform these to a single equation in x and y.

► A table of values is as follows:

t	x	y
0	-2	0
1	-1	3
2	2	6
-1	-1	-3
-2	2	-6

Plotting the values for x and y yields the curve that is shown. It may be verified that this is a parabola by solving the second equation for t to yield $t = \frac{1}{3}y$ and substituting this in the first equation to obtain $x = \frac{1}{9}y^2 - 2$ or $y^2 = 9x + 18$, which is a parabola that opens to the right and whose vertex is at $(-2, 0)$.

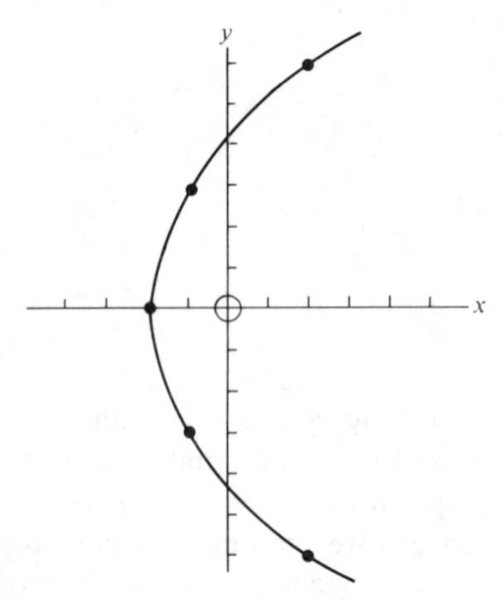

2 Plot the curve corresponding to

$$x = \cos\theta$$
$$y = \sin\theta$$

► Here the parameter is θ, and a table of values is as follows:

θ	x	y	θ	x	y
0	1	0	180	-1	0
30	0.866	0.5	210	-0.866	-0.5
60	0.5	0.866	240	-0.5	-0.866
90	0	1	270	0	-0.1
120	-0.5	0.866	300	0.5	-0.866
150	-0.866	0.5	330	0.866	-0.5
			360	1	0

The curve is plotted as shown. Here we may verify that the curve is a circle by squaring both equations and then adding to obtain

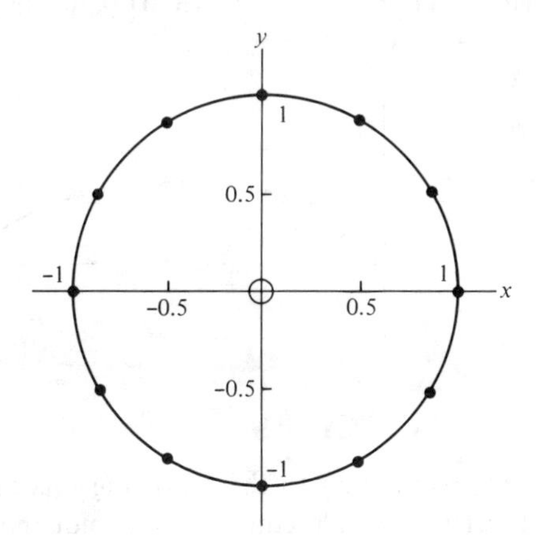

$$x^2 = \cos^2 \theta$$

$$y^2 = \sin^2 \theta$$

$$x^2 + y^2 = \cos^2 \theta + \sin^2 \theta$$

or

$$x^2 + y^2 = 1$$

(since $\cos^2 \theta + \sin^2 \theta = 1$). This is a circle of radius 1 and with center at the origin.

3 If a projectile is fired at an angle θ with the horizontal and with an initial speed V_0, its position (measured in feet) at any time t (measured in seconds from the time of firing) is given by

$$x = V_0 t \cos \theta$$

$$y = V_0 t \sin \theta - 16t^2$$

Determine the trajectory traversed by this object.

► Solving the first equation for t yields $t = x/(V_0 \cos \theta)$, and then substituting this in the second equation yields

$$y = \frac{V_0 x \sin \theta}{V_0 \cos \theta} - \frac{16x^2}{V_0^2 \cos^2 \theta}$$

or

$$y = x \tan \theta - \frac{16}{V_0^2 \cos^2 \theta} x^2$$

(since $\sin\theta/\cos\theta = \tan\theta$). This is recognized to be a parabola. As shown, it intersects the ground or x axis at (0, 0) and $[(V_0^2 \tan\theta \cos^2\theta)/16, 0]$, which may also be written $[(V_0^2 \sin\theta \cos\theta)/16, 0]$ (since $\tan\theta = \sin\theta/\cos\theta$).

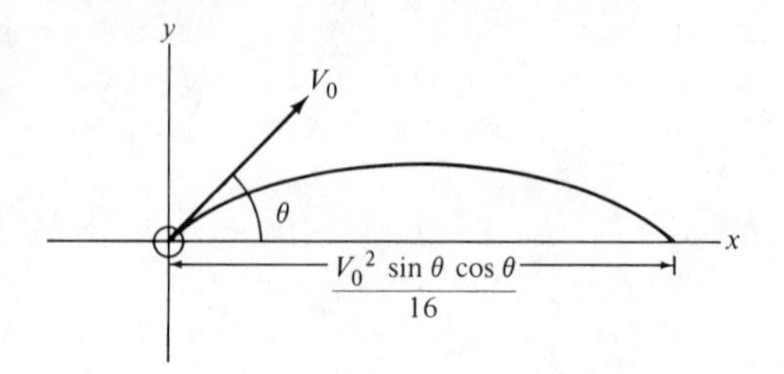

exercises 7.5

1 Plot the curves corresponding to the following sets of parametric equations. Transform each set to a single equation and plot the curve of this equation to check your answers:

(a) $x = t + 2$, $y = 2t^2$
(b) $x = 2t + 1$, $y = 5t + 2$
(c) $x = 3t^2$, $y = 2t^2 + 1$
(d) $x = t^2 - 1$, $y = 2t + 5$
(e) $x = t^2$, $y = t^3$
(f) $x = 2\cos\theta$, $y = 3\sin\theta$

2 Change the equation $y = 3x^2 + 2x$ to a set of parametric equations. (*Hint:* Let $x = t$.)

3 A projectile is fired from ground level at an angle of 40° with the horizontal and with an initial velocity of 600 ft per second. How far from the firing point will it hit the ground and what is the maximum altitude it will attain? (*Hint:* The maximum altitude is attained at the midpoint of its trajectory.)

7.6 / functions

An understanding of the notion of a function is essential for anyone who wishes to study calculus; indeed, the concept of a function is basic in most of mathematics. In most of the equations that we have studied in this chapter, you will notice that for each value we may assign to x, there is a corresponding value of y. In any situation where to each value of x there corresponds a single value of y, we say that we have defined a *function.*[1]

[1]More generally, a *function* is a correspondence between two sets A and B such that to each element of A there corresponds a unique element of B. The element b of B that corresponds to the element a of A is called the *image* of a. The set A is called the *domain* of the function, whereas the set of all images in B of elements in A is called the *range* of the function. In this section, we shall restrict ourselves to the case where the domain and the range are sets of real numbers.

When a function has been defined by means of an equation in x and y, we usually refer to x as the *independent variable* and to y as the *dependent variable.* Thus we may say that a function is a rule that assigns to each value of the independent variable a corresponding value of the dependent variable. Sometimes we have to restrict the set of values that may be assumed by the independent variable. If we limit ourselves to the real numbers, then in the equation $y = 2\sqrt{9 - x^2}$, x must be limited to values between -3 and $+3$; otherwise, there is no real value of y that corresponds to x. As mentioned in Footnote 1, the set of values that may be assumed by the independent variable is called the *domain* of the function, whereas the set of values assumed by the dependent variable is called the *range* of the function. For the function defined by the equation $y = 2\sqrt{9 - x^2}$, for example, the domain is the set of real numbers between -3 and $+3$, whereas the range is the set of real numbers between 0 and 6. In most of the functions in which we will be interested, the domain will be the set of all real numbers.

One of the properties of a function is that to each value in the domain there corresponds one and only one value of the dependent variable. Thus the equation $x^2 + y^2 = 9$ would not define a function, for if we assign the value 3 to x, we would have the two values $+3$ and -3 of y corresponding to it.

We can tell whether or not an equation in x and y defines a function by looking at the curve of the equation. *An equation in x and y defines a function if any vertical line intersects the curve of the equation in not more than one point.*

When we have a function with x as the independent variable and y as the dependent variable, we say that *y is a function of x* and we use the notation

$$y = f(x)$$

to denote this. When we say that y is a function of x, we mean that we have a relationship in which the value of y depends upon the value of x.

There are many practical situations where the notion of function enters. If a man works for \$5 per hour, then the amount of his salary depends upon the number of hours that he works. We say that his salary is a function of the number of hours that he works and write $S = f(H)$, where S denotes his salary and H denotes the number of hours worked. More specifically we have $S = 5H$ as the equation that defines this function. Here H is the independent variable and S is the dependent variable.

If a car is traveling at the rate of 40 miles per hour, then the distance, D, it travels may be given in terms of the number of hours traveled, H, by the equation $D = 40H$. Here, too, D is a function of H, and we may write $D = f(H)$.

The temperature, F, expressed on the Fahrenheit scale is a function of the temperature, C, expressed on the centigrade scale, which is defined by the function

$$F = \tfrac{9}{5}C + 32$$

If we solve this equation for C and write

$$C = \tfrac{5}{9}(F - 32)$$

then we have expressed C as a function of F. In some cases like this one, either of the variables may be considered as the independent variable, whereas the other one is considered as the dependent variable.

When the equation defining a function is written so that the dependent variable is expressed in terms of the independent variable; that is, when y has been solved in terms of x, we say that the equation is written in *explicit* form. Thus $y = x^2 + 3x - 1$ is an equation written in explicit form. When this is not the case, we say that the equation has been written in *implicit* form. Thus $xy = x^2y - 2x + 3y$ is an equation written in implicit form. It usually simplifies things if the equations defining functions are written in explicit form. When an equation is given in implicit form, it is desirable to solve for y in terms of x, when this can be done without too much difficulty, and thus express it in explicit form.

An equation is not the only way to describe a function. We may also describe a function by drawing its graph. Alternatively, we could describe a function by writing a table of values that matches the values of the dependent variable with the values of the independent variable. Finally, we could define a function in words by stating the rule that assigns to each value of the independent variable a value of the dependent variable.

These four means of defining a function are illustrated below for the function that relates the salary of a man working for $5 per hour to the number of hours that he works.

1 Equation

$$S = 5H$$

2 Graph

figure 7.28

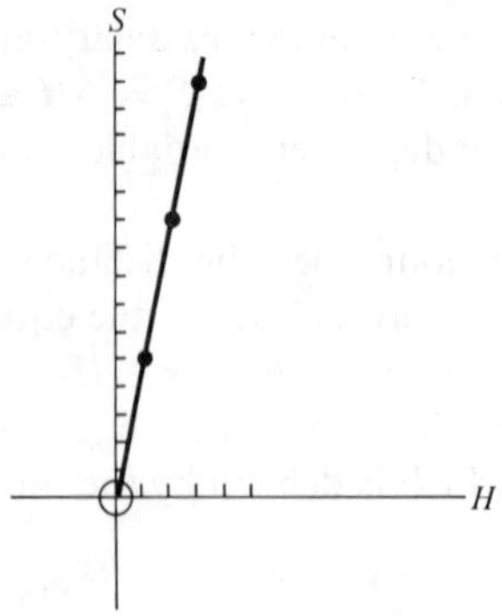

S

H

Since H (Figure 7.28) can only assume positive values, the line extends only to the right of the S axis.

3 Table

H	S	H	S
0	0	6	30
1	5	7	35
2	10	8	40
3	15	9	45
4	20	10	50
5	25		

It is of course impossible to give the entire table here, but we have given enough of it so that the other values can be determined.

4 In words

The man's salary in dollars is equal to five times the number of hours that he works.

Throughout this book we have considered many different types of equations, and thus we have really been looking at many different types of functions. To refresh our memories, we will list below some of the different types of functions that we have considered, we will include several examples for each type, and we will sketch the curves of a few of these examples. Notice that, in each of these examples, to each value of x (within the domain) there corresponds a single value of y.

1 Polynomial functions

(a) $y = 2x - 3$ (linear functions; Figure 7.29)

figure 7.29

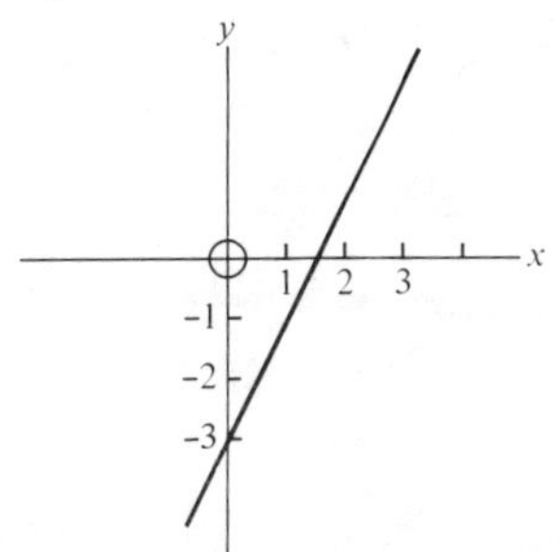

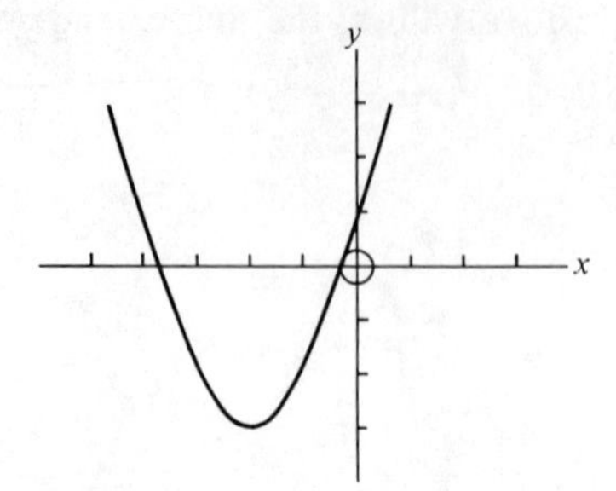

figure 7.30

(b) $y = x^2 + 4x + 1$ (quadratic functions; Figure 7.30)
(c) $y = 2x^3 - 7x^2 + 5x - 2$ (cubic functions)
(d) $y = x^4 - 7x^3 + 3x + 9$ (quartic functions)

2 *Rational functions*

(a) $y = \dfrac{x + 1}{x}$ (Figure 7.31)

(b) $y = 2x^2 - 7x + \dfrac{3}{x} - \dfrac{5}{x^3}$

(c) $y = \dfrac{x^3 - 7x^2 + 2x + 1}{3x^2 - 5x + 2}$

3 *Algebraic functions*

(a) $y = \sqrt{x}$ (Figure 7.32)

figure 7.31

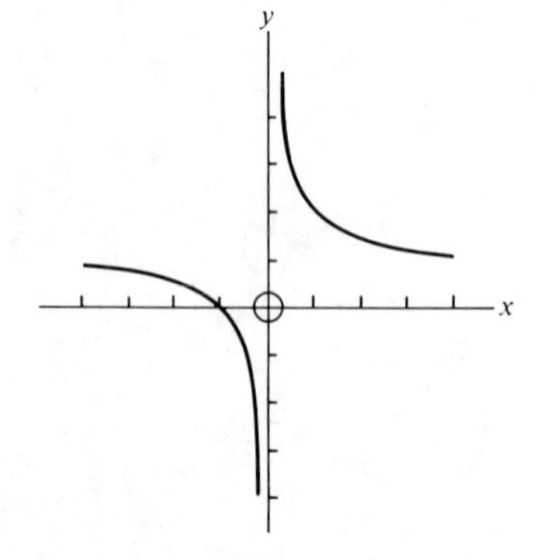

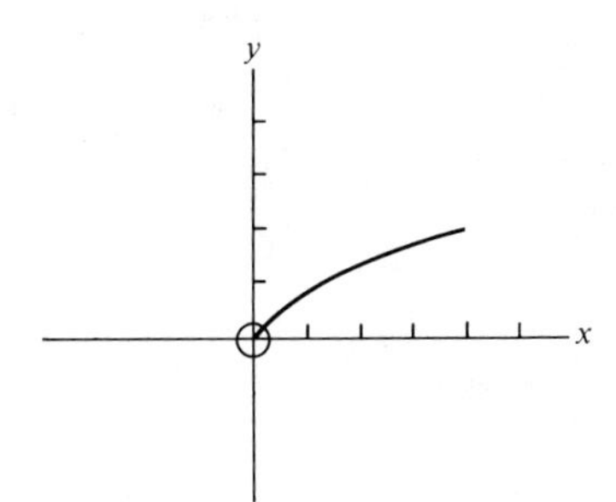

figure 7.32

(b) $y = \sqrt{9 - x^2}$ (Figure 7.33)

(c) $y = \dfrac{\sqrt{x + 1}}{x}$

(d) $y = \sqrt[3]{x^2 - 7x + 1}$

(e) $y = \sqrt[3]{x} + \sqrt[4]{x^2 + 1}$

4 *Exponential functions*

(a) $y = 10^x$ (Figure 7.34)

(b) $y = 5^{(x^2 - 7x + 1)}$

figure 7.33

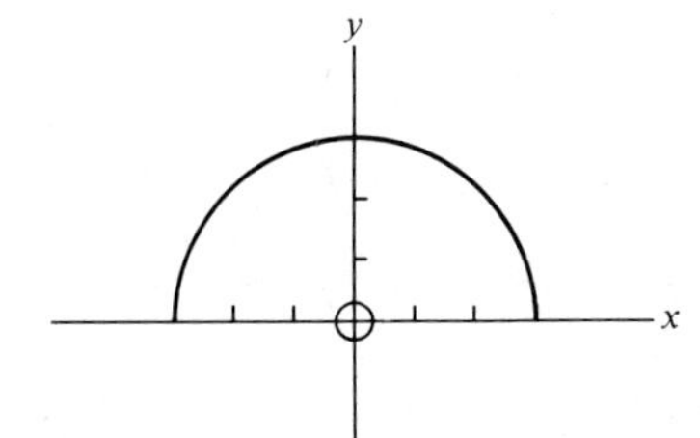

figure 7.34

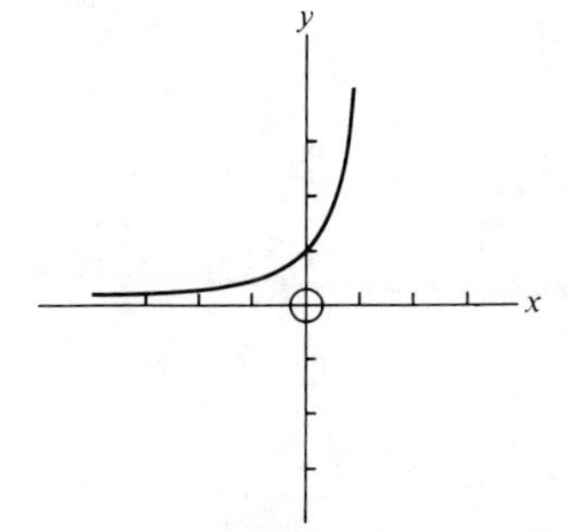

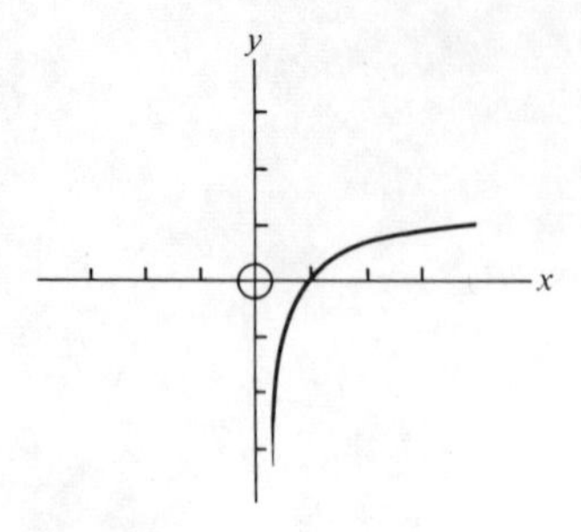

figure 7.35

5 *Logarithmic functions*

(a) $y = \log x$ (Figure 7.35)
(b) $y = (\log x)^2 + 3 \log x$

6 *Trigonometric functions*

(a) $y = \sin x$ (Figure 7.36)
(b) $y = \cos x$ (Figure 7.37)
(c) $y = \cot x$
(d) $y = \sec x + \sin x \cos x - 5$
(e) $y = \cos^2 x + 3 \sin x$
(f) $y = \dfrac{\tan x + 1}{\cos x}$

7 *Combinations of these functions*

(a) $y = x^2 + 2^x$
(b) $y = 3x - \sin x$

figure 7.36

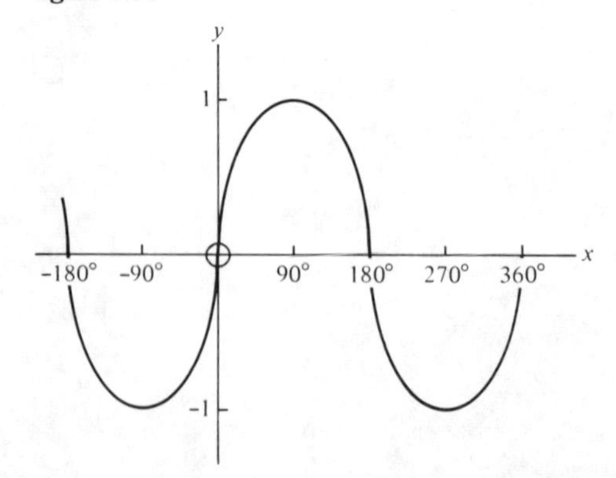

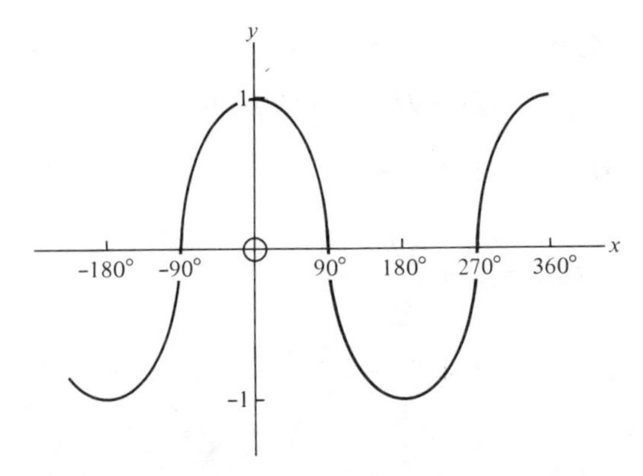

figure 7.37

(c) $y = \dfrac{\log x - x}{\cos x}$

(d) $y = 2x - \dfrac{4}{x^3} + \dfrac{1}{\log x}$

Suppose that we consider now a specific function such as[1]

$$f(x) = 3x^2 - 5x + 2$$

and ask: What is the value of this function when we substitute $x = 2$? We find that when $x = 2, f(x) = 3 \cdot 4 - 5 \cdot 2 + 2 = 4$. To indicate this we write

$$f(2) = 4$$

When we use this notation to indicate that the value of this particular function, $f(x)$, is 4 when x is 2, we are using the *functional notation.* For this same function we find, similarly, that $f(1) = 0, f(0) = 2, f(-1) = 10, f(-2) = 24$, etc. For the function

$$f(x) = \frac{2x + 1}{x}$$

we find that $f(1) = 3, f(2) = \frac{5}{2}, f(-1) = 1$, etc. To summarize the concept of functional notation, we say that for any function $f(x)$ and any number a, $f(a)$ stands for the value of $f(x)$ when we substitute a for x.

[1]We write $f(x) = 3x^2 - 5x + 2$ instead of $y = 3x^2 - 5x + 2$, because $y = f(x)$.

examples

1 If $f(x) = 1 + 1/x^2$, evaluate $f(-1)$, $f(2)$, $f(-3)$, and $f(a)$.

$$f(-1) = 1 + \frac{1}{1} = 2$$

$$f(2) = 1 + \frac{1}{4} = 1\frac{1}{4}$$

$$f(-3) = 1 + \frac{1}{9} = 1\frac{1}{9}$$

$$f(a) = 1 + \frac{1}{a^2}$$

2 If $f(x) = \sin x$, evaluate $f(0°)$, $f(90°)$, $f(45°)$, and $f(180°)$.

$$f(0°) = \sin 0° = 0 \qquad f(45°) = \sin 45° = 0.7071$$

$$f(90°) = \sin 90° = 1 \qquad f(180°) = \sin 180° = 0$$

3 If $f(x) = \sqrt{x} + 2^x$, evaluate $f(0)$, $f(1)$, and $f(4)$.

$$f(0) = \sqrt{0} + 2^0 = 0 + 1 = 1$$

$$f(1) = \sqrt{1} + 2^1 = 1 + 2 = 3$$

$$f(4) = \sqrt{4} + 2^4 = 2 + 16 = 18$$

When we wish to discuss two functions simultaneously, we cannot refer to both of them as $f(x)$, so we usually call one of them $f(x)$ and the other one $g(x)$. Suppose, for example, that

$$f(x) = x^2 + 1$$

$$g(x) = \frac{1}{x}$$

We find that $f(1) + g(2)$ would equal $(1 + 1) + \frac{1}{2} = 2\frac{1}{2}$. Similarly,

$$f(2) + g(1) = (4 + 1) + \frac{1}{1} = 6, \; f(a) + g(b) = (a^2 + 1) + \frac{1}{b},$$

$$f(-1) + g(c) = (1 + 1) + \frac{1}{c} = 2 + \frac{1}{c}, \text{ etc.}$$

Suppose that we are asked to evaluate $f(g(2))$. We find that $g(2) = \frac{1}{2}$, and so $f(g(2)) = f(\frac{1}{2}) = \frac{1}{4} + 1 = 1\frac{1}{4}$. Similarly, $g(f(2)) = g(5) = \frac{1}{5}$.

examples

4 If $f(x) = \sqrt{x+1}$ and $g(x) = x/(x+1)$, evaluate $f(0) + g(1)$, $f(1) + g(0)$, $f(a) + g(0)$, $f(x+h)$, $f(g(0))$, $g(f(3))$, $f(0)g(0)$, and $g(3)f(3)$.

$$► \quad f(0) + g(1) = \sqrt{0+1} + \frac{1}{1+1} = 1 + \frac{1}{2} = 1\frac{1}{2}$$

$$f(1) + g(0) = \sqrt{1+1} + \frac{0}{0+1} = \sqrt{2} + 0 = \sqrt{2}$$

$$f(a) + g(0) = \sqrt{a+1} + \frac{0}{0+1} = \sqrt{a+1}$$

$$f(x+h) = \sqrt{(x+h)+1} = \sqrt{x+h+1}$$

$f(x+h)$ stands for the expression that we get when we replace x by $x + h$ in the function $f(x)$.

$$f(g(0)) = f(0) = \sqrt{0+1} = 1$$

$$g(f(3)) = g(\sqrt{3+1}) = g(\sqrt{4}) = g(2) = \frac{2}{2+1} = \frac{2}{3}$$

$$f(0)g(0) = (1)(0) = 0$$

$$g(3)f(3) = \left(\frac{3}{4}\right)(2) = \frac{3}{2}$$

5 If $f(x) = x^2 - 3x + 1$ and $g(x) = 2x + 5$, evaluate $f(2) + g(1)$, $f(g(1))$, $g(f(2))$, $f(x+h)$, $g(x+h)$, $f(x+h) - f(x)$, and $g(x+h) - g(x)$.

$$► \quad f(2) + g(1) = 4 - 6 + 1 + 2 + 5 = 6$$

$$f(g(1)) = f(2+5) = f(7) = 49 - 21 + 1 = 29$$

$$g(f(2)) = g(4 - 6 + 1) = g(-1) = -2 + 5 = 3$$

$$\begin{aligned} f(x+h) &= (x+h)^2 - 3(x+h) + 1 \\ &= x^2 + 2xh + h^2 - 3x - 3h + 1 \end{aligned}$$

$$g(x+h) = 2(x+h) + 5 = 2x + 2h + 5$$

$$\begin{aligned} f(x+h) - f(x) &= x^2 + 2xh + h^2 - 3x - 3h + 1 - (x^2 - 3x + 1) \\ &= 2xh + h^2 - 3h \end{aligned}$$

$$g(x+h) - g(x) = 2x + 2h + 5 - (2x + 5) = 2h$$

6 If $f(x) = 1/x$, evaluate $f(2)$, $f(a)$, $f(x + h)$, and $f(x + h) - f(x)$.

► $$f(2) = \frac{1}{2}$$

$$f(a) = \frac{1}{a}$$

$$f(x + h) = \frac{1}{x + h}$$

$$f(x + h) - f(x) = \frac{1}{x + h} - \frac{1}{x} = \frac{x - (x + h)}{x(x + h)} = \frac{-h}{x(x + h)}$$

7 If $f(x) = 3x^2 - 5x + 1$, evaluate

$$\frac{f(x + h) - f(x)}{h}$$

► $$\frac{f(x + h) - f(x)}{h} = \frac{3(x + h)^2 - 5(x + h) + 1 - (3x^2 - 5x + 1)}{h}$$

$$= \frac{3(x^2 + 2xh + h^2) - 5(x + h) + 1 - 3x^2 + 5x - 1}{h}$$

$$= \frac{3x^2 + 6xh + 3h^2 - 5x - 5h + 1 - 3x^2 + 5x - 1}{h}$$

$$= \frac{6xh + 3h^2 - 5h}{h} = 6x + 3h - 5$$

If $f(x)$ is any function, and we draw the graph of $f(x)$, we may see that the expression $[f(x + h) - f(x)]/h$ has a certain geometrical significance, as indicated. Notice that $f(x)$ is the y coordinate of the point C, so $f(x) = AC$,

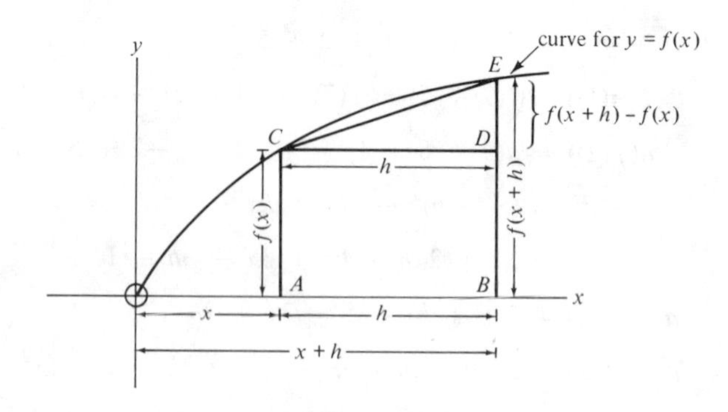

and $f(x + h)$ is the y coordinate of the point E, so $f(x + h) = BE$. Then, since $AC = BD$, $f(x + h) - f(x) = BE - AC = BE - BD = ED$, and

$$\frac{f(x + h) - f(x)}{h} = \frac{ED}{CD}$$

Thus $[f(x + h) - f(x)]/h$ is the slope of the chord CE from C to E. This is also referred to as the *average rate of change* of the function $f(x)$ from x to $x + h$.

exercises 7.6

1 Write each of the following functions in explicit form by solving for y in terms of x:

(a) $xy - y = 2x + 3$
(b) $x = \sqrt{y + 1}$
(c) $(x + y)/y = x + 1$
(d) $x^2y + 3x^2 + 1 = 2xy - y$
(e) $x = \log_{10} y$
(f) $x = \sin y$ (See the footnote of Section 7.4.)

2 If $f(x) = x^2 - 5x + 2$, evaluate $f(1)$, $f(-1)$, $f(a)$, $f(x + h)$, and

$$\frac{f(x + h) - f(x)}{h}$$

3 If $f(x) = x^2 + 1$ and $g(x) = 3x + 2$, evaluate $f(1) + g(2)$, $f(g(1))$, $g(f(1))$, $f(1) - g(0)$, $g(a) - f(b)$, and $f(1)g(2)$.

4 If $f(x) = 2x + 1$ and $g(x) = 1/(x + 1)$, evaluate $f(-3)$, $g(2)$, $f(1) - g(0)$, $g(2) - f(0)$, $g(f(1))$, $f(g(1))$, $f(1)g(1)$, and $f(1)/g(1)$.

5 For each of the following functions, find the average rate of change from $x = 1$ to $x = 3$. (*Hint:* Evaluate $[f(x + h) - f(x)]/h$; then substitute $x = 1$ and $h = 2$ (since the distance h between $x = 1$ and $x = 3$ is 2.)

(a) $f(x) = x^2 - 7x + 1$
(b) $f(x) = 4x^2 - 3x - 2$
(c) $f(x) = 1/(x + 1)$

6 For each of the following functions, find the average rate of change from $x = -1$ to $x = 2$.

(a) $f(x) = x^2 + 5x - 1$
(b) $f(x) = 3x^2 + 2x + 5$
(c) $f(x) = (x + 1)/x$

7 Determine the linear function represented by the following table:

x	y
2	-7
5	3

(*Hint:* Let $y = ax + b$, substitute the given values for x and y, and then solve the resulting two equations in two unknowns for a and b, or else, use the two-point form for the equation of a straight line.)

8 Determine the quadratic function represented by the following table:

x	y
0	-2
1	6
2	3

(*Hint:* Let $y = ax^2 + bx + c$, substitute the given values for x and y, and then solve the resulting three equations in three unknowns for a, b, and c.)

/review test 7/

NAME ____________________

DATE ____________________

1 What is the distance from the point $(2, -1)$ to the point $(3, 5)$?

2 What are the coordinates of the midpoint of the line segment joining the points $(-3, -2)$ and $(-7, 6)$?

3 What is the slope of the line through the points $(5, 0)$ and $(-3, -6)$?

4 Write the equation of the line whose slope is 2 and that passes through the point $(1, -6)$.

5 Write the equation of the line passing through the points $(4, -1)$ and $(-3, 7)$.

6 Write the equation of the line whose x and y intercepts are -3 and 1, respectively.

7 What is the slope of the line $3x - 6y = 1$?

8 What are the x and y intercepts of the line $2x - 7y = 8$?

9 Write the equation of the line parallel to the line $3x - y = 6$ and passing through the point $(-2, 5)$.

10 Write the equation of the line perpendicular to the line $6x - 5y = 1$ and passing through the point $(4, -1)$.

11 Determine the equation of the line tangent to the circle $x^2 + y^2 = 169$ at the point $(12, 5)$. (*Hint:* Use problem 14 of Exercises 6.2.)

12 Write the equation of the parabola whose vertex is at the origin and whose focus is at $(0, 2)$.

13 Write the equation of the ellipse whose foci are at $(0, 3)$ and $(0, -3)$ and whose semimajor axis is 5 units long.

14 Write the equation of the hyperbola whose foci are at $(5, 0)$ and $(-5, 0)$ and whose x intercepts are $(3, 0)$ and $(-3, 0)$.

15 Locate the focus and vertex of the parabola $(y - 2)^2 = 8(x + 1)$.

16 Locate the center and foci of the ellipse $16x^2 + 9y^2 - 64x + 18y - 71 = 0$.

17 Locate the center and foci of the hyperbola

$$9x^2 - 4y^2 + 36x + 32y - 64 = 0$$

18 Describe the curve whose equation is $x^2 - 2y^2 - 4y - 4 = 0$. Locate its focus or foci and sketch the curve.

19 How is the equation of the parabola $2x^2 - y = 7$ transformed when the origin is moved to $(3, -2)$?

20 What are the polar coordinates of the point $(\sqrt{3}, 1)$?

21 What are the rectangular coordinates of the point $(5, 30^\circ)$?

22 Change the equation $2x^2 - 5y^2 + 3y = 1$ to polar coordinates.

23 Change the equation $r = 2 \cos \theta$ to rectangular coordinates.

24 Plot the curve corresponding to the parametric equation

$$x = 3t + 1$$

$$y = 2t^2$$

Transform this set to a single equation in x and y.

25 Given the function $f(x) = 2x^2 - 3x + 1$, evaluate $f(1), f(-2)$, and $f(a + b)$.

26 If $f(x) = 1/(x + 1)$ and $g(x) = x^2 + 2$, evaluate $f(1) + g(-1), f(g(1))$, and $g(f(1))$.

27 For the function $f(x) = x^2 + 6x - 2$, evaluate $[f(x + h) - f(x)]/h$.

28 Find the average rate of change of the function $f(x) = x^2 + x - 1$ from $x = 1$ to $x = 4$.

29 Determine the linear function for which the following is a table of values:

x	y
1	4
-2	3

30 Determine the quadratic function for which the following is a table of values:

x	y
-1	2
3	0
1	-1

ANSWERS TO ODD-NUMBERED EXERCISES

Chapter 1

Exercises 1.1

1 First figure: $6 + 5 = 11$
Second figure: $3 + 8 = 11$
Third figure: $5 + 8 = 13$

3 $6 + 0 = 6$

5 Any set with five elements

7 (a) $A \cup C = \{1, 2, 3, 4, 5, 7, 8\}$ (b) $A \cap B = \{2, 4\}$
(c) $B \cup C = \{1, 2, 3, 4, 6, 7, 8\}$ (d) $B \cap C = \{4\}$

9 (a) 9 (b) 13 (c) 14 (d) 13 (e) 12 (f) 14

11 15

Exercises 1.2

1 63

3 100 miles

5 (a) 6 (b) 8 (c) 8 (d) 7 (e) 9 (f) 6

Exercises 1.3

1 (a) 5 (b) 3 (c) 7 (d) 8 (e) 12 (f) 9

3 (a) Quotient = 2 remainder = 3
(b) Quotient = 3 remainder = 2
(c) Quotient = 4 remainder = 3
(d) Quotient = 5 remainder = 2
(e) Quotient = 5 remainder = 7
(f) Quotient = 3 remainder = 7

5 (a) 3 (b) 6 (c) 6 (d) 63 (e) 40 (f) 45

7 $0.12

9 $0.33

Exercises 1.4

1 (a) 8 (b) 8 (c) 56 (d) 3 (e) 8
(f) 12 (g) 42 (h) 72 (i) 8 (j) 16

3 (a) The associative property for multiplication
(b) The commutative property for multiplication
(c) The distributive property
(d) The commutative property for addition
(e) The associative property for addition
(f) The distributive property
(g) The commutative property for addition
(h) The associative property for multiplication

5 No

7 No

Exercises 1.5

1 (a) Nine hundred and sixth-eight
(b) Two thousand, five hundred and seventy-three
(c) Eight hundred and sixteen thousand, four hundred and thirty-six

(d) Twenty-nine million, five hundred and sixty-two thousand, one hundred and eighty-five
(e) Forty thousand and forty
(f) One million, one thousand, and one

3 (a) 8861 (b) 2843

5 (a) 628,376 (b) 3,629,395

7 (a) $1318 + 3816 = 5134$ (b) $2654 + 6583 = 9237$

9 (a) $(16,827 \times 5) + 4 = 84,139$ (b) $(464 \times 628) + 90 = 291,482$

11 (a) 34 (b) 9499 (c) 8177 (d) 12 (e) 18,223

13 See Figure 1.15

Exercises 1.6

1 (a) 1142 (b) 2005 (c) 93 (d) 219 (e) 106

3 3021 (written in base 6)

5 7462 (written in base 8)

7 203331 (written in base 4)

9 24152 (written in base 6)

Exercises 1.7

1 No yes yes no yes
no yes yes no yes
no yes yes no yes

3
$58 = 2 \times 29$
$364 = 2 \times 2 \times 7 \times 13$
$96 = 2 \times 2 \times 2 \times 2 \times 2 \times 3$
$3960 = 2 \times 2 \times 2 \times 5 \times 9 \times 11$
$256 = 2 \times 2 \times 2 \times 2 \times 2 \times 2 \times 2 \times 2$
$729 = 3 \times 3 \times 3 \times 3 \times 3 \times 3$
$100 = 2 \times 2 \times 5 \times 5$
$1068 = 2 \times 2 \times 3 \times 89$
$216 = 2 \times 2 \times 2 \times 3 \times 3 \times 3$
$782 = 2 \times 17 \times 23$

5 210

7 90

9 The same number of primes

Exercises 1.8

1 In each case, name any set of members of the given set

3 (a) All girls are tall
(b) No teachers are smart
(c) No athletes are smart
(d) Bill is fast
(e) No students play ball

5 No

Chapter 2

Exercises 2.1

1 (a) -6 (b) -8 (c) 2 (d) -25
(e) 14 (f) 0 (g) $-m$ (h) m

3 (a) -8 (b) -8 (c) 2 (d) 10
(e) 4

5 (a) -7 (b) -10 (c) -4 (d) 5
(e) -2 (f) 2 (g) 11 (h) 13

Exercises 2.2

1 (a) 10 (b) -11 (c) 1 (d) -3
(e) 8 (f) -5 (g) -3 (h) -3
(i) -7 (j) 13 (k) 9 (l) 0

3 (a) -4 (b) 6 (c) -8 (d) -9 (e) -20 (f) -1

5 11 miles south

7 $72 behind

9 $1 ahead

11 $(a - b) + (b - a) = a - b + b - a = a + 0 - a = a - a = 0$. Therefore, $b - a$ is the inverse of $a - b$ and is written $-(a - b)$.

13 $(a + b)(c + d) = (a + b)c + (a + b)d = ac + bc + ad + bd$

Exercises 2.3

1 (a) 7 (b) 7 (c) -5 (d) 5

3 (a) 19 is farthest to the right and is largest
0 is farthest to the left and is smallest
(b) 0 is farthest to the right and is largest
-19 is farthest to the left and is smallest
(c) 7 is farthest to the right and is largest
-8 is farthest to the left and is smallest
(d) 9 is farthest to the right and is largest
-7 is farthest to the left and is smallest

(e) 17 is farthest to the right and is largest
-19 is farthest to the left and is smallest
(f) 14 is farthest to the right and is largest
-9 is farthest to the left and is smallest

5 11 checks for \$7 each

7 4 checks for \$8 each

9 If $a > b$, then $a - b$ is positive; therefore, $(a + c) - (b + c)$ is positive, so $(a + c) > (b + c)$.

11 Starting at 0, make 3 trips, each 4 units to the left, and you will wind up at the answer, -12.

Exercises 2.4

1 (a) $\frac{2}{3}$ (b) $\frac{1}{2}$ (c) $\frac{5}{16}$ (d) $\frac{39}{203}$ (e) $\frac{3}{5}$

3 (a) $5\frac{47}{120}$ (b) $\frac{773}{80}$ or $9\frac{53}{80}$ (c) $\frac{151}{60}$ or $2\frac{31}{60}$
(d) $\frac{51}{42}$ or $1\frac{9}{42}$ (e) $-\frac{9}{10}$ (f) $\frac{141}{28}$ or $5\frac{1}{28}$

5 $\frac{7119}{5029}$

7 GCD (198, 330) = 66; LCM (198, 330) = 990

9 $$\frac{a}{b}\left(\frac{c}{d} + \frac{e}{f}\right) = \frac{a}{b}\left(\frac{cf + de}{df}\right) = \frac{acf + ade}{bdf};$$

$$\frac{a}{b}\cdot\frac{c}{d} + \frac{a}{b}\cdot\frac{e}{f} = \frac{ac}{bd} + \frac{ae}{bf} = \frac{acbf + aebd}{bdbf} = \frac{b(acf + aed)}{bdbf} = \frac{acf + ade}{dbf}.$$

This proves that

$$\frac{a}{b}\left(\frac{c}{d} + \frac{e}{f}\right) = \frac{a}{b}\cdot\frac{c}{d} + \frac{a}{b}\cdot\frac{e}{f}$$

11 (a) $3\frac{3}{4}$ (b) $8\frac{1}{3}$ (c) $-6\frac{2}{5}$ (d) $-\frac{2}{3}$

Exercises 2.5

1 (a) Eight, and twelve hundredths
(b) One hundred and seventy-three, and eight thousandths
(c) Nineteen, and one thousand four hundred and thirty-one ten thousandths
(d) Eighty, and nine tenths

3 The following are rational numbers:
(a) $4\frac{3}{8}$ (b) 5.172 (c) $1.646464\cdots$ (d) $\sqrt{4}$ (e) -10

5 (a) $\frac{4173}{1000}$ (b) $\frac{92714}{10000}$ (c) $\frac{104}{33}$ (d) $\frac{236}{9}$ (e) $\frac{50005}{1000}$

7 (a) 10.957 (b) 13.258 (c) 0.865

9 (a) 65.6603 (b) 2.097 (c) 38,758

11 (a) 40.2375
(b) $8\frac{7}{10} \times 4\frac{625}{1000} = 8\frac{7}{10} \times 4\frac{5}{8} = \frac{87}{10} \times \frac{37}{8} = \frac{3219}{80} = 40\frac{19}{80} = 40.2375$

Exercises 2.6

1 (a) $\frac{5}{9}$ (b) $\frac{2}{3}$ (c) $\frac{6}{5}$
(d) $\frac{2}{1}$ (e) $\frac{25}{1}$ (f) $\frac{1}{4}$

3 (a) 240% (b) 2.67% (c) 0.35%
(d) 5.3% (e) 240% (f) 2.4%

5 288 : 3

7 (a) 490% (b) 5% (c) 6250%
(d) 0.03% (e) 250% (f) 62.5%

9 57

11 $2.94

13 $260

15 4.19%

17 41.67%

Chapter 3

Exercises 3.1

1 (a) 64 (b) 64 (c) $\frac{1}{16}$ (d) 729 (e) 3.24 (f) 3.539605824

3 (a) $162x^{16}y^{18}z^{8}$ (b) $\dfrac{2048x^{7}y^{8}z^{20}}{9}$

(c) $\dfrac{160x^{19}y^{8}z^{15}}{27}$ (d) $\dfrac{8x^{17}y^{3}z^{15}}{81}$

Exercises 3.2

1 (a) $\frac{1}{8}$ (b) $\frac{1}{9}$ (c) 9 (d) $\frac{1}{9}$ (e) -9 (f) $\frac{1}{9}$
(g) 9 (h) 4 (i) 4 (j) $\frac{1}{4}$

3 (a) 64 (b) 8 (c) 128 (d) $\frac{1}{128}$ (e) 243 (f) $\frac{1}{27}$

5 (a) 3.16 (b) 1.41 (c) 1.28 (d) 135.69 (e) 7.87 (f) 14.77

Exercises 3.3

1 (a) $12\sqrt{3}$ (b) $9\sqrt{7}$ (c) $3\sqrt{5}$ (d) $2\frac{1}{3}\sqrt{5}$ (e) $\left(\dfrac{1+\sqrt{2}}{2}\right)\sqrt{3}$

3 (a) $\frac{5}{4}$ (b) $\frac{4}{5}$

Exercises 3.4

1 (a) 3 (b) 4 (c) $\frac{1}{3}$ (d) $\frac{1}{2}$

3 (a) 58.1 (b) 7090 (c) 3.22 (d) 5,620,000
(e) 0.0256 (f) 0.0000249

5 (a) 2760 (b) 1.07 (c) 15,600 (d) 0.4
(e) 2.19 (f) 1.03 (g) 356,000 (h) 0.00000427

Chapter 4

Exercises 4.1

1 (a) $3x^4 - 7x^3 + x^2 - 5x - 2$ (b) $7x^3 - x^2 + 13x - 4$
(c) $3x^5 + x^4 + 2x^3 - 12x^2 + x + 5$ (d) $x^3 - 3x^2 + 10x - 3$
(e) $2\frac{1}{2}x^3 + 3x^2 + 5\frac{5}{6}x + 1\frac{1}{4}$

3 (a) $6x^3 - 31x^2 + 38x - 5$ (b) $6x^4 - 25x^3 + 22x^2 - 9$
(c) $5x^5 - 2x^4 - 6x^3 + 7x^2 - x - 1$ (d) $x^7 - 8x^5 + 10x^3 - 21x$

5 $3x^2yz^2 + 3x^3yz^4 + 2x^3yz - 2xyz$

7 $6x^2y^3z^2 + 3x^3y^4z^2 - 10x^3y^4z^3 - 5x^4y^5z^3$

Exercises 4.2

1 (a) $\dfrac{6xy - 2y^2 + x^2}{xy}$ (b) $\dfrac{xy^2 - y + 2x^3 + 2xy}{2x}$

(c) $\dfrac{8x^2 + 9xy + 2y^2}{3x^2 + xy}$ (d) $\dfrac{-xy - x^2 - y^2}{xy + y^2}$

3 (a) $\dfrac{6x^3y + 6x^2y^2 - 90x}{5xy + 5y^2}$ (b) $\dfrac{z - y}{xyz}$

(c) $\dfrac{y - x}{xyz}$ (d) $\dfrac{10x^3 + 5x^2y - 7xy^2 - y^3}{3x^3 - 3xy^2}$

(e) $\dfrac{y}{xy + 1}$ (f) $\dfrac{yz + 1}{xyz + x + z}$

Exercises 4.3

1 (a) $2x^2y(3x + 2y - 1)$ (b) $4a^2b^2(3ab - 4b + 5a)$
(c) $5p(q^2 - 3pq + 4)$ (d) $3ab(3c^2 - 5ac - 4b)$

3 (a) $(xy + z)(xy - z)$ (b) $(4x^2 + 9y^2)(2x + 3y)(2x - 3y)$
(c) $(x + 2y)(w - z)$ (d) $(2x - y)(3p + 2q)$

Exercises 4.4

1 (a) $\dfrac{14x + 7y - 8xy}{7x^2}$ (b) $\dfrac{5x - xy - y - 2}{x^2 - 1}$

(c) $\dfrac{5x^2y + 5xy^2 + 2x^2 + xy}{x^3 - xy^2}$ (d) $\dfrac{2(2p^2 + 2pq - q^2)}{2p^3 + p^2q - 2pq^2 - q^3}$

Chapter 5

Exercises 5.1

1 (a) Conditional equation
(b) Identity
(c) Identity
(d) Identity
(e) Conditional equation

3 3 and 7

Exercises 5.2

1 (a) $x = 0$ (b) $x = -\frac{1}{2}$ (c) $x = 2$ (d) $x = \frac{8}{19}$

3 (a) $x = \frac{25}{66}$ (b) $x = -21$ (c) $x = \frac{2}{51}$ (d) $x = \frac{24}{31}$

Exercises 5.3

1 Bill has \$12, and John has \$23

3 5 and 6

5 Jane is 11 years old, and Mary is 17 years old

7 11 dimes and 19 quarters

9 9 lb of the \$0.70 coffee and 3 lb of the \$0.55 coffee

11 18

Exercises 5.4

1 (a) $x = 2$ and $x = 9$ (b) $x = -\frac{1}{5}$ and $x = 5$
(c) $y = -\frac{1}{3}$ and $y = \frac{3}{2}$ (d) $s = \frac{2}{5}$ and $s = -\frac{1}{3}$
(e) $t = -\frac{1}{3}$ and $t = \frac{4}{3}$

3 (a) $x = \frac{1}{2}$ and $x = 2$ (b) $x \approx 6.7$ and $x \approx 0.3$
(c) $x \approx 0.77$ and $x \approx -0.43$ (d) $x \approx 0.62$ and $x \approx -1.62$
(e) $x \approx 7.82$ and $x \approx -0.32$ (f) $x \approx 1.19$ and $x \approx -1.69$

5 7 or -3

7 Ann is 8 years old, and Mary is 14 years old

Exercises 5.5

1 (a) $x = \frac{14}{9}, y = \frac{1}{9}$ (b) $x = \frac{22}{41}, y = \frac{23}{41}$
(c) $x = \frac{1}{2}, y = 0$ (d) $x = \frac{31}{14}, y = -\frac{5}{14}$

3 (a) $x = \frac{22}{7}, y = \frac{5}{7}$ (b) $s = \frac{32}{13}, t = -\frac{9}{13}$
(c) $p = \frac{13}{17}, q = \frac{11}{17}$ (d) $a = \frac{29}{13}, b = \frac{6}{13}$

5 $x = 2, y = 0, z = -1$

Exercises 5.6

1

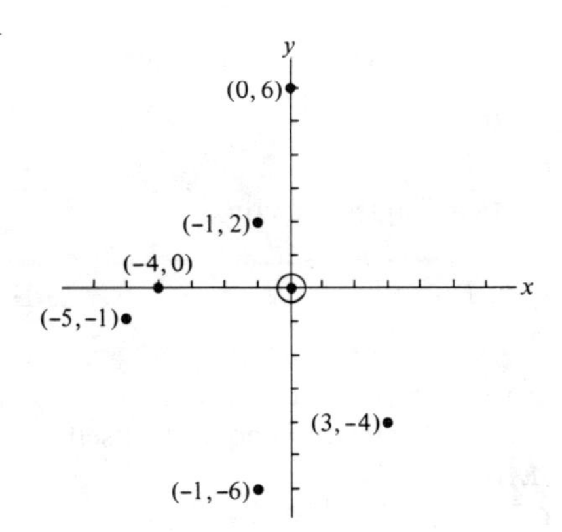

3 (a) $x = -1, y = 3$ (b) $x = 1, y = 3$
(c) $x = 4, y = -3$ (d) $x = 4, y = 0$

5

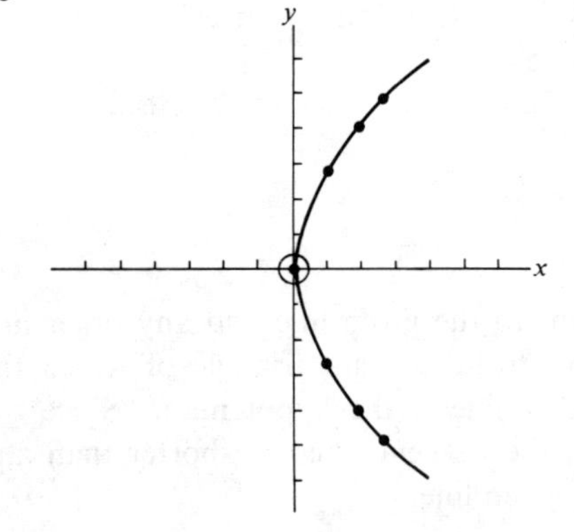

x	y
0	0
1	± 2.8
2	± 4
3	± 4.9

Exercises 5.7

1 (a) $x = 1, y = 1$ (b) $x = \frac{59}{54}, y = \frac{19}{54}$
(c) $x = \frac{22}{47}, y = -\frac{57}{47}$ (d) $x = \frac{32}{50}, y = \frac{13}{50}$

3 (a) $x = 4, y = -1$ (b) $x = \frac{38}{22}, y = \frac{13}{22}$

Chapter 6

Exercises 6.1

1 (a) 108 sq units (b) 396 sq units
(c) 179.25 sq units (d) 91.52 sq units

3 \$612.30

Exercises 6.2

1

Statement	*Reason*
1. $AB = BE, DB = BC$	1. Given
2. $\angle ABD = \angle CBE$	2. Vertical angles are equal
3. $\triangle ABD \cong \triangle CBE$	3. *SAS*
4. $\therefore \angle DAB = \angle BEC$	4. *CPCT*

3

Statement	*Reason*
1. $\angle 1 = \angle 2, \angle 3 = \angle 4$	1. Given
2. $BC = BC$	2. Any line segment is equal to itself
3. $\therefore \triangle ABC \cong \triangle BCD$	3. *AAS*
4. $\therefore AC = BD, AB = CD$	4. *CPCT*

5

Statement	*Reason*
1. $AB = AC, DB = EC$	1. Given
2. $\therefore AD = AE$	2. If equals are subtracted from equals, the results are equal
3. $\angle A = \angle A$	3. Any angle is equal to itself
4. $\therefore \triangle ADC \cong \triangle AEB$	4. *SAS*
5. $\therefore DC = BE$	5. *CPCT*

7 $x = 10\frac{4}{5}, y = 11\frac{2}{3}$

9 $AB \approx 14.1$

11 Area of triangle $ABC = 30$ sq units

13 Draw the perpendicular from the point to the given line and any other line from the point to the given line. This forms a right triangle of which the perpendicular is one side and the other line is the hypotenuse. Since the hypotenuse is longer than either side, the perpendicular is shorter than any other line from the given point to the given line.

Exercises 6.3

1 94.04 ft

3 83.88 ft

5 186.7 ft

7 $A = 56°, a \approx 17.8, c \approx 21.5$

9 $B = 23°, b \approx 2.1, c \approx 5.4$

11 See graph of $y = \sin x$ in Section 7.6

13 $y = \tan x$

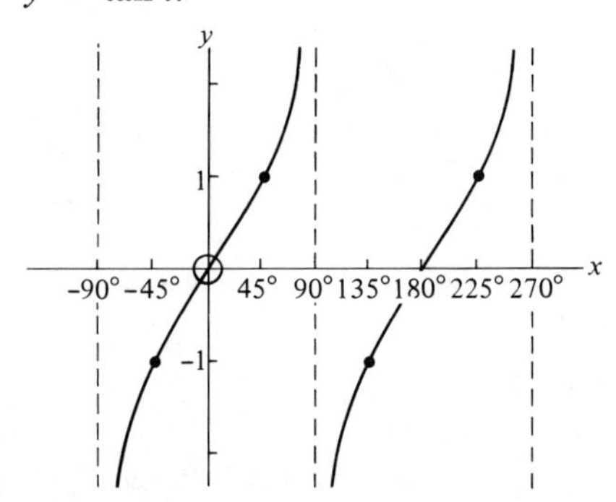

Exercises 6.4

1	(a) -0.9397	(b) -1.4281	(c) 1.494	(d) -0.9004
	(e) -1.743	(f) 0.5000	(g) 0.9848	(h) 1.1918

3 22.2 miles

Chapter 7

Exercises 7.1

1 (a) 7.81 (b) $(2, \frac{1}{2})$ (c) $-\frac{5}{6}$ (d) $5x + 6y = 13$

3 Midpoint of diagonal from $(-1, 4)$ to $(-2, -1)$ is $(-1\frac{1}{2}, 1\frac{1}{2})$. Midpoint of diagonal from $(3, 4)$ to $(-6, -1)$ is $(-1\frac{1}{2}, 1\frac{1}{2})$.

5 $2x + y = 3$

7 $5x + y = 2$

9 $2x - 5y = -17$

11 slope $= -\frac{5}{3}$, y intercept $= \frac{1}{3}$

13 $2x - 3y = -15$

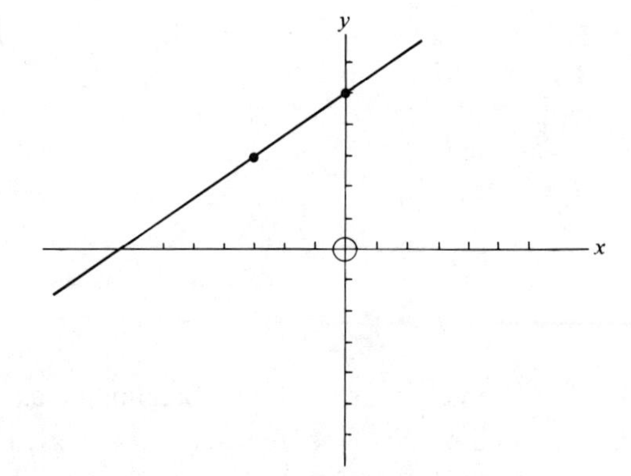

x	y
0	5
-3	3

15 $4x - 3y = 9$

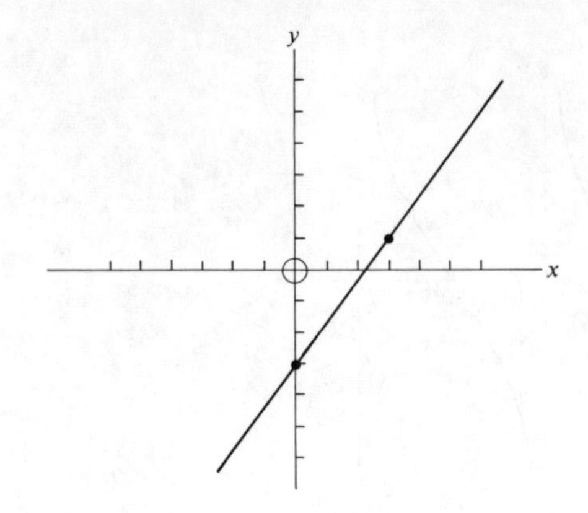

x	y
0	-3
3	1

Exercises 7.2

1 $x^2 = -16y$

3 $\dfrac{x^2}{4} + \dfrac{y^2}{100} = 1$

5 $\dfrac{x^2}{169} + \dfrac{y^2}{144} = 1$

7 $\dfrac{x^2}{16} - \dfrac{y^2}{9} = 1$

9 $\dfrac{y^2}{8} - \dfrac{x^2}{8} = 1$

11 When a becomes 0, the equation becomes $b^2x^2 = 0$; that is, the hyperbola becomes the line $x = 0$, counted twice. When b becomes 0, the equation becomes $-a^2y^2 = 0$; that is, the hyperbola becomes the line $y = 0$, counted twice.

Exercises 7.3

1 $(x - 4)^2 = -8(y + 1)$

3 $\dfrac{(x - 2)^2}{21} + \dfrac{(y + 4)^2}{25} = 1$

5 $\dfrac{(x - 2)^2}{16} - \dfrac{(y + 2)^2}{9} = 1$

7 An ellipse whose center is at $(-3, 1)$. Semimajor axis $= 5$ and is parallel to the x axis; semiminor axis $= 2$; foci at $(-3 + \sqrt{21}, 1)$ and $(-3 - \sqrt{21}, 1)$.

9 A hyperbola whose center is at $(-2, 3)$. Transverse axis is vertical, $a = 2$, $b = 3$, foci at $(-2, 3 - \sqrt{13})$ and $(-2, 3 + \sqrt{13})$.

Exercises 7.4

1 (a) $(\sqrt{20}, 333°)$ (b) $(5, 53°)$ (c) $(13, 293°)$
(d) $(10, 127°)$ (e) $(5, 0°)$ (f) $(1, 90°)$

3 $r = 3 \sin \theta$

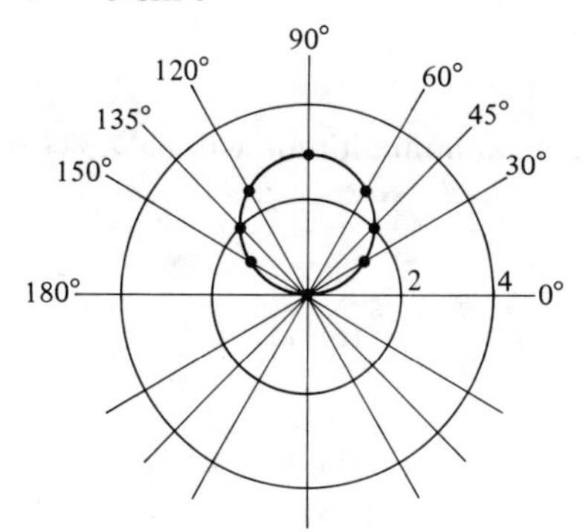

θ	r
0	0
30	1.5
45	2.12
60	2.6
90	3
135	2.12
180	0

5 $r = 2(1 - \sin \theta)$

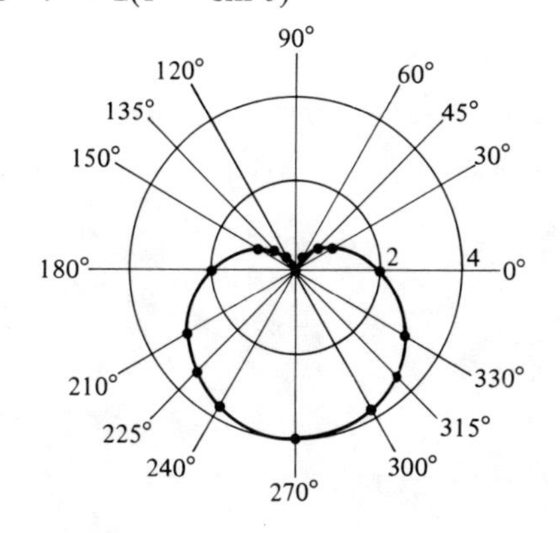

θ	r
0	2
30	1
45	0.6
60	0.3
90	0
120	0.3
135	0.6
150	1

θ	r
180	2
210	3
225	3.4
240	3.8
270	4
300	3.8
315	3.4
330	3

7 $x^2 + y^2 = 3y$

Exercises 7.5

1 (a) $y = 2(x - 2)^2$ (parabola)
(b) $5x - 2y = 1$ (straight line)
(c) $2x - 3y = -3$ (straight line)
(d) $y^2 - 4x - 10y + 21 = 0$ (parabola)
(e) $y^2 = x^3$
(f) $\frac{x^2}{4} + \frac{y^2}{9} = 1$ (ellipse)

3 It will hit the ground at 11,078 ft. The maximum altitude will be 2,383 ft.

Exercises 7.6

1 (a) $y = \frac{2x + 3}{x - 1}$ (b) $y = x^2 - 1$ (c) $y = 1$

(d) $y = \frac{3x^2 + 1}{-(x^2 - 2x + 1)}$ (e) $y = 10^x$ (f) $y = \arcsin x$

3 $f(1) + g(2) = 10$; $f(g(1)) = 26$; $g(f(1)) = 8$; $f(1) - g(0) = 0$; $g(a) - f(b) = 3a - b^2 + 1$; $f(1)g(2) = 16$

5 (a) -3 (b) 13 (c) $-\frac{1}{4}$

7 $10x - 3y = 41$

INDEX

Absolute value, 61n
Addition
  of fractions, 73
  of numbers, 8, 23–24
  of polynomials, 146–147
Addition table, 22
Adjacent side, 264
Algebraic expression, 146
  simplification of, 165–169
Alternate interior angles, 249
Analytic geometry, 216, 287
Angle, 251
  acute, 265
  base, 254
  complementary, 265
  obtuse, 265
  reflex, 265
  right, 236n
  straight, 265
  supplementary, 265n

Approximation, 123
Area, 236
  of circle, 245
  of parallelogram, 239
  of rectangle, 237
  of trapezoid, 243
  of triangle, 240
Arithmetic, 6
Associative property
  for addition, 16, 85
  for multiplication, 16, 85
Asymptotes, 303n
Average rate of change of a function, 337
Axioms, 251
Axis
  major, 300
  minor, 300
  semimajor, 300
  semiminor, 300
  transverse, 304
  x, 211
  y, 211

Base, 33
  for logarithms, 132
  of a parallelogram, 238
  of a triangle, 239
Binary number system, 39–40
Binary operation, 15
Binomial, 163
Borrowing, 26

Cancelation law, 66, 85
  for fractions, 71, 82
Cardioid, 319
Carrying, 24
Cartesian coordinate system, 211
Center
  of a circle, 243
  of an ellipse, 300
  of a hyperbola, 304
Characteristic, 132
  negative, 136
Chord, 244
Circle, 243
Circumference, 244
Closed set under an operation, 57–59
Closure properties, 85
Coefficient, 146
Commutative property
  for addition, 15, 85
  for multiplication, 15, 85
Completeness property, 97
Complex numbers, 197n
Composite number, 41
Congruent triangles, 248
  criteria for, 251
Conic sections, 297, 306
Connected curves, 305
Coordinate, 212
  r, 315
  x, 212
  y, 212
  θ, 315
Cosecant, 264
Cosine, 264
Cosine law, 276
Cotangent, 264
Counting number, 6
Cross multiplication, 183

Decimal, 88
  repeating, 88, 96
  terminating, 88, 96
Decimal system, 9, 19–31
Deductive method, 251
Degree of a polynomial, 146
Denominator, 70
Density property, 84, 86
Determinant of coefficients, 220
Determinants, 219–226
Difference of two squares, 163
Digits, 20
Directrix, 297
Distance formula, 288
Distributive property, 17, 85
Dividend, 28
Divisor, 28, 41
Division
  of fractions, 72
  of numbers, 13–14, 27–31
  of polynomials, 150–151
  by zero prohibited, 70
Domain, 326n, 327

Elimination, method of, 201–203, 201n
Ellipse, 297, 299–301
Empty set, 7
Equality of fractions, rule for, 71
Equation, 177
conditional, 178
linear, 178, 180–183
quadratic, 178, 191–195
Euclid's theorem on prime numbers, 43
Euclidean algorithm, 42
Explicit form, 328
Exponent, 109, 110
fractional, 116–117
negative, 115
Exponents, rules for computation, 113

Factor, 41, 149n
common, of polynomials, 160
Factoring rule, 123
Factorization of polynomials, 158–165
Focal distance, 299, 301, 304
Focus, 297, 299
Fractions, 70
improper, 76
Function, 326n, 327
types, 329–333
Functional notation, 333
Fundamental theorem of arithmetic, 41

General form of an equation of a conic, 311
General form of a quadratic equation, 192
Generalized distributive rule, 149
Greatest common divisor, 42

Height
of a parallelogram, 238
of a triangle, 240
Hyperbola, 297, 302–305
equilateral, 304
Hypotenuse, 257, 264
Hypotheses, 251

Identity, 178
trigonometric, 269
Identity element
for addition, 61, 74, 85
for multiplication, 70n, 85
Implicit form, 328
Indeterminate, 152n, 178
Integers, 60
negative, 61
positive, 61
rules for computation, 61–62, 65
Intercept form, 293
Intercepts of a line, 293
Intersection
of lines, 215
of sets, 7
Inverse of a number
additive, 58, 85
multiplicative, 70n, 86
Inverse operations, 13
Irrational numbers, 96

Least common multiple, 42
Leg of a right triangle, 257
Lemniscate, 320
Length of a rectangle, 236
Line segment, 288
Logarithms, 129, 132
rules for computation, 134
table of, 130–131
Logic, 44–47
Lowest common denominator, 78

Mantissa, 132
Midpoint formulas, 290
Mixed number, 76
Multiple, 14
Multiplication
of fractions, 72
of numbers, 10–11, 24–25
of polynomials, 149
Multiplication table, 22

Natural numbers, 6, 41n
Negative numbers, 58
Nondecimal number systems, 33–40
*n*th root, 116
Number line, 62
Numbers, 5, 6

Numeral, 21
Numerator, 70

Operations of arithmetic, 6
Opposite side, 264
Ordered pair, 201n
Order properties of the integers, 67–69
Origin, 211

Parabola, 297–299
Parallel lines, 216, 238, 249, 252
  slopes of, 295
Parallelogram, 238
Parameter, 322
Parametric equations, 322
Percent, 99–100
Perpendicular lines, 236, 252
  of slopes of, 295
π (ratio of circumference of circle to diameter), 244
Place-value system of numeration, 21
Plane figure, 235
Plotting
  curves, 216–217, 316, 322–323
  lines, 213
  points, 212
Point-slope form, 291
Polar coordinates, 315–316
  relation to rectangular coordinates, 321
Polar coordinate system, 315
Polynomial, 146
  cubic, 147
  irreducible, 159
  linear, 147
  quadratic, 147
  quartic, 147
  quintic, 147
  rules for computation, 152
Postulates, 251
Power, 110
Prime number, 41
Product set, 11
Pythagoras, theorem of, 257–258

Quadrant, 272
Quadratic formula, 195
Quotient, 13, 150

Radical, 116
Radius, 243
Radius vector, 315
Range, 326n, 327
Ratio, 99
Rational expression, 154
  rules for computation, 154–155
Rationalizing the denominator, 125–127
Rational number, 70
Rational point, 84
r coordinate, 315
Real number line, 97
Real numbers, 96
Reciprocal, 72
Rectangle, 236
Rectangular coordinate system, 211
Reduction to lowest terms, 71
Regrouping, 24
Relatively prime numbers, 43
Remainder, 13, 150
Root of an equation, 178
Rotation of axes, 310
Rounding off, 125

Scientific notation, 119
Secant, 264
Semicircle, 244n
Sets, 6
  equal, 8
  equivalent, 7
  disjoint, 7
  intersection of, 7
  union, 7
  use in logic, 44–47
Similar triangles, 256
Simple closed curve, 236
Sine, 264
Sine law, 275–276
Slope, 290
Slope-intercept form, 291–292
Solving a right triangle, 267
Solving a triangle, procedures for, 277
Spiral of Archimedes, 320
Square root, 116, 117
  computation of, 119–121
Starting point, 132
Substitution, method of, 201, 205–207, 209–210

Subtraction
of fractions, 74
of numbers, 12–13, 25–27
of polynomials, 148
System of linear equations, 201
System of numeration, 9

Tangent, 264
to a circle, 261
Term, 149n
Theorem, 41n, 251
θ coordinate, 315
Three-leaved rose, 318
Transitive property, 68
Translation of axes, 307–308
formulas for, 308
Transposing terms, 179
Trapezoid, 242n
Trial divisor, 120
Triangle, 239, 251
isosceles, 254
right, 257
Trichotomy, law of, 67
Trigonometric functions, 262
table of, 263–264
Trigonometry, 261
Trionomial, 160
factorization of, 160–162
Two-dimensional figure, 236
Two-point form, 292

Union of sets, 7
Unit interval, 84
Unknown, 178

Variable, 152n
dependent, 327
independent, 327
Vectorial angle, 315
Venn diagrams, 44
Vertex, 238n
of a parabola, 298

Width of a rectangle, 236
Word problems, 185
rules for solution, 186

x axis, 211
x coordinate, 212

y axis, 211
y coordinate, 212